THE CONQUEST OF BREAD

Also by Richard A. Walker

The Capitalist Imperative: Territory, Technology, and Industrial Growth
(with Michael Storper)
The New Social Economy: Reworking the Division of Labor
(with Andrew Sayer)

THE CONQUEST OF BREAD

150 Years of Agribusiness in California

Richard A. Walker

NEW YORK
LONDON

for

ZIA

the sweetest flower in my garden

and

ANNIE

the introduction who took root

Published in the United States by The New Press, New York, 2004
Distributed by W. W. Norton & Company, Inc., New York

LIBRARY OF CONGRESS CATALOGING-IN-PUBLICATION DATA

Walker, Richard, 1947-
The conquest of bread : 150 years of agribusiness in California / Richard A. Walker.
p. cm.
Includes bibliographical references and index.
ISBN 1-56584-877-2
1. Agricultural industries—California—History. 2. Agriculture—Economic aspects—California. 3. Agricultural industries—Environmental aspects—California. 4. Agricultural industries—Social aspects—California. 5. Agricultural laborers—California—History. I. Title: Agribusiness in California. II. Title.
HD9007.C2W35 2004
338.1'09794—dc22 2004042631

The New Press was established in 1990 as a not-for-profit alternative to the large, commercial publishing houses currently dominating the book publishing industry. The New Press operates in the public interest rather than for private gain, and is committed to publishing, in innovative ways, works of educational, cultural, and community value that are often deemed insufficiently profitable.

www.thenewpress.com

Composition by dix!

Printed in the United States of America

2 4 6 8 10 9 7 5 3 1

Contents

Illustration Credits

Thanks to the following people and institutions for providing images and graphics:

Darin Jensen for drawing the frontispiece maps and all graphs.

Donald Bain for getting all the photographs in shape as digital images.

Bob Dawson for the following photographs:
Plowed field, San Joaquin Valley, 1985
Furrow irrigation, San Joaquin Valley, 1984
Cotton harvesting by machine, Los Banos, 1998
Irrigation well and pump, San Joaquin Valley, 2003

Don Bain for the following photographs from his GeoImages project:
Orange groves, Ojai Valley, 1993
Tule Lake bed, drained for agriculture, 1992
Almonds in blossom with beehives, San Joaquin Valley, 1985
Aerial view of the California aqueduct and the Delta-Mendota canal, 1980
Bunching broccoli, Los Osos, 1974 (taken by Naomi Shibata)

The Department of Geography, UC Berkeley, for the following lantern slides:
Sheep grazing in Central Coast region, circa 1900
Grain harvest, San Joaquin Valley, circa 1900
Gravenstein apple orchards, Sonoma County, circa 1900

Asparagus cutting, El Centro, Imperial Valley, circa 1910
Caterpillar tractor in the Delta marshes, circa 1910
Chickens and sheds, near Petaluma, circa 1910
Drying fruit, Turlock, Stanislaus County, circa 1900

The Bancroft Library for the following photographs from its collection:
"California, the Cornucopia of the World," California Immigration Commission, 1885
Frank Namimatsu on his farm, San Jose, 1945
Luther Burbank and Atlee Burpee, Santa Rosa, early 1900s
Freight boats on the Sacramento, circa 1915
Cutting room, George Hyde cannery, San Jose, 1920
Giant wine cask, Lachman Cellars, 1898
Sunnyside Fruit Salad label, circa 1920

The San Francisco Public Library History Room for the photograph of the Safeway store interior, circa 1954.

The family of Richard V. (Dick) Coryell and Lincoln Cushing of the Institute of Industrial Relations Library, UC Berkeley, for the woodcut "Vineyard March," 1970.

The Palace of Horticulture photograph from Todd 1921 is in the public domain.

The following photographs are by the author:
Lettuce rows, Watsonville, 2003
"Rain for Rent" sign, east of Stockton, 2003
Brussels sprouts, Watsonville, 2003
Vineyards in winter, Napa Valley, 2002
Field covered in plastic, near Watsonville, 2003
Dairy farm near Linden, 2003
Silos, near Colusa, 2002
"God Bless America," Marysville 2002

Dot-maps are reproduced from the U.S. Bureau of Agricultural Economics, 1940.

Preface

I have been fascinated by California agriculture all of my life. As a child, I lived close by the orchards draping the Santa Clara Valley foothills, and on summer vacations I inhaled the vast fields of the Great Central Valley without air-conditioning. In those days, before freeways, bugs would be plastered across the grill and windshield, until the pesticides erased the bugs—and much of the rest of the wildlife. I began thinking about agriculture seriously in the 1970s as a consequence of the labor and water politics of the time. That upsurge of activism captured many people's imagination besides mine, and led to a spate of fine books touching on California agriculture by the likes of Donald Worster, Lawrence Jelinek, Clete Daniels, and Marc Reisner. I also had the good fortune to meet some like-minded people with an interest in agriculture, such as Paul Taylor (at the end of his life), Merrill Goodall, Don Villarejo, Frank Bardacke, and Isao Fujimoto.

Bill Friedland has always been an inspiration, both as a scholar and as an adaptable but true-hearted survivor of the Old Left; he also gave me valuable support as a young fellow finding my way in academe. I owe special thanks to Jeff Lustig, who led me to many of the old farmhands, and who founded the California Studies Association and started its series of incredibly educational conferences. I also thank Ellen Widess for bringing me into the circles of pesticide activists and introducing me to the likes of Don Dahlsten (who recently passed away) and Michael Storper, who sucked me into the Peripheral Canal controversy and led me to know such activists as David Brower, Phil Williams, and Bob Gottlieb.

On the Berkeley campus, I have had the good fortune to be a colleague of Michael Watts and Gillian Hart and to have shared many wonderful students with them. Through long discussions, oral exam preparations, and dissertations on distant places, agrarian studies and political ecology became lodged in my bones. Bringing it all back home were students working on North American agriculture and resources, such as Brian Page, Scott Prudham, David Igler, Jake Kosek, Jim Buckley, Greig Guthey, Kate Davis, Ellen Liebman, James McCarthy, Geoff Mann, Daryl Birkenfeld, Paul Sabin, Joan Cardellino, and Jen Sokolove. Jason Moore and Kathy Johnson have been an inspiration in other regards. They have been the best students money couldn't buy, and have taught the teacher more than they realize. I count them all as the best of friends.

Geographers seem to have a special obsession with California agriculture, probably because it is such a perfect example of regional difference, the intersection of nature and culture, and the mobilization of natural resources on a vast scale. I owe a begrudging nod of thanks to my late colleague Jim Parsons in this regard—since he and I disagreed on almost everything except a mutual fascination with our home state. Don Mitchell's eloquent and clear-headed assault on injustice in the agrarian landscape has certainly been a beacon. But two former students, George Henderson and Julie Guthman, bear special responsibility here, because I never intended to write this book and they caused me to do it by force of inspiration—George by doing capital theory better than I ever have and Julie by dragging me out to look at the real world of farms.

Today California agriculture has come back into focus for a different reason: the compelling movement for better, safer, healthier food. Even if agribusiness has captured much of the organic market and consumer politics are sorely limited, there is something irresistible about the demand for good food as it runs into the wall of unnatural practices that today dominate agriculture in America. Clearly, present-day food politics have influenced my approach to the subject. Some parts of the California agriculture story are well known and have been well told, like irrigation, farm size, the role of the university, and the mechanization of tomato harvests. Yet many other crucial facets, such as consumption habits and the role of supermarkets and restaurants, have barely been touched by historians. Nor has the biological foundation of farming in California been sufficiently considered in most critical studies, so I have included plant breeding, animal feeding, biotechnology, and bees to rectify this omission.

I believe this sketch of California's agrarian history and geography has the virtue at least of giving form to an immense amount of material, which might start another scholar on his or her way to filling the still-yawning gaps in our understanding of an amazing place—and its astonishing run of fortune. I don't think I've gotten too many things wrong, even if I haven't always got it right. I am well aware of the shortcomings of my own comparative knowledge, as of the pitfalls of local history. It is so easy to get wrapped up in the twists and turns of fate and the facts at hand, and to start to believe the exaggerations—both good and bad—about the place. Writing on California is replete with those. Hyperbole is never more than a stone's throw away. I've tried to temper such proclivities while still emphasizing what is, in fact, distinctive, unique, or advanced in the case of California agribusiness.

I also want to pay special tribute to two friends and inspirational thinkers, Mike Davis and Gray Brechin. While they have been known to take a wrong turn on the Highway of Hyperbole themselves, they have utterly changed the way thinking people view California. No one since Carey McWilliams has so nimbly straddled the line between journalism and scholarship on the Golden State, or given the Left a greater voice in the politics of this amazing, galling place. I am, as always, in debt to my mentor, David Harvey, for teaching me to get my head around Marx, philosophy, and geography when I was a student at Johns Hopkins.

As an academic, I am constantly on notice about the need to do comparative studies and to theorize at the highest levels of abstraction. But I took a fateful turn a while back by becoming mired in California studies, and trying to do serious historical geography and political economy on the place. I am painfully aware of how difficult it is to do regional studies as comparative social science because of long-standing schisms in history and geography, in which the local is seen as the province of the idiosyncratic, and the national or international as the arena of true social science. Even where the writer has used a local "case study," s/he always shies away from seeing it as distinctive in its own right, rather than as symptomatic of some greater truth. Andrew Sayer and Doreen Massey led me to understand that the two things, locality and abstraction, are not mutually antagonistic, but reinforcing.

I want to give special thanks to those who read the first draft and gave me valuable suggestions for revision: Michael Watts, Michael Johns, Frank Bardacke, Charles Post, George Henderson, Julie Guthman, and Bill Friedland. That's real friendship in an overworked academia, let me tell you. I should acknowledge the wonderful assistance of the Bancroft Library staff and the staff of the Baker book search system in the UCB library. I received terrific research help along the way from Anchi Mei, Madeline Solomon, Julie Guthman, Kate Davis, and Sebastian Petty. I have only nice things to say about my publishing staff at The New Press: Colin Robinson, Abby Aguirre, Sarah Fan, and Lizzie Seidlin-Bernstein. I want, as well, to thank those special friends who have been a constant source of encouragement: Joe Matthews, Iain Boal, Tim Clark, Anne Wagner and others in the Berkeley Retort Group, Erica Schoenberger, Bob Brenner, and Annie Girard-Ducasse.

THE CONQUEST OF BREAD

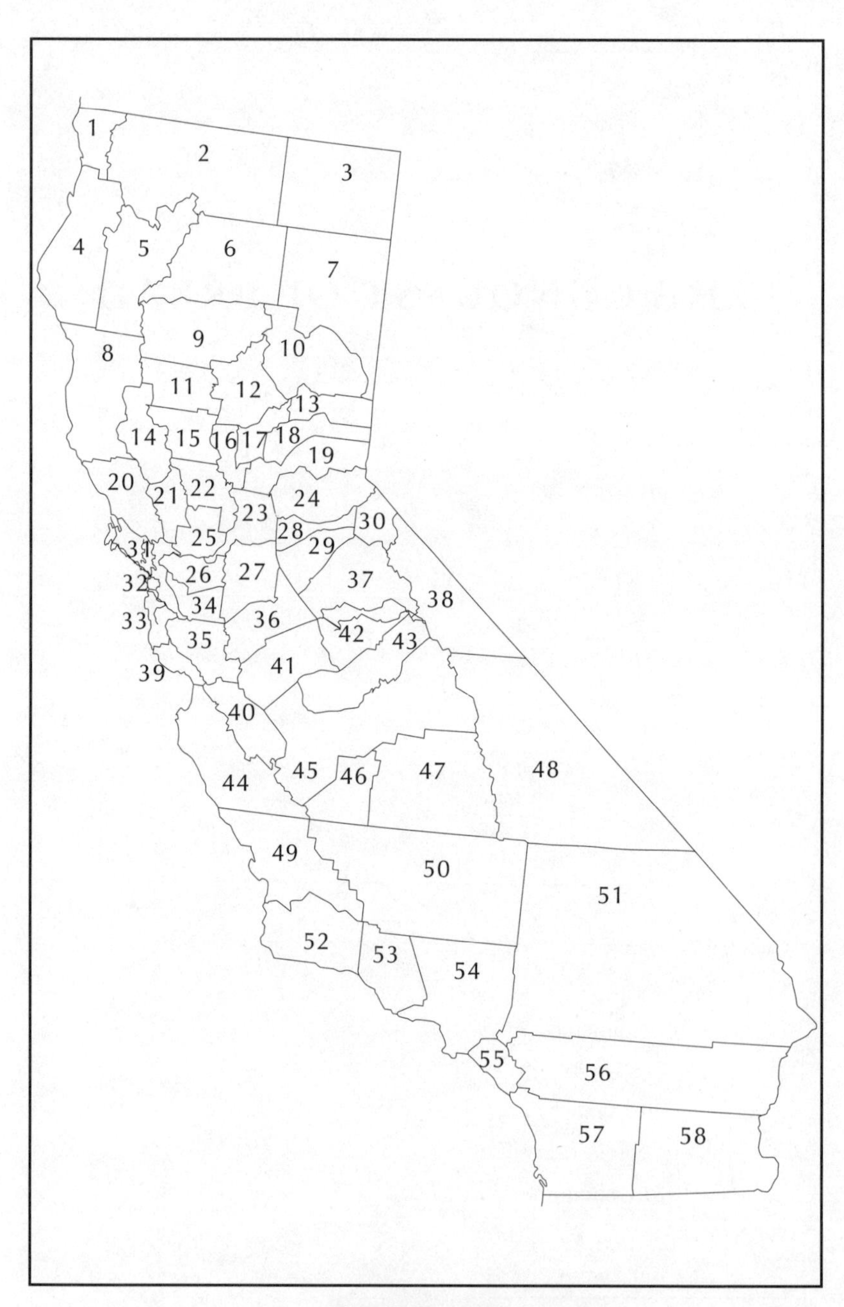

California counties

County Names

1	Del Norte	30	Alpine
2	Siskiyou	31	Marin
3	Modoc	32	San Francisco
4	Humboldt	33	San Mateo
5	Trinity	34	Alameda
6	Shasta	35	Santa Clara
7	Lassen	36	Stanislaus
8	Mendocino	37	Tuolumne
9	Tehama	38	Mono
10	Plumas	39	Santa Cruz
11	Glenn	40	San Benito
12	Butte	41	Merced
13	Sierra	42	Mariposa
14	Lake	43	Madera
15	Colusa	44	Monterey
16	Sutter	45	Fresno
17	Yuba	46	Kings
18	Nevada	47	Tulare
19	Placer	48	Inyo
20	Sonoma	49	San Luis Obispo
21	Napa	50	Kern
22	Yolo	51	San Bernardino
23	Sacramento	52	Santa Barbara
24	El Dorado	53	Ventura
25	Solano	54	Los Angeles
26	Contra Costa	55	Orange
27	San Joaquin	56	Riverside
28	Amador	57	San Diego
29	Calaveras	58	Imperial

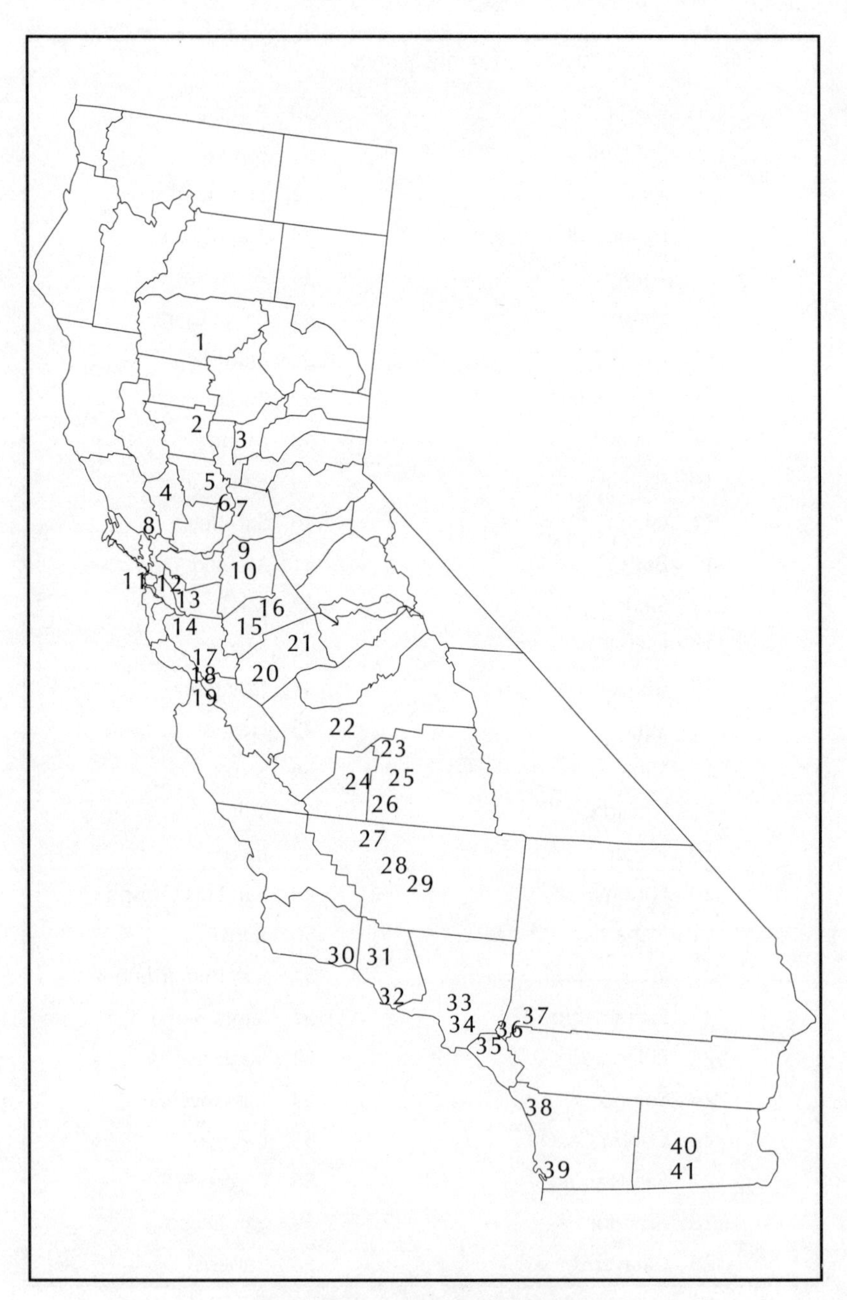

Some California cities and towns

City Names

1	Corning	22	Fresno
2	Colusa	23	Dinuba
3	Marysville	24	Corcoran
4	St. Helena	25	Visalia
5	Woodland	26	Pixley
6	Davis	27	Wasco
7	Sacramento	28	Bakersfield
8	Petaluma	29	Arvin
9	Lodi	30	Santa Barbara
10	Stockton	31	Ojai
11	San Francisco	32	Oxnard
12	Oakland	33	Pasadena
13	Hayward	34	Los Angeles
14	San Jose	35	Anaheim
15	Patterson	36	Chino
16	Modesto	37	San Bernardino
17	Gilroy	38	Oceanside
18	Watsonville	39	San Diego
19	Salinas	40	Brawley
20	Los Banos	41	El Centro
21	Merced		

California's agricultural valleys

Introduction

Unlocking the Secrets of Agribusiness

California agriculture is an extraordinary thing. California is probably the most intensely farmed landscape in the world outside the tropical rice zones and the most prosperous agricultural region of the advanced industrial nations. It accounts for roughly 12 percent of agricultural output in the United States and supplies one-third of the table food consumed by Americans. By 1870 the value of California's farm products, $50 million, exceeded that of gold. By 1925 California passed up Iowa and briefly had the highest value of farm output of any state, $87 million. Over the last half century, no other part of the country even came close. In 2000 the value of California's farm output was $24.6 billion; Texas was second at $13.2 billion and Iowa third at $10.9 billion. Fresno County alone would rank fourth among the *states.* Farm income differentials are even higher: in 1930, for example, California farms generated roughly three times as much output value as the national average. As geographer James Parsons once put it, California agriculture is "one of the wonders of the world."[1] In food processing, California has also led the United States for over half a century and its lead is even more astonishing than in farming; in 1997, the state shipped over $50 billion in food products, while Illinois, ranked second, shipped only $10 billion.

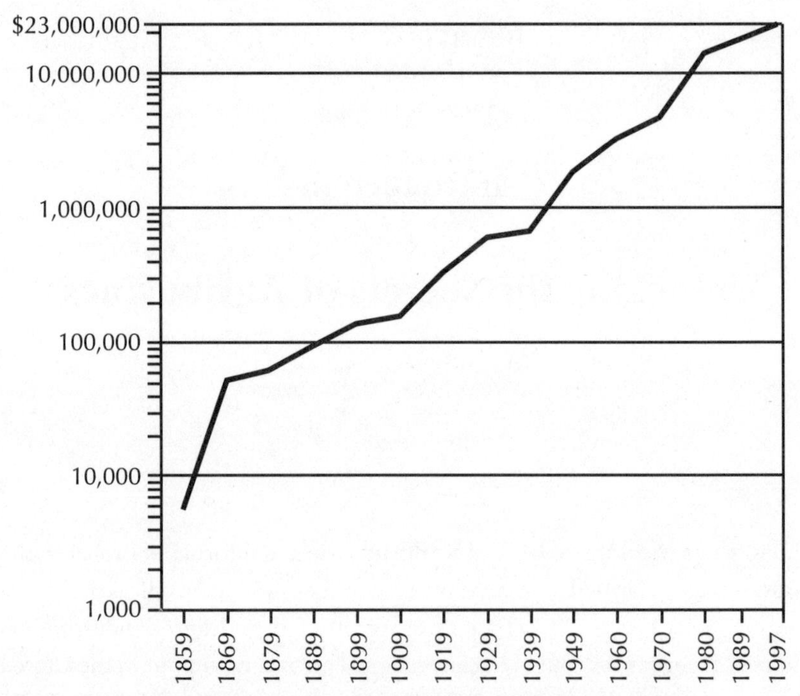

Note: 1859–1949 Value of Farm Production, 1960–1997 Value of Agricultural Goods Sold.
Sources: U.S. Census of Agriculture (various years), Historical Statistics of the United States.

California agricultural production, 1859–1997

One might think that California agriculture would therefore be a fixture in the panorama of agrarian studies, yet it has not been. Most attention has been fixed on the temperate granaries, ranches, and feedlots of the Midwest and Northern Europe. On a global scale, it is a neglected subject compared to the plantation zones of the tropics and subtropics, the long-suffering peasants of the underdeveloped world, and today's worldwide subcontracting networks that supply flowers and fresh fruit to the global north. In the great debates over modernization, capitalism, and agriculture from V. I. Lenin to Susan Mann, California barely rates a mention. Even the major theoretical works of the last generation on agro-industrialization and global agriculture, such as those of

Top Ten Farm States 1900–1997 (in value of output)

1900	1910	1920	1930*	1940
Iowa	Illinois	Texas	Texas	Iowa
Illinois	Iowa	Iowa	Iowa	Texas
Ohio	Texas	Illinois	California	California
Texas	Ohio	Ohio	Illinois	Illinois
Missouri	Missouri	California	Kansas	Minnesota
Kansas	Kansas	Kansas	Nebraska	Ohio
Nebraska	Nebraska	Missouri	Minnesota	Wisconsin
Minnesota	Minnesota	Nebraska	Wisconsin	New York
Wisconsin	California	Minnesota	Missouri	North Carolina
California	Wisconsin	North Carolina	Ohio	Missouri

1950*	1960	1970	1987	1997
Texas	California	California	California	California
California	Iowa	Iowa	Texas	Texas
Iowa	Texas	Texas	Iowa	Iowa
Illinois	Illinois	Illinois	Kansas	Nebraska
Minnesota	Minnesota	Nebraska	Nebraska	Kansas
Nebraska	Nebraska	Kansas	Illinois	Illinois
Kansas	Kansas	Minnesota	Minnesota	Minnesota
Wisconsin	Missouri	Missouri	Florida	N. Carolina
Ohio	Wisconsin	Wisconsin	Wisconsin	Florida
Missouri	Ohio	Ohio	N. Carolina	Wisconsin

* California was briefly #1 in 1927 and 1947

Source: U.S. Census of Agriculture, various years.

David Goodman, Harriet Friedmann, and Philip McMichael, give California short shrift. So do such popular writers on food as Eric Schlosser and Michael Pollan.[2] It is a curious neglect, given that many of the principle features of 21st-century agro-food systems, such as subcontracting, brand names, petro-farming, feedlots, biotechnology, and concrete dams, were pioneered here.

There is, of course, a rich literature on California agriculture, much of it path breaking, written almost exclusively by people who have lived here at one time or another. Henry George, Carey McWilliams, and Frank Norris gained national acclaim, and live on in the annals of American literature, journalism, and social criticism. More recently, California agribusiness has inspired exceptional studies by scholars such as Donald Worster, Miriam Wells, and Vickie Ruiz, books that are highly regarded in academic circles, and sometimes reach larger audiences. These authors have shed a great deal of light on California's agrarian order at different times and in various dimensions. They have written with great eloquence on labor, irrigation, finance, canning, and other aspects of the agribusiness world.[3] Like any other author, I stand on many shoulders to get a better view of my subject.

So why revisit California agribusiness? Above all, because these studies, for all their excellence, focus on only one or two aspects of the agribusiness complex; no one since Lawrence Jelinek has attempted a systematic overview with a unifying vision.[4] In addition, there are several dimensions of the problem that scholars persistently get wrong—for example, by placing undue emphasis on large land holdings, farm co-ops, and irrigation—while others, like supermarkets, farm management, and nurseries, remain neglected. In this book, I hope to broaden and enrich our view of what constitutes the domain of agribusiness, and tie together the disparate parts of a sweeping, turbulent history. Furthermore, my approach will draw on the theoretical toolbox assembled by recent agrarian studies, making use of such concepts as commodity chains, nature-based production, circulation of capital, and agro-industrialization. I will, however, push the theoretical envelope in arguing that the one unified idea capable of encompassing the breadth, depth, and dynamics of California agribusiness is *agrarian capitalism*.

We shall commence, in chapter 1, with the most striking and self-evident part of California agriculture: the profusion of crops. This cornucopia has been synonymous with the state's image for over a century. The mind boggles at the plenitude coming off the land, and at the rapidity with which the fertile possibilities of the state have been exploited. Cropland and grazing land expanded at a remarkable rate from the outset, and agriculture soon spread across the valleys and hills of the whole state. It is a modern wealth of nations manifest in fruits, eggs, and vegetables. This fecundity sets the table for everything else to follow, but explains nothing by itself.

The crops in all their worldly riches are cyphers for something else. Adam Smith would have said that they represent the power and profusion of markets. Commercial farming begins and ends with the market, and presses forward the general commodification of food, land, and labor. The farmer buys land and hires wage labor (commodity inputs) and turns around to sell a crop (commodity output). The commodification of farming is a basic point of reference in contemporary agrarian studies. As Jack Kloppenburg points out, the long-run trend has been to bring all farming under the rhythms of the market, even seeds, which today constitute a multibillion-dollar business worldwide and entrap farmers in a commercial net of payments and debt.[5]

As we shall see in chapter 2, California not only harvested grain and fruit, it quickly ripened the markets in land and labor. Land had to be brought into the commodity nexus after the Gold Rush, wrested away from the rancheros of the Mexican era and privatized by the state and federal governments. This transition marked the early years of the state and its social order, provoking critiques from Henry George to Tomás Almaguer. The commodification of land continued through several generations of drainage, subdivision, and concentration. Labor, too, was rapidly commodified, and California agriculture became principally a hired-labor system, not a family-farming nexus. This is a crucial divide in agrarian history that would only become more dramatic over time. California amassed the country's largest agricultural wage-labor force, a motley crew of immigrants, tramps, women, and children, into a huge, mobile army of harvest workers, whose exploitation has shocked observers from the days of John Steinbeck to the era of Cesar Chavez. Farmworkers, too, would be continually recommodified in a way that left workers naked before the market and subject to extreme levels of exploitation. What is striking about California agriculture is how the processes of commodification in crops, land, and labor continues unabated to this day. What Adam Smith first called "primitive accumulation" is not confined to the distant past.[6]

Still, it takes more than apples and oranges, land and labor, to make modern agribusiness in California. So we proceed in chapter 3 to the site of commodity production, the farm. The farm is fundamental to agrarian studies because on-farm production appears as something vastly different from modern industry. In place of the factory and the industrialist, one normally finds farmers with a clear sense of class separation from the city capitalists. The emergence of capitalist out of peasant (or family) farming is the oldest obsession of agrarian theory, going back to the Lenin-Kautsky-Chayanov debates of a century ago. It has sometimes been held that farming virtually demands precapitalist relations of production—small holders, family labor, and self-exploitation—by virtue of the barriers to productivity growth and labor displacement created by the exigencies of natural rhythms of planting and growing. Echoes of this debate still ring today in discussions of agricultural transformations around the world. But the class position of farmers is less controversial in California because they have mostly been fully market-oriented employers of wage labor, as observers from R.L. Adams to Stephen Stoll have born witness. They are, in short, business people, or what in California are known as *growers*—not farmers.[7]

Another magnificent obsession of agrarian theory has been the importance of nature in agricultural production. The direct confrontation with nature makes agriculture different from all other sectors of the modern economy, as argued most tellingly by Mann and Dickinson. Abstractions of the market, price and marginal cost, production functions, and easy substitution of inputs just don't wash out in the hot sun and along dusty country roads. Farm production is where biology, geology, and labor come together in fruitful ways, or wither on the vine under the force of nature's vicissitudes. This is as true in California as anywhere, as seen by writers such as Richard Sawyer and Julie Guthman.[8] So, in chapter 4, we shall engage the fundamentals of soil, plant and animal stocks, water, and pests. This story begins with a long process of discovery of the natural parameters of farming in California's environments, followed by increasingly urgent attempts to raise farm productivity. Making biological organisms jump through productive hoops has not always been easy, however, and the growers have often had to turn to modern science in order to unlock nature's secrets—a process that continues in today's biotechnology revolution.

In popular imagery, California's abundance is attributed directly to the munificence of nature: sunny Mediterranean summers, deep alluvial soils, and the Sierra snowpack. In agrarian studies, on the other hand, nature and the biological pivot of farming are usually portrayed as a problem, not a benefit. Nurturing biological organisms—plants and animals—through long growth cycles is particularly difficult, and not amenable to factory-like production methods.[9] Neither view is correct. The abundance of California agriculture has issued from two human sources. On the one hand, the artful manipulation of organisms, soil, and water has made it possible to convert natural potential into a wealth of human-made commodities, bearers of prodigious market value. On the other hand, the wealth of production has only been possible because of an abundance of cheap human labor, worked hard under dire conditions and kept in a permanently degraded state. The farm labor process occupies the last part of the fourth chapter.

One cannot stop with the farm. One must follow the commodity chain forward and back in order to locate the main redoubts of modern agribusiness. After all, on-farm production today accounts for only about one-tenth of the value added to finished agricultural products in the advanced economies; one-quarter comes from inputs, and two-thirds from postfarm processing. Bill Friedland deserves credit for being the first student of food systems, a generation ago, to drag commodity chains out of the storage shed in order to link up the fields with the factories and distributors in a systematic way. This idea has been greatly expanded upon by David Goodman, Bernardo Sorj, and John Wilkinson, who argue that the central dynamic of modern agriculture has been an ever-widening social division of labor. This insight enlarges, enhances, and complicates the simple picture of on-site production that has prevailed since Adam Smith, and it underlies the two longest chapters of this book.[10]

Chapter 5 treats the upstream side of the commodity circuit, or the provision of inputs to farming. This includes the supply of plants and animals, water, machinery, feed, and fertilizer. It introduces a set of inputs that are themselves industrial commodities. That is, agribusiness is not only a commodity chain but an extended production system. Inputs are usually produced somewhere away from the farm, and involve several stages of production and commodity flow (not to mention scientific input). Such production systems stretch the term "industry" in ways that defy normal classifications, since some portion of the chemical and biotechnology industries, for example, are integrated with agriculture. The term is bent even farther because the production of water or seed is, by their natures, not carried out through conventional manufacturing processes. California has developed the provision of such modern agrarian inputs to a high degree, in a system that reaches far beyond the farm and into the mountains, oceans, deserts, and cities. The massive reconfiguration of the natural landscape to suit the needs of agribusiness for profit has been the subject of trenchant study and criticism from Robert van den Bosch to Donald Worster.[11]

The even-more-extensive downstream—or post-farming—side of the commodity circuit will be investigated in chapter 6. California agriculture has flourished through the development of the downstream stages, from coveys of canneries to swarms of supermarkets. Moving crops to market was the first step, and scholars such as Stoll and Victoria Woeste have devoted a great deal of attention to the reworking of wholesale marketing through growers' cooperatives. They have paid much less attention to three other steps, however. The first is processing, which has become more and more elaborate—and synthetic—with time. The second is retail distribution, by which the final commodity, fresh or processed, is sold in grocery stores, supermarkets, and farmers' markets. Today, moreover, the retail end is commonly a restaurant and fast-food outlet, where food undergoes yet another processing stage, the manufacture of meals. At the end of the chain stands the consumer, and California agriculture has prospered by the conquest of national and international consumer markets. This achievement raises all manner of thorny questions about consumer sovereignty and its manipulation that have long beset economics, and recently embroiled cultural studies.[12]

The production chain and its extended division of labor constitute the larger domain of what may be called *agro-industry.* As Goodman, Sorj, and Wilkinson observe, large portions of the agrarian labor process have been shifted off the farm and into the factory. The whole of the agro-production complex employs ten times the number of people as farming. While only a minuscule one-fiftieth of Americans work on farms today, over one-fifth work in the food system as a whole. This agro-industrial complex provides the key to industrialized agriculture in California. As Jelinek remarks, "California became the first and most complete example of industrialized agriculture in the nation." Yet this is less a matter of factories *in* the fields, as Carey McWilliams called them, as of factories and fields *working together.*[13]

Agro-industrialization is central to any explanation of the immense dynamism of modern agriculture. As Charles Post has shown, the development of the United States in the 19th century cannot be understood apart from this vital interplay of agriculture and industry. With the help of plows, harvesters, railroads, canneries, and the rest, farm productivity has been raised much faster than would have been possible by better farming practices alone. This has been true in California, in spades. Alan Olmstead and Paul Rhode argue convincingly that technological change has been at the root of California's agricultural fecundity, not just sunshine and good soil. This technology has come from all points of the industrial compass (especially from local suppliers and processors), as well as out of research laboratories. California has, not surprisingly, led the way nationally (and globally) in several key arenas of agro-industrial innovation, such fruit drying, harvesting, animal feeding and breeding, irrigation, and canning.[14]

A puzzling side of California agribusiness and its immense division of labor is the relatively modest role of large corporations in knitting it all together. DiGiorgio, Safeway, and Del Monte are all big players, but none dominates in the manner of Cargill in midwestern agriculture, let alone General Motors and Ford in automobiles. The story of California agriculture is not a recapitulation of the rise of large corporations chronicled by Alfred Chandler, although the first large agribusiness corporation in history, Miller and Lux, arose here. Since the 1980s, there has been a major rethinking of organizational theory in business and economics. This has been sparked by the resurgence of small firms in high tech, expansion of global subcontracting, and the rediscovery of industrial districts. Out of this organizational revolution comes a new and more open language of business networking, contracting, and alliance that suits California agriculture very well.[15] Agribusiness had developed flexible production before flexibility became a buzzword of the New Economy, as first seen by Margaret FitzSimmons. This has been apparent for some time in global studies of food systems, as well. In short, the search for business concentration and monopoly power that has entranced so many observers of California agriculture turns out to be less interesting than a host of interactions between parts of the agro-industrial complex, city and country, farm and factory.[16]

There remains a missing term in our presentation of commodity chains and production systems in California agriculture. In and through the whole runs an invisible thread, capital, which weaves together all the elements of the agribusiness system. How does capital perform this magic? It does so, principally, by *circulating* through all the elements of purchase, production, and sale, looking for profit (and interest). George Henderson did the great service of introducing the analysis of finance capital into agrarian studies in California. We shall explore the role of money and finance capital in the development of California agriculture in chapter 7. There we shall find that agribusiness is not just about a cornucopia of crops, commodity production, business farmers, or agro-industries; it is, finally, about the mobilization, circulation, and accumulation of capital.[17]

Capital is more than money, however. It is a class phenomenon, based on the ownership of the means of production. If we are to speak of capital in the countryside in chapter 7, then we must also talk about the men and women who acted as the capitalists, and how they came together as a group to get things done, to form partnerships and political alliances. Agricultural florescence in California is thus more than crops, factories, and money: it is the collective achievement of a class of people, reaching across the boundaries between city and countryside, who made it their business to promote agribusiness and its interest statewide. Not that this agrarian capitalist class is self-evident: it has to be sought out through the organizational and personal traces it leaves, and through its collective actions.[18] This necessarily has involved government—state, federal, and local—as a handmaiden of capital. Such a claim raises ghosts of Max Weber, Karl Marx, and Karl Polanyi, and a century of debate over the relation between capital and the state in market societies. We shall come down solidly on the American side of the issue, because of the distinctive intertwining of local and state governments (the decentralized federalist system) with business and growth (the unrivaled national pastime).[19]

Human labor, too, runs all through the agro-industrial system. It has all too often been a missing term in popular presentations of California's agrarian splendors. Hence the phrase "California—the Beautiful and the Damned," used by IWW organizers in the early 20th century and picked up by Don Mitchell in his brilliant excoriation of labor's invisibility.[20] Farmworkers and other workers in the long chain of production have tried to oppose the class power of capital in the countryside, but without much success, as will be seen in the last section of chapter 7. Field workers' struggles for unionization have been particularly heroic, and make for a long-running California tragedy. Cannery and grocery workers have done better, but their voices have been little heard in the corridors of power relative to the chorus of the growers and their helpmates. Exposés of the labor system in books and movies such as *The Grapes of Wrath* have done more than anything else to publicize the dark underside of agribusiness and its ruthless application of class power, and they still hold pride of place in scholarly critiques of California agriculture. Nonetheless, even though the prosperity of agribusiness has been built on workers' backs, their story has to be seen in view of the vast structure as a whole, which occupies most of this book.

Many of the tools for analyzing California agriculture can be found in the literature on agrarian systems, business organization, theories of the state, and the like. But when all is said and done, there is still no unifying explanatory framework for California agribusiness that both pulls together the disparate parts and gives life to the dynamic forces behind agrarian development. Instead, we find widely divergent fixations: Henry George denounced the "land monopoly," Carey McWilliams the "factories in the fields," and Lawrence Jelinek "the empire of agribusiness." Donald Worster rejects these classic interpretations to highlight state-driven "rivers of empire." Bill Friedland refers to "manufacturing green gold," and Margaret FitzSimmons and Stephen Stoll speak more broadly of "industrialized agriculture." Alan Olmstead and Paul Rhode see technology as the fulcrum of agrarian progress, while George Henderson highlights the circulation of capital through the countryside. David Vaught hails the small farmers as the pivot of agrarian development, while Don Mitchell and Miriam Wells revive a vision of common labor as the moving force in the landscape.[21] What to do in the face of this brilliant cacophony?

In order to embrace the whole of the agrarian system in California, I have seized upon a single name, *agribusiness.* It is, on first glance, the best term to weave together all these disparate parts and to capture the moving force behind California agriculture, useful because it's popularly accepted, brings farming into juxtaposition with the rest of American business, and connotes the scope and scale of the agrarian enterprise.[22] But what do we mean by it? Agribusiness is handy, but it lacks a deeper analytical content. On that score, it would be well to recall a more classic term, *agrarian capitalism.* The countryside here is and has always been capitalist in modern times, as critical observers such as Friedland, Wells, Stoll, and Henderson have recognized. But the state's agribusiness has too rarely been designated as, simply, capitalist. Indeed, in official circles one goes to great lengths to make sure not to breathe a word of it. Here's how the editor of the official history from the University of California tiptoes around the question:

> California represents in microcosm the evolution of agriculture from pastoral grazing through extensive dryland grain farming into today's complex, intensive specialty cropping with irrigation . . . The history of California agriculture

> is the development of an enormously promising but challenging land by the restless sons and daughters of a pragmatic culture. Certainly the state's farm scene is a product of a unique historical past as well as a rare combination of topographical and climatic conditions. California was a mecca for dreamers and the dispossessed from the days of the Gold Rush onward. The state was open for settlement during the beginnings of an era of revolutionary technological change, and agricultural expansion took place in an opportunistic and nontraditional political and social climate.[23]

So there we have it in a nutshell: pragmatism, restlessness, uniqueness, topography, dreamers, the dispossessed, technology, modernism. Every evasion, just to avoid calling it what it is: agrarian capitalism.

I hope to establish, once and for all, that California agriculture is capitalist through and through, and has been from the outset of the American era. Indeed, it is one of the purest cases of capitalist agriculture in the world, inserted quickly into the world market at a most auspicious moment: the conjunction of the California Gold Rush, rapid antebellum expansion in the United States (faster even than the post–World War II spurt), and the Victorian boom across the Atlantic. It has only become more thoroughly capitalist with time, and has grown, divided, expanded, and flourished at an astonishing rate for a century and a half. Full-throated capitalism in the countryside made California the richest agricultural region of the 20th century. A fully evolved capitalism embraces all the elements so far enumerated as the key dimensions of agribusiness.

Why call it capitalist—what difference does that make?[24] To begin with, California agribusiness shares the elemental features of the capitalist mode of production: a small class owning the land (the means of production), the exploitation of wage labor, and production for profit. This I call the Basic Triad of capitalist production, consisting of capitalists, workers, and profit making. Capitalists in today's economy appear in the guise of companies, usually limited liability corporations; but behind all companies are a class of owners of stock and managers of the enterprise. But the Basic Triad is only a starting point, and not a satisfactory characterization of a full-blown capitalist system.

For that we need to add a second layer of concepts, which I will call the Basic Model of the capitalist economy. This is an expanded account of both the features of capitalism and the propulsive forces it unleashes.

The Basic Model starts with *commodities,* or marketable goods and services, and the tendency of capitalism to turn anything and everything into something to sell because in commodity form goods have measurable *value,* the source of capital and profit. Commodities must be sold, however, to return *money* (the universal form of value) to their owners. Commodification begins with outputs but comes around to inputs, particularly land and workers, which come to be freely traded in property and labor markets even though they are not produced commodities at all, but the basis of life; capitalism is extremely corrosive of all preexisting social relations over land and labor, in a process called *primitive accumulation.* The third element in this mix is the use of money gained from the sale of commodities to expand the means of production and labor force, that is, investment. The return of profit as investment, in an ever-growing spiral, is the sine qua non of capital: *accumulation.* Accumulation is the overarching purpose of all capital and capitalists. In order to accumulate, however, capitalists need to produce commodities—in ever-increasing number, flooding the world with goods and services.

Continuing with the Basic Model, we find the mobilization of human labor in production by the capitalists in order to make commodities. This is the meeting place of work and nature, in what is usually called *the labor process.* Under the spur of capitalist profit seeking and competition among capitalists, there is a constant pressure to improve upon the labor process, whether by minimizing the use of materials or by producing more output in the same period of time. This improvement shows up principally as *rising labor productivity,* but also as higher productivity of capital in the form of plant and equipment. Altogether, this process of increasing production has come to be known as *industrialization* (or industrial revolutions). Such industrial advance rests on a number of things characteristic of modern industry. One is the bringing together and direction of collective labor, usually in factories and offices but equally at construction sites and in agricultural fields. Another is the *rationalization of the labor process,* or what is often called the detail division of labor but includes the whole range of task assignment, sequencing, timing, layout, and so on at the work site. Yet another aspect of modern industry, much ballyhooed in the history of industrial revolutions, is the application of mechanical aids to human labor, or *mechanization,* which has become (with advances in regulation and feedback in machines) a system of *automation.*

This much is well known by students of modern industry, or manufacturing, and shows up clearly in the progress of modern agriculture, as well. A few supplementary elements should be added, a third layer of concepts making up an Advanced Model of capitalism. These pertain to the whole of the industrial system. To begin with, there is a *social division of labor,* which extends across factories and workplaces, and, indeed, across companies and industrial sectors. Modern industry is an immense weave of inputs and outputs constituting complex commodity chains and production systems. Furthermore, the sum of the industrial arts, commonly known as *technology,* consists of more than factories, machines, and divisions of labor. It requires, above all, human knowledge and practical labor skills, and advances in technology are commonly made through human *learning,* not just better engineering. It further demands inputs of abstract knowledge, or *science,* converted to practical ends through what is known as *research and development.* One should not forget, however, that just as factory buildings and equipment demand investment, so do science, learning, and innovation; and they, too, are driven by flows of capital.

Finally, the ever-greater output of commodities in the system needs to move smoothly from step to step, place to place, beginning to end, and around again. That is, an Advanced Model must include *circulation* as well as production. That circulation embraces commodities chains, commodity transport and flows, and the institutions of wholesale and retail commerce. This circulation intersects production along the way: selling targets not just final consumers but intermediate producers (like farmers); transport involves not just final markets but railheads and warehouses; and retailing may involve making additional changes to the product (like cooking dinners). But most of all, circulation involves money. Circulation through the (agro)industrial system begins and ends with money. Commodities exchange for money. Production requires money. Workers are paid in money. Above all, the investment of capital sets in motion the buying, producing, and selling, all for the purpose of making profit. Successful circulation is, then, the accumulation of capital. The preeminence of money and the flow of value from investment to profit is what makes the system not just a set of commodity chains and production systems, but a tapestry of *capital chains.* And the great weaver of capital flows is the financial system.

Putting this all in a diagrammatic form, we can reconstruct the outline of the book, chapter by chapter, in the following manner. Readers whose eyes glaze over at the sight of symbols, even pseudo-mathematical ones like these, may want to skip ahead.

We begin with the crops, $\mathbf{C_o}$, and the basic inputs, labor and land, $\mathbf{C_i}$. These flow in and out of the farm, $\mathbf{P_f}$, thus:

$$\mathbf{C_i - P_f - C_o}$$

The diagram above represents the simple circuit of agricultural production. But this can be amplified to include the upstream part of the chain of inputs and input production, $\mathbf{P_i}$ - $\mathbf{C_i}$, and the downstream chain of outputs and processing, $\mathbf{P_o}$ - $\mathbf{C_o}$, making the circuit of agro-industrial production

$$\mathbf{P_i - C_i - P_f - P_o - C_o}$$

That circuit requires that the final output be sold to realize its value in money, **M,** by the uptake of household consumption, $\mathbf{C_h}$, thus:

$$\mathbf{P_i - C_i - P_f - P_o - C_o - M - C_h}$$

But the point of the chain is not just production for use, or exchange for sale, but investment of capital to make more capital, to wit:

$$\mathbf{M_k - P_i - C_i - P_f - P_c - C_o - M'}$$

where $\mathbf{M' > M_k}$ (on an annualized or other amortized basis).

In fact, the simple chain depicted is more like a matrix or network of intersecting chains, with capital flowing in and out of every crossing.

Our modeling exercise does not end with the circuits of capital. It has a final layer, going beyond the merely economic and making for a Social Model of capitalism. That last layer is no mere frosting. It demands the reintroduction of *human action,* individually and collectively, to make the wheels turn (whether the actors understand the full effects or not). So any number of key names and events must be added to our barebones outline of California's agrarian development. Moreover, as a model of capitalism, it requires that these people form into *classes* around their position in production, property holding, and labor markets, and that they come into conflict over their differing needs and interests. *Class formation* and *class struggle* are basic components of a dynamic theory of capitalist society. Finally, these social relations, actions, and mobilizations inevitably bring into play the state and its many arms, as the cyborg of capital—a powerful extension and hybrid of class power.[25]

The question asked at the beginning still hangs in the air: why study California agribusiness? Is this simply a local preoccupation, or does it have something to tell the wider world? At the very least, a thorough treatment of California should be useful to anyone thinking about modern agriculture in advanced economies. It provides a systematic overview of an agro-industrial complex and its dynamics over many decades, and that should assist others in working out the dimensions of the agrarian economy and the basic capitalist logic of it. California, in this sense, is just a good example of an advanced agrarian production system, and one that rests on a remarkably pure form of the capitalist system. I would go further, however, by saying that because of the early installation of capitalist production—i.e., a commercial, businesslike, and industrial agriculture—California represents an historical vanguard of sorts, in which many features of agribusiness now found around the world were originally worked out (or very nearly so), from the irrigation district to scientific agronomy to mass marketing. In many ways, the rest of the country and much of the world have caught up with California agribusiness and its practices, but California still offers an amazing record of agrarian development.

I make an additional claim, however, concerning the geography of place and the theory of agrarian capitalism. There is inevitably another layer of analysis beyond a "pure theory" of agrarian capitalism, one that attends to the way history and geography unfold together. It pertains to the peculiarities of place and the need for comparative analysis of variations in actually existing capitalisms, something that has become a necessary part of economic geography and international studies in the wake of the competition among the United States, Asia, and Europe in our time.[26] It is readily apparent that things like financial systems, labor relations, and government involvement in the economy vary considerably from nation to nation, and even from region to region, or continent to continent. China is decidedly not like Spain. But there is more than variation involved here: there is the accumulated weight of history, class struggles, and political decisions. That is, capitalist development has followed different trajectories, which are not easily undone, and this is no less true of agrarian capitalisms.[27] Telling those stories involves something more creative than a stripped-down modeling exercise; it means painting a picture on the broad canvas, a panorama of historical development. This is what I hope to do in this book.

1

Cornucopia

The wealth of modern nations presents itself as an abundance of goods, as Adam Smith—himself one of the first modern farmers—demonstrated in *The Wealth of Nations,* considered by many the bible of political economy. Karl Marx starts his investigations on the opening page of *Capital* with the same image: the wealth of nations is, in the first instance, "an immense accumulation of commodities." So it is with the wealth of California agriculture. It shows up as a phenomenal quantity, variety, and value of crops and food products. California has long grown the greatest variety of crops by far of anywhere in the United States. Commentators are in the habit of saying that the state grows everything from A (avocados) to Z (zucchini); a book-length report was even prepared at the University of California, Davis, just to describe the state's two hundred–plus crops. No other state can boast more than a couple dozen. Moreover, it has been the largest producer of canned goods of any state for over a century, and the biggest generator of processed food for half a century.[1]

California has, through most of its history, been represented by fantastic images of agricultural abundance. One is a color lithograph poster, used to draw settlers in the 1880s, of California as the Cornucopia—the legendary

Horn of Plenty. Another popular image is the delirious fruit crate label of the early 20th century, of which there were thousands. Yet another is the Edenic vision of the southern San Joaquin Valley that was seen by the Joad family as traveled from Oklahoma to California in the 1940 film *The Grapes of Wrath*.[2] Most people experience the power of agribusiness more simply, through the everyday abundance of California products on supermarket shelves or restaurant menus, along with delightful labels on cans and boxes or imaginative names for the chef's specialties. Nonetheless, the profusion of crops is even greater than most people realize.

The sheer variety of farm output has led many to refer to California agriculture as a separate species of farming: "specialty cropping." This term, which implies niche marketing, does not do justice to the vast quantity of every commodity coming from the state's farms. If this variety of goods were the product of modern manufacturing, no one would be surprised by either the number of goods, their various qualities, or the sheer volume produced. Indeed, if one were referring to processed foods, such as cans of corn or cartons of milk, industrial images would make perfect sense. Why the surprise when such profusion comes from the land instead?

The marvelous produce of California agriculture would sit comfortably at any of the great modern industrial expositions from the Crystal Palace exhibit in London in 1851 to present-day trade fairs like the annual Comdex Convention in Las Vegas and the Paris Air Show. In fact, the wonders of California agribusiness have been put on exhibit in just such a way, at several world's fairs. One of the first such displays occurred in San Francisco in 1915, to celebrate the opening of the Panama Canal and the rebuilding of the city following the great earthquake of 1906. The Panama-Pacific International Exposition drew hundreds of thousands of visitors from around the state, the country, and the world. What would they have seen that bore witness to California's wondrous agrarianism?

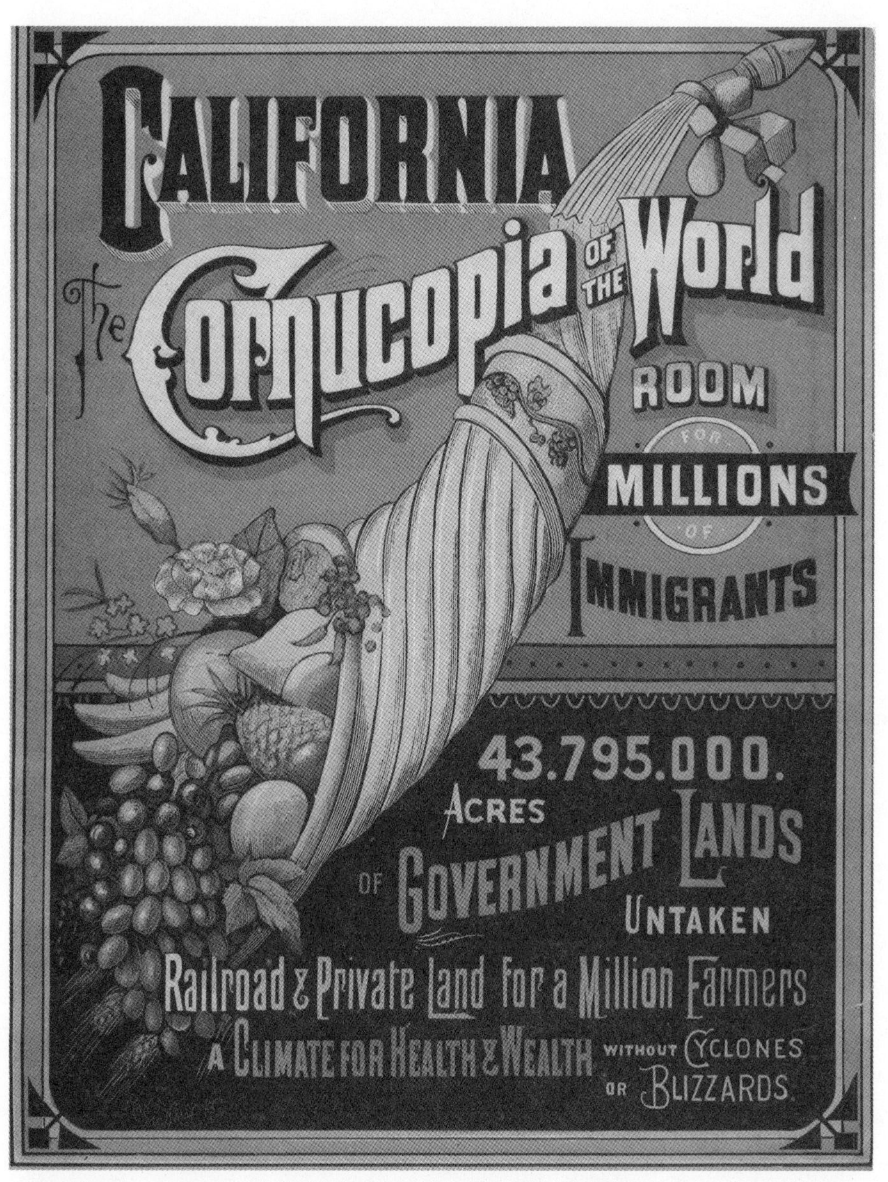

"California, the Cornucopia of the World," California Immigration Commission, 1885

ALL THE WORLD'S A STATE FAIR

A visitor entering the grounds of the Panama-Pacific International Exposition would have walked into a fantasyland of immense proportions: hundreds of acres of buildings and grounds where San Francisco's Marina District now stands. It was like stepping right into the Cornucopia poster or the lithograph off a Sunsweet orange crate—an astonishing array of taffeta domes, columned walls, *grand allées,* and magnificent flower beds. At center stood the 430-foot-high Tower of Jewels, a calico festival of illumination meant to be San Francisco's Eiffel Tower and to manifest California's preeminence in the new world of hydroelectricity. This was the first world's fair to be powered by electricity and gas rather than steam, and electric lights played on every surface and shown through every glass skylight. It was also the first to feature motion pictures throughout the exhibits. Most fantastic of all was the Palace of Fine Arts, standing off at the west end of the fairgrounds, Bernard Maybeck's monumental homage to romanticism, built on a scale to match the Pantheon in Rome.[3]

The exposition was a giant conceit for the city of San Francisco, to show the world that the ruined city was back from the abyss and ready again to play its appointed role on the Pacific stage. But it was also a classic world's fair and hence an international undertaking, drawing on the resources of many countries and putting on display the wonders of modern industry from all corners of the earth. Bourgeois civilization was celebrated in all its grandeur, as it had been at the Crystal Palace and the great Paris expositions. There was a sheen of high culture put on the fair in the exhibits of the Palace of Fine Arts, the Palace of Education, and the Palace of Liberal Arts. But as at any world's fair, the real sinew and bone was to be found under the lofty roofs of the commercial exhibitions in the Palaces of Machinery, Mines and Metallurgy, and Transportation, and their ilk. The fair was an unalloyed success for the city and the state of California—which also boasted a companion fair in San Diego at the same time and the newly completed El Camino Real running between the two sites, the longest paved road in the world.[4]

Palace of Horticulture, Panama-Pacific Exposition, 1915

Agriculture found a place of honor in the pantheon of modern arts on show. A Palace of Agriculture, second only in bulk to the Palace of Industry, lay across the *allée* from the Palace of Fine Arts. Visitors to this cavernous hall found it mostly devoted to offerings from the cornfields and granaries of the eastern states, and to the North Coast lumber industry. Apparently, wood and wheat constituted *real* agriculture in the American mind. California agri-business was not made of the same stuff as the others, however, and it would get to strut before the captivated audience in other venues: the Palace of Horticulture, the Palace of Food Products, and the California Building.

The Palace of Horticulture flanked the Tower of Jewels on the west, with a companion Palace of Festivals on the east. A separate horticulture department was something entirely new to such fairs, and symbolic of the transformation of the California agrarian landscape that had taken place over the previous generation. The visitor walking into its conservatory looked up at a gigantic greenhouse dome, 185 feet high and 152 feet in diameter, designed by the same architects, Bakewell and Brown, who had built the magnificent new San Francisco City Hall. Under the dome was a sumptuous display of palms, cycads, and other tropical plants from Cuba, and on the sides aquaria, pools, and a collection of orchids installed by John McLaren, head gardener at Golden Gate Park. Outside lay almost ten acres of ornamental gardens, featuring a stunning array of flowers, trees, and plants, many from local nurseries and specialists. C.C. Morse Seeds of San Francisco presented a floral calendar that metamorphosed over the course of the year. All this prettiness was but window dressing for the real meat of the palace, the economical horticulture section in the west hall; here were displayed the industrial arts of viticulture, pomology, and fruit preserving.

Many states, and some nations, had exhibits in the Horticultural Palace, but the biggest space of all was devoted to California. Over two dozen counties contributed exhibits, featuring the best of their fruits and nuts: pomegranates and grapefruit from southern California, dried apples and sugar plums from the North Coast, Kelsey plums and Vicar of Wakefield pears from the Sierra foothills, almonds and cherries from the Bay Area. Also on display were exotica such as *feijoas* and seedless avocados, which would not seem out of place in today's farmers' markets. The great hybridizers and nurserymen of the time, who were responsible for much of the florescence of horticulture across the state, had a place in the hall, too, with the limelight occupied by Santa Rosa's Luther Burbank, the country's most famous plantsman. His latest introductions were laid out in the central booth of the state's exhibit: Santa Rosa plums, spineless cactus, thornless blackberries, and many more. Off to the side was an amazing array of strawberries developed by Burbank's lesser-known comrade in cross-pollination, Albert Etter of Humboldt County.

A prime expression of the modern food industry of the times was the sparkling displays of glass jars of fruit by the California Fruit Canners' Association and Hunt Brothers, and the fresh-looking dried fruit packages of Castle Brothers and Winters' Dried Fruit Company. Holding court over all was Mrs. Freda Ehmann, queen of the canned olive—then considered a remarkable delicacy, and exported around the world. As a backdrop to these products, the National Canners' Association, big brother to California's canning companies, had installed a full, working canning line in the hall, chiefly to demonstrate the cleanliness of the process for the visitor. Anderson-Barngrover of San Jose set up the canning machinery, most of which it had under patent; and American Can Company provided the cans from its Oakland plant, where it probably employed the can-making machines on show from Berger & Carter of San Francisco.

California agricultural machinery could be found elsewhere in the fair, as well. In the gigantic Palace of Machinery at the east end of the exposition grounds, Parker Machine Works of Riverside showed off its wooden box–making machines used for crating oranges. Krogh Manufacturing of San Francisco and Layne & Bowler of Los Angeles exhibited their deep-well centrifugal pumps. The Pelton Water Wheel Company of the city displayed the Pelton-Francis turbine for hydroelectric generation (without which the state's thousands of irrigation pumps would not run). The Palace of Agriculture had the largest exhibition of farm equipment ever brought under one roof, much of it hailing from the Midwest. Nonetheless, California companies weighed in with the distinctive Holt motor-driven combine harvester, a highly maneuverable Yuba Ball Tread Tractor for working in orchards, the Wallace Egg Carrier, John Bean's array of spray pumps, and the Victor Water-Heated Incubator. The most amazing of all was the recently invented Holt Caterpillar Tractor from Stockton, which was kept busy chugging up and down a 45-degree incline.

In the Food Products Palace, the most spellbinding exhibit was a fully operative flour mill installed by the Sperry Flour Company of Vallejo. Oakland's Shredded Wheat Company put up machinery to demonstrate how its breakfast cereal was manufactured. Albers Milling of Oakland also had a booth to show off its morning mush. A big attraction was P.S. Luttrell's demonstration of his Reminder Timer for the harried cook. California Central Creameries made a splash with its flowers carved from butter. Alaska Packers displayed models of its ships and canneries for packing salmon. The wine industry put on a comprehensive show of California's fifty wine grapes and more than two dozen wineries, including Italian-Swiss Colony, Gundlach-Bundschu, and Cresta Blanca. The State Board of Viticulture gave out information about grape growing and wine making.

Over in the California Building, lying between the Palace of Fine Arts and the Bay, was space for every county up and down the state to show its many assets. The exhibits employed classic forms of the time: displays, dioramas, taxidermy, and paintings. Mining and nature's glories figured large in the overall panorama, but agriculture was featured everywhere. The most impressive canvas was said to be "Santa Clara Valley in Blossom Time," representing the county's 7 million fruit-bearing trees, mostly prune and apricot. Sonoma County showed off its Gravenstein apples and White Leghorn hens on a miniature Ferris wheel; Napa its olive oil, nuts, cider, and leather; Alameda its fruits in jars; and Contra Costa its refined sugar and canned fish. The Central Valley counties furnished an octagonal tableau of immense, sparkling jars of fruit of every kind, along with shows of dairy products, grains, and wine. Southern California counties put on displays of citrus, dates, nuts, and cotton—the latter just introduced to Imperial County. The North Coast made much of its redwoods, of course, but also its dairies and wineries. New and delightful for the curiosity seeker were the films shown in little movie theaters around the hall, accompanied by lectures, on irrigation, beekeeping, stock raising, fruit packing, and the rest of California's agricultural and industrial arts.

One area where California agriculture made a relatively poor showing was livestock; but some compensation was found in hosting the largest poultry show ever seen on the Pacific Coast. There were, of course, things of passing fancy here and there around the exposition that spoke volumes about farmers' experimentation in a state still trying out its legs; these included silkworms from Napa and ostrich plumes from Imperial, both of which would be abandoned. The wine and beer on exhibit would also face the chill winds of Prohibition blowing out of the Midwest, as the 18th Amendment went into effect four years later.

At the end of the yearlong fair the throngs passed out of the fairgrounds, the lights went off, and the whole magical kingdom disappeared in a matter of weeks. The exposition was torn down to make way for great new tracts of housing, still standing as today's Marina District. The fairgrounds had, after all, been part of the San Francisco Bay only a year before that, and this was prime bayside property. (Much of the land was owned by one of the city's richest families, the aptly named Fairs, descendants of James the Silver King of the Comstock Lode.) Like a Hollywood movie set, the palaces were built to suit the occasion—gossamer showcases of metal, glass, and plaster never meant to last. In this sense, too, such world's fairs were the perfect embodiment of the age of capital in which, as Marx famously said, "All that's solid melts into air."

The Panama-Pacific International Exposition was a wonderful showcase for California's burgeoning agriculture and industrial enterprise. It must have convinced tens of thousands of happy strollers that this was a state on the move, a business culture to be reckoned with. And, indeed, California would overtake every other state in the union in agricultural output within a decade after the fair closed. But the exposition itself would prove no more permanent than a California wheat field or vineyard, uprooted to make way for the next most profitable crop, and the herds of visitors to the fair were as transient as migrant farmworkers. Only the Palace of Fine Arts escaped the wrecking ball to stand sentinel over history, a landscape oddity much like the grand valley oak—a species once regnant over the Sacramento Valley that one still spies from time to time standing watch over an aging farmhouse.[5]

WHEAT SHALL OVERCOME

The Panama-Pacific International Exposition halls gave the viewer a good introduction to California agriculture as it stood in 1915. But it was only one frame in a moving panorama. We need something more like John Banvard's 500-foot painted diorama of the Mississippi Valley of the 1840s, rolled and unrolled before astonished viewers to portray the immense sweep of the mighty river over hundreds of miles, or perhaps the "Soaring over California" attraction installed at Disneyland in 2001. What we have, instead, is words. So we will walk into the Cornucopia using two resources: the first is a short guidebook to crop types and histories that reminds us of C.C. Morse's and Luther Burbank's catalogues; the second is an unfolding diorama that shows how the crops spread out across the state, found their geographic niches, and jumped from place to place over time. The reader is forewarned, however: if you are one of those people who finds driving down the Westside of the Central Valley on Interstate 5 to be boring, you may want to skim over the details. But if you are someone who thinks the same view is endlessly fascinating, then tarry awhile over this staggering panorama of expansion and change.

Livestock

Livestock have too often been treated as a sidebar to California agriculture, yet they are central to the state's agrarian development. Cattle raising, for example, exploded in the 1850s under the impetus of Gold Rush demand, making beef the most important agricultural product of the time. But it collapsed under the strain of statewide recession, drought, and flooding in the early 1860s, and the number of cattle fell precipitously from 3 million to .5 million by 1870. Afterward, the cattle industry rebuilt on a new basis, and California climbed back among the top ten states in beef production by the turn of the next century. Sheep numbers soared just as cattle collapsed, thanks to the demand for wool during the Civil War, and they remained the state's dominant livestock for the rest of the century. They were also less susceptible to the vagaries of climate and problematic forage, although wool markets were variable. California jumped to sixth among the states in wool production by 1860, second by 1870, and first by 1889, with 13 percent of national output. Sheep populations rose from 1 million in 1860 to peak at around 6.5 million in the 1870s, then tailed off as the state began exporting stock to the Rocky Mountain region.[6]

Dairying, overlooked in the dramatic stories of cattle and sheep booms, be-

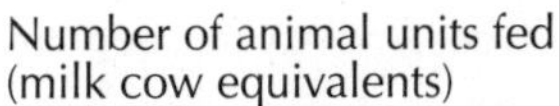

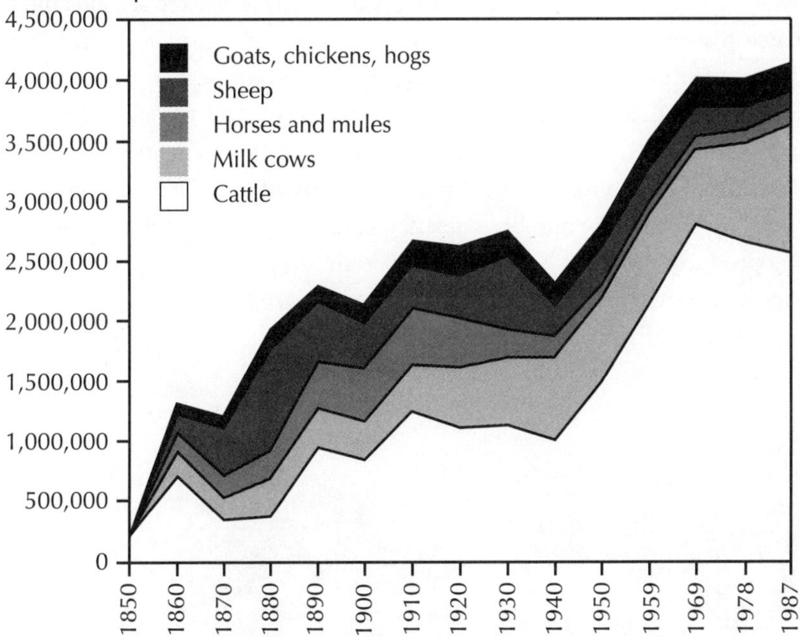

After: Olmstead & Rhode 1997, fig. 3.
Source: U.S. Bureau of the Census, Census of Agriculture, 1959 and 1987.

Livestock in California, 1850–1987

came a major activity thanks to urban demand for milk, butter, and cheese. Local dairying had to start from scratch, but caught up to demand by the end of the Civil War. Dairy cows favored the cool pastures of the North Coast and Bay Area, and Marin became the leading milkshed of the 19th century. The number of cows rose to 200,000 by 1870, and the output of butter and cheese reached 10 million pounds by 1870. Dairying was second only to wheat in value of output in 1894. Dairying grew even more rapidly in the 20th century, with 500,000 milk cows by 1919, generating $63 million in milk products.[7]

An extraordinary poultry industry arose in the early 20th century, shipping millions of eggs eastward. Dairy products, eggs, and beef were exceeded only by oranges in commodity value in 1930. California still ranked second among states in sheep raising, as it does to this day. Livestock expanded markedly in the postwar era. After slipping during the Depression, cattle numbers rose briskly, more than doubling to 25 million head by 1975. Cattle and dairy products topped the list of the most valuable commodities in the state from 1950 to 1975. Sheep also revived, hitting 1 million head by 1980. California finally overtook Wisconsin as the number-one dairy state in the 1990s and remains the top egg producer. Turkeys became an important crop, as well. Only cattle has slipped in recent years, although there are still over 50 million head and beef output ranks fifth among farm products in value.[8]

Sheep grazing in Central Coast region, circa 1900

Grains

California did not have to wait long for grains to flourish. The first to do so was not wheat, but barley—a lowly feed grain easily overlooked. The Gold Rush came astride pack trains of thousands of animals, and California soon had the largest stock of horses and mules in the country. The state quickly led the country in barley production, generating almost 30 percent of national output by 1870 and remaining a top producer up until 1900, when animal power began to give way to electricity and petroleum. Barley covered roughly half a million acres during this time. Corn, oats, and sorghum were other significant feed crops. All the feed grains went into beer and whiskey, as well.

Then the glamour grain, wheat, stole the show. Farmers began cultivating wheat in the 1850s to feed the miners' taste for bread. It passed cattle as the

Grain harvest, San Joaquin Valley, circa 1900

number-one crop in the mid-1860s, and was shipped in quantity to Europe and China. Output grew exponentially, hitting 7.5 million bushels by 1860 and 20 million bushels by 1870; production peaked in 1888 at just under 40 million bushels grown on 3 million acres, putting California second among the states. Grains made up almost 80 percent of California's agricultural output value in 1890. But the end of the era came hard after that, just as it had to the first cattle boom. Acreage and output fell an astonishing 75 percent between 1900 and 1910. The great wheat boom had lasted forty years, not a bad record of longevity by capitalist standards; so the common idea that wheat had a transitory role in California's development doesn't pass muster.[9]

Rice was the last of the grains to arrive in California. Rice cultivation took off in 1912, expanded to 125,000 acres by 1918, then continued to grow in the 1920s. Rice continued strong right through the Korean War, then slumped in the 1950s. It came back with the Vietnam War and subsequently entered into Asian markets, exceeding 500,000 acres by 1980. About 40 percent of the California rice crop is exported today, principally to Japan. An ironic coda to the grain story is that wheat enjoyed an amazing comeback in the mid-20th century, covering 1.2 million acres in 1979, with an output greater than at any previous time—although it has since fallen away along with cattle.[10]

Fruits and Nuts

Fruits were in high demand from the moment the Gold Rush began, especially the durable apple. Apples and grapes went to slake the miners' thirst for alcohol. Almost 30 million vines were in place by 1870 and 2 million apple trees, representing half the orchards in the state. That was only a runup to the horticultural boom to come. Fruit output grew tenfold in the 1870s, then fiftyfold in the 1880s as trees and vines were planted on hundreds of thousands of acres. Over the latter decade fruits grew from 4 percent to 20 percent of the value of all farm output, becoming California's leading crop group by 1890. The expansion of the late 19th century was led by soft fruits—peaches, pears, apricots, plums, and cherries—of which there were some 25 million trees by 1900.

Horticulture hit a bump in the depression of 1893–95, then took off again, growing at an annual rate of 5 percent right through to 1930. Land in trees and vines expanded to over 300,000 acres by 1893, 650,000 acres by 1909, and 3.1 million acres by 1929. Horticulture's share of California's farm produce surpassed one-third by 1909. By this time, citrus had taken the lead, with the number of trees hitting 6 million in 1910, mostly oranges but also lemons and grapefruit. Surprisingly, Prohibition brought continued prosperity to vineyardists, who sold grapes and juice in bulk for home use; vineyards doubled their acreage from 300,000 in 1919 to 650,000 in 1928 (not surpassed for another fifty years). For comparison, citrus acreage was only 300,000 in 1938.[11]

In 1889, William Mills, land agent for the Southern Pacific, proclaimed before the State Agricultural Society that it was possible for California to become "the orchard for the whole world," a vision that very nearly came true. In 1900

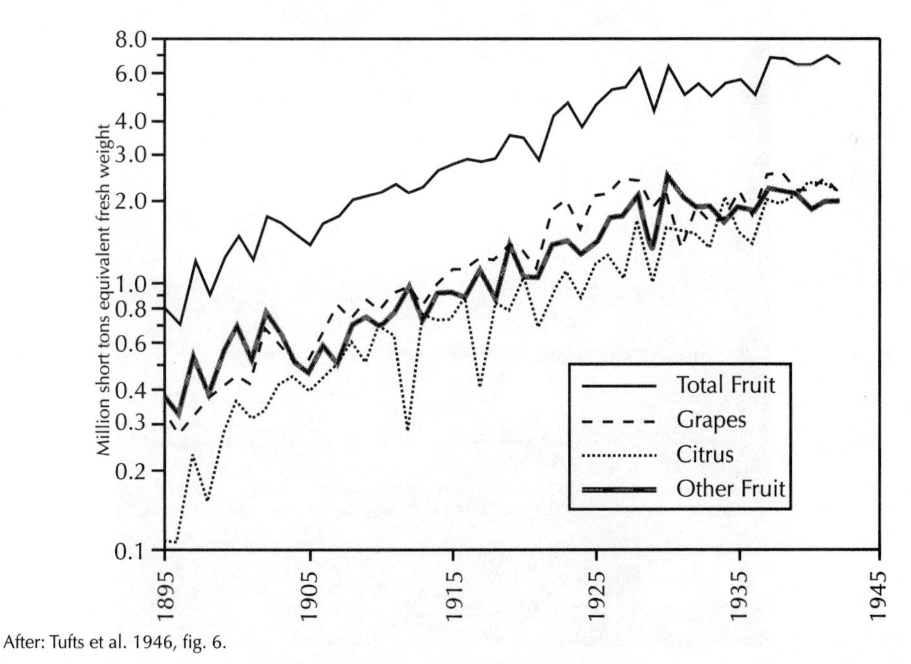

After: Tufts et al. 1946, fig. 6.

California fruit production, 1895–1945

the state was growing more than a quarter of all the fruit in the United States. By 1919 the state produced over half the country's oranges and all its lemons, two-thirds of the plums, four-fifths of the grapes and figs, and almost all the apricots, almonds, walnuts, and olives. In 1930, California was producing almost half the nation's fruit and nut crop.[12]

Expansion after World War II came chiefly in walnuts and almonds, as orchard lands went from 1.3 million to 1.8 million acres from 1950 to 1980.

Gravenstein apple orchards, Sonoma County, circa 1900

The war ushered in the age of the table grape, especially Thompson Seedless. Avocado production expanded hugely after 1930, and nectarines finally took off as a commercial fruit. By the 1980s something new was on the table, with California consumers looking for new and exotic fruits, such as clementines, persimmons, and pistachios. The great kiwi boom of the eighties was symptomatic. The number of varieties of citrus and stone fruit continued to increase with new breeds, such as pluots and Japanese apple-pears, and the reintroduction of heirloom varieties and organically grown fruits. Olives and olive oil returned as more than just pizza topping. Most valuable was the vastly expanded array of premium wine grapes, which shot up from 300,000 to 500,000 acres during the 1990s.[13]

Field Crops

Sugar beets were the first of the non-grain field crops to burst on the scene, as demand for canning sugar expanded in the late 19th century. Beet tonnage rose around three-hundred-fold between 1880 and 1910, as Claus Spreckels shifted his operations from Hawaii to California, and California became the nation's leading sugar producer. After a slump in output in the 1920s, beet production surged to new heights in the 1940s and '50s.[14] Sugar beets continued to do well up to the 1970s, thanks to the boycott of Cuban sugar, but persistently depressed world prices have relegated them to a secondary crop in recent years.

Cotton was introduced to California in 1906. Production rose tenfold during World War I, and expansion continued through the 1920s and '30s, exceeding 500,000 acres. Cotton was the glamour crop of the interwar period; hence its prominence in the public imagination, as portrayed in *The Grapes of Wrath*. Cotton peaked at over 1 million acres in the early 1950s and became the state's most valuable crop (only Texas surpassed California in production). It fell back by half in the 1960s and then surged again to 1.5 million acres by 1980. J.G. Boswell is the world's largest single producer.[15]

Alfalfa and hay became prominent field crops in the 20th century by virtue of the demand for intensive animal feeding. Alfalfa is the more important be-

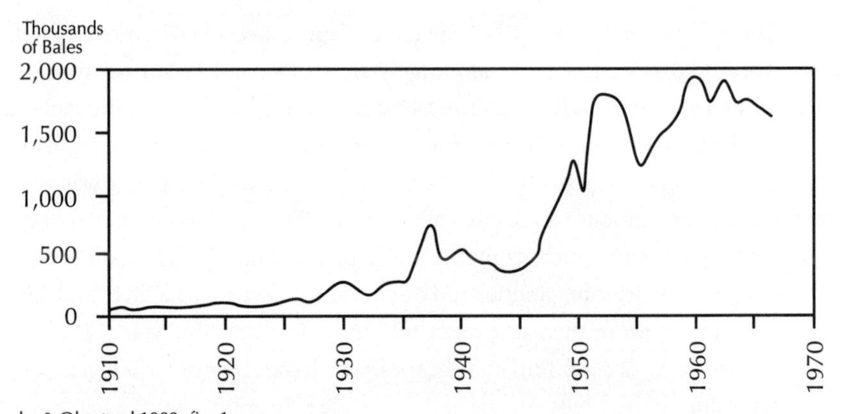

After: Musoke & Olmstead 1982, fig. 1.
Sources: U.S. Department of Agriculture, *Statistics on Cotton and Related Data*, Statistical Bulletin 99 (Washington, D.C., June 1951), p. 38, and *Statistics on Cotton and Related Data: 1920–73*, Statistical Bulletin 535 (Washington, D.C., October 1974), p. 67.

California cotton production, 1910–1970

cause of its value as a rotation crop to maintain soil fertility. Increased irrigation and feeding of livestock pushed pasturage up to a generous 1.5 million acres in 1980, making it the single largest type of cropland—quite the opposite of one's vision of California specialty cropping.[16] Alfalfa and hay have retreated somewhat over the last generation, because of rising competition from higher-value crops and the decline of cattle numbers. Late in the day, oilseeds have been planted in some quantity, but soybeans almost never.

Sturdy vegetables, such as lima beans and peas, and root crops, such as carrots, beets, and potatoes, are sometimes treated as row crops rather than as vegetables. These developed earlier than the perishable vegetables and berries, and already occupied some 50,000 acres, yielding $6 million, in 1900. Beans were mostly dried, and they could be exported. After that, beans and potatoes leveled off in the face of a rapid expansion in garden vegetables and berries.

Vegetables and Berries

This diverse group, sometimes called garden or truck crops, is made up of green vegetables (green beans, peas, celery, asparagus, and artichokes), field

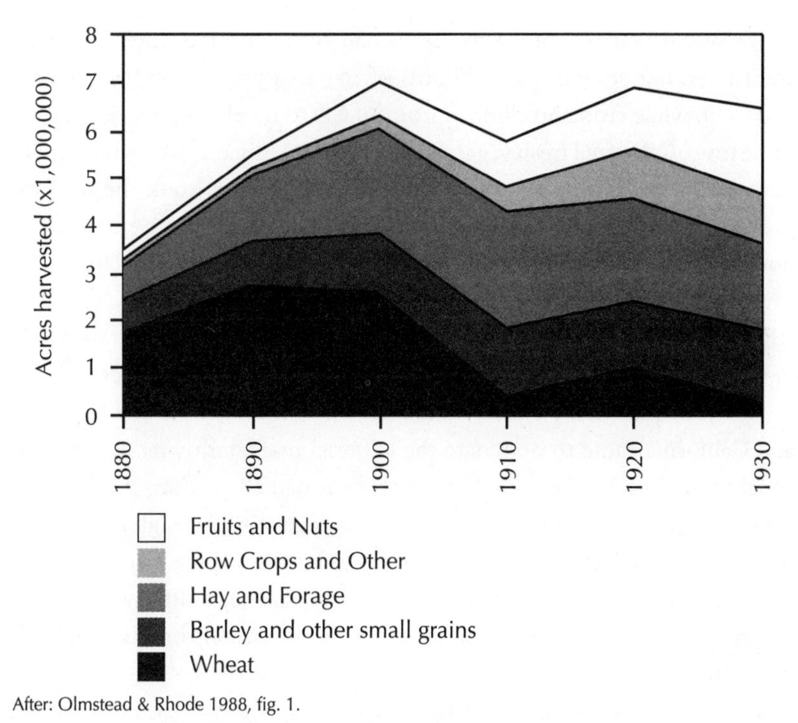

After: Olmstead & Rhode 1988, fig. 1.

California farm acreage by crop group, 1880–1930

berries (strawberries, raspberries, and blackberries), and vine fruits (tomatoes, cucumbers, squash, and melons). Garden crops were in place from the earliest days of the Gold Rush to feed the miners, but only became large-scale specialty crops after 1910. By 1919 vegetable farms exceeded 200,000 acres (10 percent of intensive cropland), with a value of output nearing $50 million. By that year, California was the foremost state in vegetable output other than potatoes. By 1939, vegetables covered over 500,000 acres and sales exceeded $100 million. The leading fresh crops in the interwar period were lettuce, potatoes, melons, celery, and cauliflower; the leading vegetables for canning were tomatoes, asparagus, and spinach.[17]

By World War II canning tomatoes had come into their own. California went from one-seventh to two-thirds of national production between 1930 and 1970, while cropland climbed from 50,000 to nearly 150,000 acres. At the same time, California fresh vegetables, especially lettuce, came to monopolize national markets, and for the first time fresh produce took the lead over processed in value of sales. Vegetable acreage held steady at just under .7 million acres, then jumped to over .9 million in the late 1960s; this land yielded over 13 million tons of produce in 1979. At the time, however, "fresh produce" still meant a few mass-produced, conventional, and hardy varieties, such as iceberg lettuce, Better Boy tomatoes, russet potatoes, and cantaloupes.[18]

Berry output took off after World War II. Strawberries came into vogue and California came to dominate the national market, driving out local seasonal producers; by the end of the century, it had a four-fifths market share. By the 1980s fresh vegetables gained a new edge with the high-price market for organics, delicacies, and exotics created by the yuppie food revolution. Suddenly there was a flourishing of species on offer that had never been seen before in this country, like arugula and radicchio. Asian immigrants added bitter melon, chinese cabbage, and lemongrass to the increasingly diverse mix. At first these came out of the fringe areas of the coast, but they soon spread around the state. Vegetable production rose by 75 percent in the 1980s and continued upward in the 1990s. California at the end of the century was growing about half the nation's output of vegetables (fresh and processed), by value.[19]

Ornamentals

A last family of crops—barely considered in most accounts of California agribusiness—is ornamentals for home and garden. These usher in a direct form of aesthetic, leisure, and conspicuous consumption of plants. Ornamental gardening has a robust history in California, nurtured by grand estates, public gardens, and enthusiasts since the Civil War. But this branch of nursery agriculture only took off as a mass production and consumption arena after World War II, thanks especially to the emergence of inexpensive air shipping.

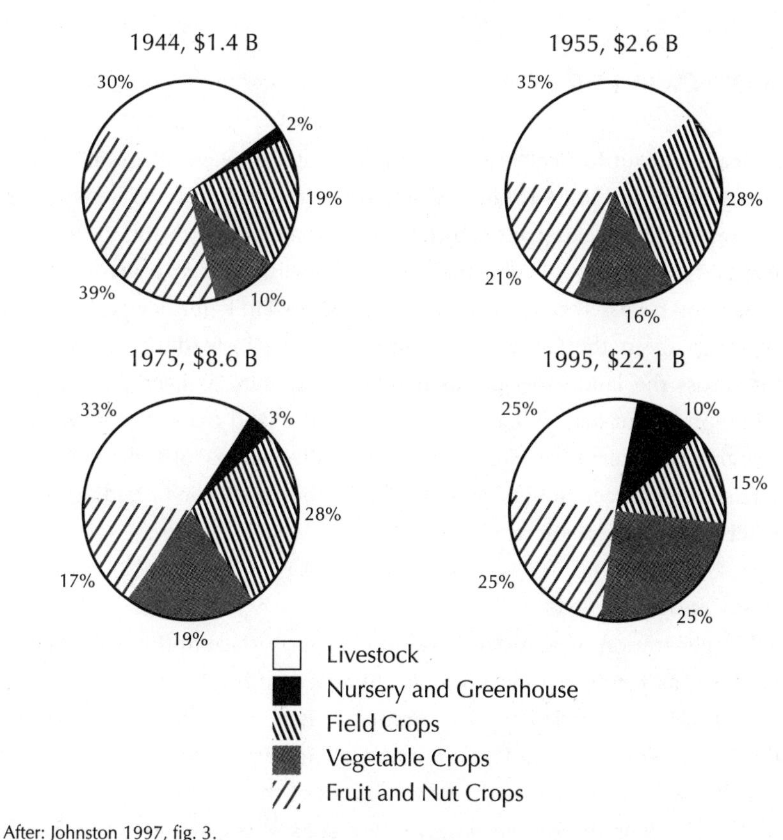

After: Johnston 1997, fig. 3.
Sources: California Department of Food & Agriculture and U.S. Census of Agriculture.

California farm output by crop group, 1944, 1955, 1975, 1995

California became a national supplier of cut flowers, potted plants, and nursery stock, and its output of ornamentals soared from about $50 million in 1959 to $140 million in 1965 to over $800 million in 1980—by which time this segment had become the fourth-largest crop type in the state. In the last twenty years, there has been a vigorous revival of landscape gardening using Mediterranean and native plants, a herbaceous parallel to the arrival of organic produce.[20]

GOLDEN WAVES OF GAIN

"America the Beautiful" celebrates in song the "amber waves of grain" across the heartland of the United States, which Americans readily associate with the vast sweep of grain fields planted by farm settlers along the advancing frontier. Where is the song about California's orange groves, vineyards, lettuce fields, and peach orchards? Too many verses, perhaps, so Tin Pan Alley preferred to write songs about the Golden Gate and cable cars instead. Yet agriculture swept across the landscape just as magnificently here as over the prairies. Nowhere in recent history has production on the land magnified so rapidly over such a large territory. Hence, the commodity history of California agriculture should be seen not just as a list of crops but also as an unfolding tapestry of cropping patterns across the landscape.

The Bay and River Axis

The first generation of agrarian development took place in the wake of the Gold Rush, as farming grew up along the axis from San Francisco to the Mother Lode. It encircled the San Francisco Bay and followed the course of the Sacramento and San Joaquin rivers; all around the fringes it followed the smaller watercourses into the fertile valleys of the coast ranges and Sierra foothills. This was the region with money, consumers, settlers, merchants, and transport; hence, as Rodman Paul observes, "When a new American agriculture began, its birthplace was not in Southern California, where Hispanic settlement had started, but rather in the central part of the state."[21] Most treatments of California history get this geography wrong because the San Joaquin and Imperial valleys are the heart of agribusiness today. It wasn't always so.

Grains thrived on the prime valley bottoms, pushing cattle into the hills. The wheat boom began in the Santa Clara Valley, then spread up the Petaluma River in Sonoma County, over the Montezuma Hills into Solano County, and thence up the west side of the Sacramento Valley. Milch cows were thick along the cool coastal valleys north of the Golden Gate, all the way up to Humboldt

County; cattle pushed south into the lower Santa Clara Valley around Gilroy. Garden farming for city tables took hold under the stimulus of high food prices, tucked around the bay flats and even in San Francisco itself. As horticulture spread, apples thrived in western Sonoma County, plums and olives in the Napa and Sonoma valleys, plums and cherries around San Jose, and apricots and vegetables in Alameda County. Orchards continued up the Sacramento River and into the foothills of Placer County. Eastern Contra Costa County grew walnuts and apricots; Solano nurtured soft fruits in its northern hills; Sacramento, Yuba, and Yolo led the state in peaches and pears.

In the census of 1869, the leading agricultural counties in the state—measured in wheat, fruit, and dairy production; value of total output; and land value—were almost all along the Bay-Delta axis. The region represented two-thirds of the state's wheat and dairy output, two-thirds its land value, and over half of total farm production and orchards. The high rankings of Alameda and Contra Costa counties might surprise people today, as would one contemporary's declaration that Solano County was "one of the most wealthy, populous and large productive agricultural counties in California."[22] Farm centers sprang up all around the Bay and Delta, such as San Jose, Petaluma, Napa City, Vallejo, Fairfield, Rio Vista, Walnut Creek, Davisville, Winters, Locke, and Walnut Grove.

By the turn of the century, areal specialization became more marked. The Santa Clara Valley was by then the prune capital of America, Sonoma top ranked in vineyards and wine, and Yolo the leader in almonds. Petaluma would soon be the largest poultry center in the United States. The flatlands along the southeastern rim of the Bay were developing into a flower, seed, and nursery zone of national prominence. Berries became significant north of San Jose. Beans and potatoes were being grown as row crops in southern Alameda County and upriver in Sacramento and San Joaquin counties. Delta agriculture grew apace as the islands were reclaimed, with asparagus a valuable specialty grown in soft peat soils. San Joaquin County, on the southern flank of the Delta, became the center of processing-tomato cultivation, as this crop grew into one of the major moneymakers in the state.[23]

Bay and Delta agriculture peaked in the 1920s, and then began to decline. Alameda would slip down in the interwar period, as vegetables and potatoes moved elsewhere, the East Bay urbanized, and floriculture shifted to the San Mateo coast. Dairies in Marin would lose their leadership to mass-production operations in the San Joaquin Valley. The wineries would largely die out during Prohibition, to be revived in the 1930s and '40s. Silicon Valley swept away the orchards of the Santa Clara Valley within a generation, as urban expansion is now doing to the southern Santa Clara Valley and eastern Contra Costa County. Cattle have been driven out of the hills by urbanization. The apple orchards and chicken coops are almost all gone from Sonoma. The Delta still has tomatoes, but not much asparagus or other green vegetables anymore; and pears are a declining fruit. Wine grapes are the only crop for which the Bay Area is still known, thanks to the reflorescence of wine making in the Napa Valley and the North Bay. Otherwise, the action in California agriculture has long since moved south.

Into the San Joaquin Valley

The San Joaquin Valley is the largest agricultural zone of the state, and has been the most important over the last fifty years; indeed, it is the richest in the United States. But it wasn't always so. Although the east side of the valley down to Fresno was inviting to farmers, the southern and western portions offered daunting obstacles of aridity, alkaline soils, perched water tables, and periodic lake beds in the Tulare Basin. Grain cultivation spread down the San Joaquin in the 1870s, and new towns such as Fresno, Visalia, and Modesto sprang up along the river and the railroad. The Fresno colonies of the 1870s settled the region, and they throve. Orchards of various kinds began to sprout up in the 1880s, but the largest acreages went to grapes for raisins. The west side of the San Joaquin River and the Tulare Basin were not settled by small farmers: they were the domain of the cattle kings Miller and Lux, whose stock were raised in the valley, then sent over the San Felipe Pass to the Santa Clara Valley, and thence to San Francisco and Oakland for slaughter.

By the turn of the century, Fresno County had leapt to second place in the

Top Ten Farm Counties in California 1869–1997

1869	1899	1919	1939
San Joaquin	Los Angeles	Los Angeles	Los Angeles
Solano	Fresno	Fresno	Tulare
Yolo	Santa Clara	San Joaquin	Fresno
Santa Clara	San Joaquin	Tulare	San Joaquin
Stanislaus	Sonoma	Sonoma	Kern
Sacramento	Sacramento	San Bernardino	Stanislaus
Sonoma	Alameda	Orange	Sonoma
Alameda	Solano	Santa Clara	Santa Clara
Colusa	Yolo	Stanislaus	San Bernardino
Monterey	Tulare	Sacramento	Monterey

1950	1959	1979	1997
Los Angeles	Fresno	Fresno	Fresno
Fresno	Imperial	Kern	Tulare
Kern	Kern	Imperial	Kern
San Joaquin	Los Angeles	Tulare	Monterey
Imperial	Riverside	Monterey	Merced
Stanislaus	San Joaquin	Merced	San Joaquin
Monterey	Stanislaus	Riverside	Stanislaus
Orange	Tulare	San Joaquin	Riverside
Sonoma	Merced	Stanislaus	San Diego
Merced	Ventura	Ventura	Imperial

Source: U.S. Census of Agriculture, various years.

state in value of farm output, and Tulare County to the south was rising fast (it would hit fourth place by 1919). Table and wine grapes had been introduced by this time, along with a mix of field and orchard crops—all of which yielded prodigiously in the valley heat. The northern San Joaquin (Merced and Stanislaus counties) was thriving on dairying and processed milk products (not fresh milk). Cotton arrived in the early 1920s, and exploded across the former cattle lands from western Fresno County down to Kern County around Bakersfield; it even replaced orchards in some places, as around Patterson. Potatoes, sugar beets, and carrots moved into Kern County in the 1930s.[24]

After World War II, the San Joaquin region's agriculture grew prodigiously, thanks to unlimited water supplies and urban pressure pushing crops out of the coastal areas. It moved to the forefront in state livestock operations, adding cattle feedlots and poultry to the already expansive dairy operations in the north. Table grape and bulk wine production surged in Fresno and Modesto counties, while cotton remained king in the southern realms. Apricots and cherries moved to the eastern slopes from the Bay Area, while citrus crops were driven over the Tejon Pass into Kern County. Almonds and walnuts appeared in western Fresno County, with many other tree crops, such as olives and pistachios, besides.

Southern California in Bloom

In the Mexican era, Southern California was the heart of settlement and the ranchero country. It fared well only too briefly in the Gold Rush before the drought of the 1860s killed off the cattle. Vineyards were more extensive here than anywhere in the state, until they were decimated by disease in the 1860s. Sheep replaced cattle and wheat replaced grapes, but the southland fell far behind the north. Southern California was given a welcome boost by the arrival of the Southern Pacific and Santa Fe railroads in the 1880s, which brought a host of settlers, but its economic salvation lay in citrus. A million orange trees were in place by the end of the 1880s, from Riverside to Ventura, across the San Gabriel Valley and into the Los Angeles plain. Less dramatically, the coastal plains grew ample crops of beans, potatoes, and sugar beets.[25] Olives and walnuts were also grown in quantity along the southern coast until swept aside by the citrus mania.

By 1900, Los Angeles was already number one among counties in California—indeed, in the United States—in total value of agricultural output. It would hold that position for almost half a century, until the sprawling cities would pave over the coastal plains and river valleys. The citrus boom continued unimpeded for fifty years, and brought Riverside and San Bernardino counties up the ranks, as well. The Los Angeles basin and other coastal zones, such as the Oxnard Plain in Ventura County, developed into important garden farming areas as the population of Los Angeles exploded. Growers of vegetables, berries, and root crops did a land office business. So did dairies around Artesia and Chino, propelling Los Angeles to the top region of the country in the production of fresh milk. Fresh tomatoes were so prolific that they were shipped east along with the oranges. Los Angeles supplied the West with ornamental plants, as well—a niche still filled today from San Diego and Ventura counties, beyond the urban fringe. Avocados cropped up around Santa Barbara and San Diego, but did not become significant until the 1970s.

In the inland desert region of the Imperial and Coachella Valleys, agriculture arrived late. It only took off after 1910, and then rocketed up the ranks of California production. Cotton was introduced to California in the Imperial Valley, before insect damage drove it north to the San Joaquin. After that, Imperial switched to melons, winter vegetables, dairy, and cattle feeding. It also grew some exotic desert plants, such as dates, of some commercial value.

As urbanization gobbled up the valleys and plains of the southern coast after 1950, Los Angeles declined as an agrarian realm, followed by Orange County and the inland empire of Riverside and San Bernardino counties. The Imperial and Coachella valleys emerged, stronger than ever, as the heartland of Southern California agriculture. Along with lettuce and melons, postwar growth was propelled by sugar beets, cattle, alfalfa, and cotton (which had been reintroduced). Imperial County vied with Fresno County as the richest farm area in the nation. Just as suddenly, cattle and the big field crops declined, and in the 1990s, Imperial joined the shift into specialty vegetables and melons.

The Sacramento Valley and Central Coast

Although Frank Norris made the town of Mussel Slough on the San Joaquin River the site of his famous novel of the Granger Wars, *The Octopus,* the greatest wheat region of California was the Sacramento Valley. Wheat stretched up the west side of the river from the Delta all the way to Red Bluff. Glenn, Colusa, and Yolo counties produced 30 percent of the wheat crop in 1880. When wheat died out, the valley suffered mightily. It was only suitable for horticulture in the alluvial deposits along the Delta, Cache Creek, Putah Creek, and the eastern margins of the Sierra tributaries. Pears and almonds did well in the southern end, as we've seen; peaches and hops prospered farther north around Marysville; and olives were abundant up at Oroville; but the central river corridor was of limited use because of its perennial marshlands and flooding. For years, it languished compared to the rest of the Central Valley, until it was saved by rice, which thrived in the watery paradise of the unruly river.[26]

The Central Coast was mostly a sheep and cattle run in the 19th century, and its hidden valleys deep in the coast ranges in San Benito County remain remarkably unchanged to this day. The coastal valleys are quite the opposite. Sugar beets changed them profoundly in the 1890s, as Claus Spreckels shifted his operations to the Salinas Valley and the Oxnard brothers moved into Ventura County. That was succeeded, in turn, by vegetables. The Monterey Bay area, from Watsonville in Santa Cruz County south past King City in Monterey County, began to flourish by specializing in lettuce, artichokes, and broccoli. At the same time, coastal sites in northern Santa Barbara and San Luis Obispo counties developed into the preeminent commercial seed center of the Americas for both vegetables and flowers.

Monterey County moved into the upper ranks of California agricultural counties by the 1930s. After the war, it became one of the wonders of the agrarian world with its absolute dominance in fresh lettuce, broccoli, and strawberry production—exceeding four-fifths of American production on only a few square miles. That concentration has solidified in recent years, though the boom in specialty crops over the last generation has spread the wealth of celery, cauliflower, and the like down the coast. Meanwhile, vineyards have come to the Central Coast in a big way. Once confined to the Santa Cruz Mountains, they now cover the flanks of the coastal river valleys down to Santa Maria in Santa Barbara County—where some of the world's finest pinot noir is bottled.[27]

This gives us a bird's-eye view of California croplands and circumnavigation of the state agricultural regions. But what have we learned of agribusiness? Not much, as yet. A munificent landscape of fields, orchards, vineyards, and pastures presents itself, to the eye's delight and the stomach's delectation. But are the crops there only to feed and fill the senses? Before we can partake of the movable feast, we had best get out our wallets and be ready to pay. California's millions of acres and vast array of crops are grown to sell, to make money; that is the nature and purpose of commodities. So put away the cameras, set aside the ravishing posters of plenitude, leave the world's fair stage setting behind, and turn back to the black-and-white landscape of economics. On the immense chessboard where the strategic moves of growers are made—where to put this thousand-acre orchard? when to take away that 500 acres of sugar beets?—the colors don't matter, just the market positions and the takings. In this game, the crops are only pawns, the game begins with the chief commodity inputs—land and labor—and profit is king.

2

A Landscape of Commodities

California agribusiness has generated an immense flow of commodities because that is the point of capitalism: production for the market. Capitalism in the countryside presupposes a well-established market system, in which crops are not for sustenance but for sale. A proliferation of goods is as much a part of such agriculture as it is of modern industry, where it has led to millions of new commodities on the market over the course of the last two centuries. That logic is as relentless today as it ever was, and California goes on adding to its product mix. But something even more profound has been going on behind the showy cornucopia of crop markets.

The growth of commercial farming has been, at the same time, the commercialization of farm *inputs.* Commodification has meant the turning of the two elemental conditions of farm production, land and labor, into things for sale on the market. This is what Marx, parroting Adam Smith, called "the so-called primitive accumulation," including the transfer of land into private hands and the mustering of human beings as hired hands. If Marx begins *Capital* with the accumulation of commodities, he ends volume I with primitive accumulation, on the initial assemblage of land, labor, and capital.[1]

The coming of capitalism into the countryside through the freeing up of land and labor took place rapidly in California after the Gold Rush. Land was transferred by the millions of acres from Mexican and government ownership to private hands, ready to be put to use growing wheat and other crops. This process took only a generation. Equally fast was the mobilization of labor. By 1870, more people were working in agricultural occupations than in mining, 48,000 versus 36,000. From there to the end of the 19th century, agriculture employed roughly 30 percent of the state's labor force and generated about 35 percent of income (a higher share than it would achieve in the 20th century). The common belief that agriculture arrived *after* mining—Gerald Nash's "stage" theory of California's economic evolution—is simply wrong. Agriculture was there from the outset, and at full throttle by the end of the first decade of the mining era.[2] But the capitalist transformation of rural land and labor was not instantaneous, nor was it once and for all. Indeed, the processes of primitive accumulation continue to the present day.

A PROFUSION OF PRODUCE

California's cornucopia of crops has been much remarked upon, but always attributed to the natural environment, as if the land determined what should be done with it. This argument may work for ancient agriculturists, but will not suffice for explaining the nature of modern agribusiness. To be sure, agricultural profusion has been made possible by California's unique climate and topography. It is one of only five Mediterranean climate zones in the world, which are noted for their winter rain, summer drought, generally mild temperatures, moderating coastal influences, and highly variable annual precipitation. California is largely an "island on the land" by virtue of its surrounding mountains and deserts. It harbors, moreover, an amazing diversity of microclimates and soils that make possible a wide variety of growing regimens, including some of the longest growing seasons outside the tropics.[3]

On this score, California trumps Kansas, but California does not exceed the rest of North America in diversity. Most of its crops compete head to head with those from the rest of the United States: cotton, citrus, tomatoes, sugar beets, avocados, peaches, plums, apples, berries, grains, poultry, cattle, and dairy are all examples. Just a handful are nearly exclusive to the Mediterranean climate, such as apricots, olives, pistachios, walnuts, and figs. Wine grapes are being successfully grown from Oregon to Ontario, and even in Kansas! Nonetheless, California agribusiness pushed ahead of the rest of the country, not to mention Canada and Mexico, and drove out competitors that were well established elsewhere. It became America's greenhouse. Today it leads the nation in the production of seventy-six kinds of crops.[4]

In the process, California became the king of specialty crops. The process of product proliferation did not begin or end with apples, plums, and oranges. Sugar beets, barley, beef, milk, wool, grapes, and potatoes were already in production by 1900; eggs, berries, tomatoes, rice, and cotton joined them in the early 20th century; lettuce, watermelons, carrots, and asparagus were in profusion by mid-century; and diversification continues right down to the purple cauliflower, kale, and Sharlin melons of today. Some estimates run as high as 300 crop varieties.

Nature provides the potential to do this. But who really knows for sure what can be grown in this state? That depends on much more than climate. What is striking is the willingness to try anything and everything, and even to bring back a crop like wheat from oblivion. This sense of wild experimentation will become apparent when we explore horticultural introductions and breeding in later chapters. Some of it has been propelled by the gathering of plant materials from around the world; some by the gathering of *people* from around the world who brought with them plants, know-how, and tastes. This continues today with the introduction of Asian vegetables like bok choy and long beans into the market by immigrant Vietnamese, Hmong, and Cambodian farmers.

But behind all the human culture and horticulture is an economic logic that has run full throttle through California agribusiness for 150 years. The profusion of crops derives from a prime logic of capitalism: the multiplication of commodities. New commodities mean new sources of value for the producer. Better yet, they usually mean an extra measure of surplus value (profit) while

California Leads the Nation
in seventy-six crops, 1997

Alfalfa
Almonds
Apricots
Artichokes
Asparagus
Avocados
Beans, Blackeye
Beans, Dry Lima
Beans, Dry Kidney
Beans, Garbanzo
Beans, Green Lima
Bedding Plants
Bell Peppers
Broccoli
Brussels Sprouts
Bulbs
Cantaloupes
Carrots
Cauliflower
Celery
Chicory
Chinchillas
Chinese Cabbage
Chinese Snow Peas
Currants
Dates
Eggs
Figs
Flowers, Cut
Garlic
Goat's Milk
Grapes, Table
Grapes, Wine
Herbs
Honey
House Plants
Kiwi
Lemons
Lettuce, Iceberg
Lettuce, Leaf
Lettuce, Romaine
Melon, Casaba
Melon, Crenshaw
Melon, Honeydew
Melon, Persian
Nectarines
Olives
Onions, Dry
Onions, Green
Parsley
Peaches
Pears, Bartlett
Persimmons
Pigeons & Squabs
Pistachios
Plums
Pomegranates
Prunes
Rabbits
Raisins
Rice, Sweet
Rice, Wild
Safflower
Seed, Alfalfa
Seed, Bermuda Grass
Seed, Ladino Clover
Seed, Veg. & Flower
Sod
Spinach
Strawberries
Sudan Grass
Sweet Rice
Tomatoes, Processing
Vegetables, Asian
Vegetables, Greenhouse
Walnuts

Source: California Department of Food and Agriculture, *California Agricultural Statistics 1998.*

the product is scarce and/or has the cachet of novelty, like star fruit and lychees do today. Just as in manufacturing, agrarian capitalism inherited a set of products from the past and has gone on to spin out more and more new ones over time. What General Motors did for automobiles by assembling a line of

Orange groves, Ojai Valley, 1993

different products and changing models (besting Henry Ford's plain black Model T in the process), California agribusiness has done for fruits and vegetables. In our time, Hollywood and Silicon Valley put out a huge and ever-changing variety of entertainment and computer products, and no one thinks this the least bit surprising. So why all the amazement at California's specialty crops?[5]

Much the same sense of amazement has gripped observers since the 19th century when they have beheld the vast spaces of California's fields and vineyards. California agriculture is based on *monoculture.* This was crystal clear from the outset of the wheat boom, and could be found just as surely on hop farms or sheep ranches of the time. The horticultural era brought smaller farms on average, but they were no less relentlessly monocultural. The vast swaths of orange groves that covered Riverside County and prunes that swept across the Santa Clara Valley were as immense and monochromic as the fields of grain that preceded them—even if their spring blossoms fooled people into thinking the latter was the Valley of the Heart's Delight rather than a major agribusiness center. The cotton fields that inspired McWilliams's shocking phrase "factories in the fields" were enormous clusters of specialized production, pure stands of cotton as far as the eye could see. As the director of the Agricultural Extension Service remarked in the 1920s, "There is a general dislike of [California] farmers for diversified farming." In 1940 only 6 percent of holdings in the state were classified by the U.S. census as general farms. Don

California dairy farms, 1930

California poultry farms, 1930

Mitchell aptly calls California "an agricultural landscape of both stunning monotony . . . and stunning diversity."[6]

Here again, the differences between agribusiness and any other business are slight. Manufactured products such as Chevy Impalas and audio CDs are regularly produced in large batches and mass production runs. All capitalist production involves specialization and monoculture; that is the nature of the factory, modern industry, and the division of labor. California farming was never a mixed farming system such as the ones found in Europe, in early New England, or across the corn-hog belt of the Midwest. Farmers here could not "abandon diversification" in the 20th century, as Lawrence Jelinek

California fruit farms, 1930

puts it, because they had hardly ever (or had rarely) practiced it. That charge would only have purchase elsewhere in the country, where agriculture has followed California's lead to become almost exclusively specialist and monoculture today.[7] Postwar petro-farming meant an even more perfect monoculture because it ended traditional crop rotation with legumes and allowed repeat planting of the same crop. There has, however, been some retreat from absolute monoculture with the spread of organic farming in recent years, but less than might be imagined, since vast fields of carrots or potatoes can be claimed as "organic" so long as a few restrictions on inputs are observed.

A clear dynamic of the cropping system is the way new plantings have repeatedly swept over the face of California. Commodity proliferation has come in waves with every ten-to-twenty-year period of economic growth—a kind of monoculture of the moment. These waves typically lead to crises of overproduction after a time, as demand saturates and the economy turns down. Growers jump in to plant the latest high-value crop, acreage expands rapidly, and then oversupply drives down prices. Every bubble that bursts leads inexorably to the introduction of other crops with more promising markets. These waves of expansion can be seen clearly in the history of single California crops, or in the patterns exhibited by particular places, like the Salinas Valley. Pulling together the various crop histories, we get a rich overlay of introductions and expansions. This, too, echoes the experience other lines of modern business, from the bicycle mania of the 1880s to cell phones in the 1990s.[8]

The first great crop bubble was the Mexican cattle boom of the 1850s, which ended in 1863. Sheep ranching benefited from the same intensive enthusiasm before it, too, peaked circa 1875. Wheat experienced its first local glut in the recession of the late 1850s, then roared ahead until meeting the general crisis of the mid-1870s. It hit a new peak in 1888, only to collapse dramatically in 1893.[9] Fruit producers behaved very much like wheat growers. While returns in the mining era were exceptionally high, especially on prime alluvial lands, most took little notice of quality and rushed to market to keep their cash flowing in the go-go years. Fruit met its first overproduction crisis in 1858. The Civil War era saw another takeoff, followed by a second glut in 1868 before the transcontinental railroad opened up new markets. The horticultural boom of the 1880s rocketed upward with the same abandon as wheat, only to hit the wall in 1893.[10]

The 20th century opened with two great waves of growth, the first ending in 1914 and the next in 1929. In the first, fruit acreage continued to expand up to the sales slump in 1913–14 (a general downturn in the national business cycle, often overlooked). After the turn of the century, citrus took the lead over deciduous fruits. Mass production of eggs came into the picture along with the cultivation of sturdy vegetables, and canned fruit and vegetable production increased steadily. From World War I through the 1920s, California prospered despite the severe downturn in farming elsewhere in the country. Cotton and canning tomatoes took off, joined by a growing array of fresh vegetables, berries, and nuts. Dairying did particularly well in supplying the booming populace of Los Angeles. Wine grapes surged, despite the prohibition on wine making. This new wave of crops contributed to the continued rise of the share of "intensive crops" in output value to 80 percent by 1929. By the end of the 1920s, these, too, had hit the wall, thanks to overplanting and falling prices, together with rising mortgage debt.[11] The output slump of the 1930s shows the baleful effects of the Great Depression on most every crop.

From World War II onward, the big field crops led the way, with cotton, rice, and alfalfa enjoying huge waves of expansion. The postwar period was the golden age of mass-produced and standardized iceberg lettuce, almonds, and table grapes, as well as feedlot beef, milk in cartons, and butterball turkeys. By acreage, for 1975, this breaks down into 73 percent field crops, 17 percent fruits and nuts, and 10 percent vegetables, worth $2.7 billion, $1.9 billion, and $1.0 billion, respectively. In addition, cattle and dairy products generated over $1 billion apiece.[12] After the severe farm crisis of the early 1980s, a new wave hit, bringing with it the proliferation of new kinds of specialty crops. These include exotic vegetables, such as Thai chilies and Chinese broccoli; fruit delicacies, such as Champagne grapes and orange raspberries; and flowers and landscaping plants. In the wake of these high-value niche products, the production of field crops fell. Nonetheless, dairies, grapes, cattle, cotton, almonds, and hay remain among the top ten crops produced in the state.[13]

Overall, there is no evidence for the old truism that California agriculture experienced a once-and-for-all shift from "extensive" to "intensive" farming with the arrival of horticulture circa 1880–1900. Olmstead and Rhode call this the "great transformation," but it's more of a great mirage.[14] California farming has been nothing if not intense from the beginning—intensively commercial, opportunistic, and enthusiastic in its search for profit. It has been perfectly capitalist in that regard, and prone to the same rapid shifts in fashionable prod-

California's Top Fifteen Crops 1910–1996

(by value)

	1996		1980		1970	
1	Dairy	3,717,116	Dairy	1,771,383	Cattle/Calves	----
2	Grapes	2,158,543	Cattle/Calves	1,438,667	Dairy	----
3	Nursery	1,610,257	Cotton	1,389,342	Grapes	235,923
4	Cattle/Calves	1,151,419	Grapes	1,215,585	Hay	233,220
5	Cotton	1,073,986	Hay	723,316	Eggs	----
6	Almonds	1,008,576	Nursery	498,005	Tomatoes	165,250
7	Hay/Alfalfa	812,588	Tomatoes	490,310	Lettuce	149,410
8	Lettuce	735,088	Almonds	487,320	Cotton	137,215
9	Flowers/Foliage	702,842	Rice	423,612	Nursery	----
10	Proc. Tomatoes	664,167	Flowers/Foliage	395,907	Sugar Beets	121,688
11	Strawberries	584,860	Lettuce	382,563	Oranges	108,268
12	Oranges	532,140	Eggs	370,165	Rice	89,205
13	Chickens	457,502	Wheat	357,945	Potatoes	86,889
14	Broccoli	368,990	Chickens	229,177	Flowers/Foliage	----
15	Eggs	367,317	Oranges	224,548	Almonds	80,104

	1960		1950		1940	
1	Cattle/Calves	514,721	Cattle/Calves	363,198	Cattle/Calves	102,014
2	Dairy	382,711	Dairy	219,304	Dairy	67,128
3	Cotton	311,056	Cotton	195,549	Hay	40,164
4	Hay	177,086	Hay	111,021	Oranges	39,353
5	Eggs	166,875	Oranges	91,144	Grapes	30,357
6	Grapes	131,761	Grapes	86,922	Cotton	26,519
7	Oranges	110,453	Eggs	83,904	Plums/Prunes	15,662
8	Tomatoes	108,008	Potatoes	56,531	Dry Beans	14,171
9	Potatoes	79,988	Barley	55,808	Barley	13,895
10	Lettuce	76,554	Lettuce	42,741	Sugar Beets	11,778
11	Turkeys	71,078	Dry Beans	40,993	Walnuts	10,909
12	Barley	70,258	Sheep/Wool	39,070	Peaches	10,639
13	Rice	57,528	Walnuts	36,733	Sheep/Wool	10,411
14	Prunes	53,515	Chickens	36,634	Chickens	10,071
15	Sugar Beets	----	Rice	33,892	Potatoes	9,090

	1930		1920		1910	
1	Cattle/Calves	126,070	Cattle/Calves	120,681	Hay	42,187
2	Oranges	119,112	Hay	90,751	Dairy	19,083
3	Dairy	95,413	Oranges	67,071	Cattle/Calves	18,589
4	Hay	65,204	Grapes	65,781	Barley	17,185
5	Eggs	45,799	Dairy	52,510	Oranges	12,952
6	Grapes	43,113	Wheat	36,938	Cotton	11,744
7	Sheep/Wool	28,875	Barley	35,036	Grapes	10,847
8	Dry Beans	27,002	Dry Beans	31,448	Eggs	6,717
9	Cotton	26,207	Peaches	29,543	Dry Beans	6,517
10	Barley	20,179	Plums/Prunes	28,382	Wheat	6,324
11	Peaches	15,712	Sheep/Wool	25,906	Plums/Prunes	5,474
12	Chickens	14,699	Eggs	22,278	Potatoes	4,879
13	Wheat	12,830	Rice	20,433	Peaches	4,574
14	Pears	12,322	Potatoes	18,901	Sugar Beets	4,321
15	Walnuts	12,077	Walnuts	17,727	Walnuts	2,250

NB: The top fifteen crops represent between two-thirds and three-fourths of total production throughout.

Sources: CDFA Statistical Review 1996, California Statistical Abstracts 1981, 1961, and 1910 and 1950 U.S. Census of Agriculture.

ucts and speculative excesses as is all market-driven enterprise. All in all, the crop history of California is a record of restless expansion, diversification, and plunges into uncharted waters, or what Henderson calls "commodity revolutions."[15] To this day, there is no sign of a letup in the tendency of agribusiness to generate new products and find new enthusiasms. But cropping patterns only begin to tell the story of commodification.

BACK TO THE LAND

When California entered the United States in 1850, its lands were not yet freed up for the market. Millions of acres of prime farmland were under the dominion of Mexican rancheros, native tribes, and the U.S. government, and would have to be transformed into modern private property under the new American regime. This would require a generation, and until then full-fledged capitalist agriculture and high-value cropping could not be fully established. The native peoples' land presented the least obstacle. The indigenes had no system of property in land, no legal standing, and few weapons. They were summarily removed from any land the Americans desired. California's Indian removal was draconian, even by the brute standards of the United States up to that time. The state even refused to establish federal reservations for a generation.[16]

Mexican land, covering around 8 million acres, presented a thornier problem. The Californio rancheros did, at least, have a system of private property, and their rights were officially recognized by the Treaty of Guadalupe-Hidalgo, which concluded the U.S.-Mexican war of 1844–46. The Spanish-Mexican idea of land grants was far different from the Anglo-American law of real property, however, and most of the rights were poorly documented and registered in a form acceptable by U.S. courts. Although the rancheros were commercial operators and economic liberals in Mexican terms, this was not yet a world of fee simple property and free exchange of land. On top of that, Anglo settlers found it easier to squat on disputed claims and fight for the right to remain, rather than to purchase or lease from the Mexican owner. A forty-year process of squatting, litigation, and occasional murder and mayhem ensued. But the main mechanism transferring Mexican ranch land to Anglo suzerainty was the failure of the ranchero cattle operations in the face of natural disasters and plunging prices in the 1860s.[17]

Another 8 million acres, conveyed by the federal government to the state of California as part of its entry into the union, passed rapidly into private hands by the end of the 1860s. The federal government then undertook its own direct land disposal between 1870 and 1918. There were a bevy of programs to effect this, but most of the 30 million acres dispensed went under three headings: cash sales, homesteading, and railroad grants. Federal lands in California were privatized at upwards of a million acres per year, the amount tracking the ups and downs of the business cycle. Confusion, conflicting claims, speculation, fraud, and litigation were rife all through the process of public land disposal.[18]

Because the transformation of land into private property took a whole generation, for a long time agriculture rested on uneasy foundations. During the period of transition, grains were the most rational crop because of their low investment and quick returns; indeed, they were often grown on land leased from claimants. Wheat needed little preparation to succeed; all one had to do was drop in seed, wait, and harvest. It was agriculture without barriers and without regrets. Conversely, the risk of losing title and capital investment acted as a restraint on planting trees or making other improvements. Wheat pushed grazing out of the valley bottoms and into the foothills (this was not the Argentine pampas, where cattle would reign supreme over field agriculture). Therefore, wheat served as midwife to millions of acres of land privatization and as a decisive step in the development of California agriculture, contrary to most opinion that it was ephemeral.[19] Livestock played a similar role in the transformation of property rights in the hill country *after* the bust-up of the initial cattle and sheep booms, when grazing moved from the open range to demarcated ownership, water rights, and pasturage.[20]

The process of primitive accumulation of land during the first quarter-century of modern California history is justly notorious for making this a region of large land holdings. In 1880 California's average farm size was 460 acres, compared to 134 nationally, and many properties ran into the thousands of acres. Miller and Lux put together an empire of around 1 million acres up and down California. Grain production and grazing played an essential part in this, since both were easily undertaken on a large scale. The government facilitated consolidation by giving away the public domain with almost no restrictions. State lands, in particular, were decisive to the concentration of property in the early years—much more so than Mexican ranchos, contrary to popular opinion. Most important, the land accumulated in the hands of capitalists made wealthy by the gold and silver rushes, such as Ben Ali Haggin, Lloyd Tevis, the Stephens brothers, Hugh Glenn, and William Chapman.[21]

That was only the first wave of land entry, however. The story does not end there. Once title was settled, many large land holdings began to be subdivided for more productive purposes, especially as the horticultural era took off. Indeed, roughly a third of the largest acquisitions in the first wave had simply been speculative investments to cream off the profits from resale, and a huge amount of railroad land was redistributed in this manner. Big holdings were broken up into tens of thousands of fruit farms from the 1870s to the 1910s. As a result, land ownership became much more dispersed from 1875 to 1925, as Ellen Liebman has shown in an exhaustive survey; it fell to an average size of 224 acres in 1930, which was much closer to the national average of 157 acres. In the process of this massive sell-off, land was further transformed from mere private property to "real estate." It meant that titles passed freely through a wide-open land market, a necessary and quite general step to the capitalization of California agriculture, as George Henderson has so clearly described. The horticultural era was crucial to the commodification of land in this sense, especially in the making of modern Southern California in the 1880s. As Stephen Stoll puts it, "The landscape of the industrial countryside was founded by developers."[22]

Total land in farms and ranches in California shot up through the second half of the 19th century. It hit 29 million acres by 1899, with improved land encompassing almost 12 million acres and harvested farmland 6.4 million acres. Those totals are surprisingly close to the figures today (see table on page 87). It is significant that improved farmland had already topped out by the 1890s, although harvest-acreage would expand further in the 20th century to a peak of roughly 9 million acres. Once land was securely within the capitalist orbit, intensification could become the main route to further growth.

A third wave of land entry came in the early 20th century with the opening of new lands by drainage, irrigation, and oil discovery. The floodplain of the Sacramento River only came into production when it was converted to rice growing and when the river's flooding was controlled by levees, floodways, and finally massive dams. The Delta was gradually reclaimed from 1870 to 1920 and leased for row crops—mostly through the labor of Chinese workers. The lands of the Tulare Basin (southern San Joaquin) were progressively

Lettuce rows, Watsonville, 2003

drained for cotton from the 1920s to the 1950s. Imperial Valley only became significant as a winter crop zone with the stabilization of the Colorado River after the flood of 1905–07, and only flourished after control of the river in the 1930s. The dry west side of Fresno County came under irrigation late, through federal and state water projects, and grazing land was sold for cropping. Southern San Joaquin Valley lands were privatized en masse in the search for oil, then subsequently irrigated.[23] The Klamath-Tule lake basin was massively dewatered after World War II.

Most of these newly reclaimed lands, perhaps 2 million acres in all, were acquired by large, absentee owners, speculating in agriculture. They recapitulated early land entry in other regards as well. Where water control or soil quality was uncertain for many years, as in the Delta and Imperial Valley, leasing was a favored strategy to avoid risky commitment to ownership by farm operators. Similarly, transient crops like barley, rice, and alfalfa were planted to minimize fixed investments. Unlike earlier entries, these reclaimed lands were subsidized by government water projects.

Another reshuffling of landed property occurred after World War II, one conducive to the new production regime of the petrochemical age. There was considerable land concentration, as smaller growers sold out to more efficient and financially well-heeled competitors. Just as striking was the extraordinary shift toward leasing as a strategy of farm operation. By 1978, half of California cropland was leased, usually on three-year contracts, and that percentage has held pretty much steady to the present day. This marked a major, if unheralded, step in the devolution of landed property as the cornerstone of agriculture and the perfection of land as commodity and capital.[24]

Total acreage (including grazing land) and harvested cropland hit new highs in the 1970s (36 million and almost 8 million, respectively). Ranchland started a long decline while harvested fields only peaked in the 1970s, then fell back in the 1980s, thanks to low global prices, growing competition in specialty crop markets, high mortgage debt, and drought (plus cutbacks in cheap irrigation water). This was the first absolute pullback in California agricultural land use in over a century. It was also another signal that land, per se, had become less important in the high-productivity world of modern agro-industry.[25]

Because land continues to change hands through sale and leases, it is forever recommodified and priced in the open market at a value that reflects the profitability of farming on it. This will vary with crop prices and the quality of the land, as well as variable input costs (especially labor). Over time, there has been an almost inexorable rise in land values, because California agriculture is remunerative. Crop prices and productivity have, by and large, kept up with input costs, or even moved ahead. Any difference is quickly capitalized into land rents and land sales. Further pressure on agricultural land values come from the cities, which demand more land for urban development, often very far out. As a result of these considerations, farm land prices are higher in California than most other parts of the United States.

Land prices then feed back on agricultural production. They exert a con-

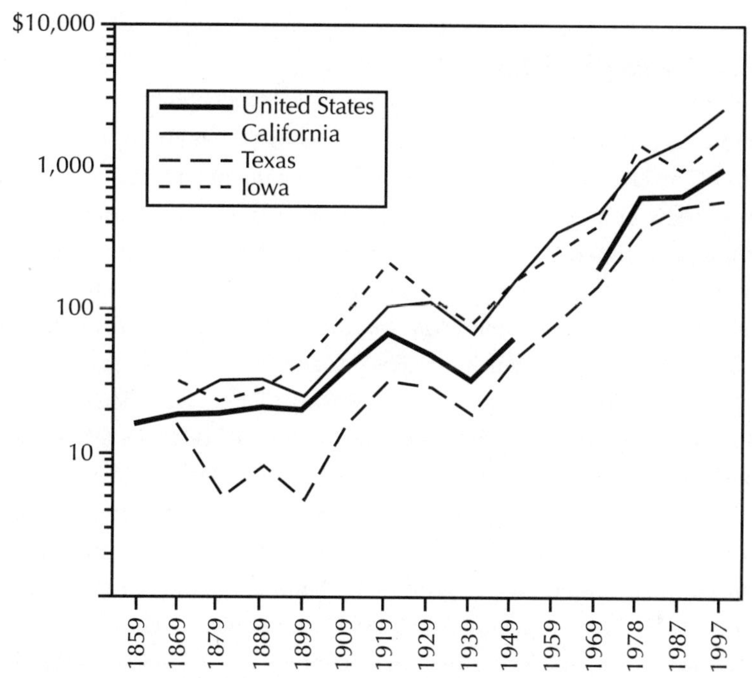

Note: Data missing for U.S. total in 1959 and state values for 1859.
Source: U.S. Census of Agriculture, various years.

Land and buildings per acre: California, United States, Texas, and Iowa, 1859–1997

stant pressure on growers, who feel the market individually as an external force, even if it is measuring collective performance. Rising prices press growers to cash out on rising land values (common in the case of urban pressures) or to increase profits, which they can do either by growing higher-value crops, raising crop yields, or by being more efficient. Efficiency means cutting unit costs by operating at a greater scale, reducing input costs (especially labor), or raising productivity through the introduction of new farming technologies. Competition means that growers are trapped in a "production treadmill" from which there is no escape, and the land commodity is imbricated in this dilemma they all face.[26]

AMBER WAVES OF LABOR

California agriculture has used hired labor from the outset of the American period. The employment of wage labor is fundamental to capitalist production systems, and the commodification of labor is the most basic element of primitive accumulation. The use of wage labor distinguished the Yankee era from the Mexican one as much as the revolution in landed property did. Ranchos had certainly been commercial, but they had mostly used the indentured labor of native peoples. Rancheros like Mariano Vallejo would go out on raids into Indian territory to bring back new workers. There were nominal wages, but in no sense was there a free labor market. In the face of the footloose and monied economy of Gold Rush California, the old system dissolved.[27]

The most striking thing about the labor system in California farming—and the one most remarked on by observers—is not just that it involves hired labor but that it has used one group after another, in a vast, repetitive cycle of recruitment, employment, exploitation, and expulsion. The cycle has been all the more noteworthy in that the majority of farmworkers have been nonwhites and noncitizens: California indigenes, Chinese, Japanese, Filipinos, Mexicans, Punjabis, Oaxacans, and Vietnamese. In this, California has been absolutely marked off from the mainstream of American farm history (though more closely linked to the rest of the Southwest and Florida). For geographer James Parsons, writing in the 1950s, "One of the distinguishing features of California agriculture, which for the most part lies outside of the rural farm tradition of the rest of the country, has been its dependence on a pool of mobile, foreign-born farm labor." Jelinek describes the endless labor turnaround as a "race and ethnic rotation" to match the rotation of the crops. Henderson calls this condition of production, succinctly, "The Ever-New, Ever-Same."[28]

But the ever-new quality of the workforce profoundly affects the ever-same quantity of labor. There is something more afoot in the fields of California than wage relations alone. This has been a peculiar variant of agrarian capitalism in which labor is commodified not once, but again and again. Labor has been repeatedly commodified in the most minimal sense of work for hire, but has never able to establish a solid foothold in the agrarian order. Hence it has never been able to make legitimate claims for good wages, political rights, or economic justice. This is the nearest thing one can imagine to the naked exposure to the market denounced by Karl Polanyi in his study of the British agrarian revolution, *The Great Transformation*. But it has not been a passing phase in California agriculture; rather, it has been the permanent condition of farm labor.[29]

As with the land, the commodification of human labor was not a once-and-for-all thing after the Gold Rush. After all, most forty-niners were in the goldfields and not looking to work for someone else. Wheat and cattle were once again the starting point for the process of primitive accumulation in agriculture. Their great virtue was a relatively low demand for labor at a time when there were few wage workers of any kind available (just as they were helpful in not demanding clear property rights in land). Large expanses of grain could be sown and harvested with little land preparation, and cattle could wander free with little supervision. The first hired hands were a mix of Chinese, whites, Indians, and Mexicans. Shepherds working the mountains were typically Basque. The Chinese were highly regarded as an agrarian labor force, but constituted perhaps as little as one-tenth of agricultural labor before 1870—because they were in the goldfields, too.[30]

As the agricultural economy took off during and after the Civil War, California farm labor was relatively well paid—receiving the highest farm wages in the country—because of its scarcity. Wages peaked just before the depression that hit in 1875–77.[31] That fact will surprise most people, because even the finest observers of farm labor have been vague about when the mass labor system came into being. This was not yet the age of dirt-cheap harvest labor that we now associate with California agribusiness. That can be understood in the same terms as the commodification of land: it was not an instantaneous transition. The agrarian labor market would have to evolve over decades—and it is never a finished product.

The expansion of horticulture after 1870 ushered in the modern harvest labor system. The age of fruit could not have proceeded without a larger agrarian labor force than had previously existed, because fruit picking is so labor intensive (plus ancillary labor demand for packing, canning, and shipping fruit). California's harvest-labor markets had to be filled up with new recruits, and they were. The post–Civil War workforce was a mishmash of U.S.-born whites and immigrant Irish, Germans, Chinese, and Portuguese, many of them ex-miners, and quite a few footloose migrants referred to as "tramps." The Chinese played a leading part in this period, their numbers totaling perhaps 20,000 workers, and ranging from 5 to 50 percent of the farmworkers in the northern farm counties.

A common myth of the time, widely disseminated in the declarations of Henry George, was that large landholders gained a decisive advantage over family farmers because of their access to plentiful wage labor, particularly the "coolie" labor of Chinese. But that is simply wrong, a view distorted by fears of land monopoly and anti-Chinese prejudice. Sucheng Chan has empirically refuted the blame put on the Chinese: not only were they not decisive in working large land holdings, they were critical in establishing many of the apple and pear orchards in the Delta and foothill counties, getting the wine industry off the ground in Sonoma, and clearing the foothills in the Santa Clara Valley of chaparral.[32] It is even more fantastic to maintain, in the manner of David Vaught, that horticulturalists were not heavy users and exploiters of wage labor. Indeed, as horticulture overtook grain and cattle as the advancing front of agribusiness, the new generation of fruit growers was the leading force refining the pool of cheap, mobile labor—without which they would have been hard-pressed to make the subdivision of large holdings a workable proposition.[33]

Chinese labor supply dwindled in the face of the racially motivated Chinese Exclusion Acts of 1882, 1892, and 1902 (passed despite opposition from many growers who valued Chinese labor). New bodies had to be found. Japanese were the next large group of recruits, beginning in the 1890s. Most made their way to California via the Hawaiian sugar plantations. Some 30,000 *Issei* were at work in the fields by 1909. Japanese workers were instrumental in the burgeoning cultivation of berries and fruits, and established vigorous communities in Sonoma, Santa Clara, and Alameda counties, as well as around Los Angeles. But the Japanese wanted to be farmers themselves, and began to move out of the labor force into farm tenancy and ownership. Furthermore, racial antagonism to Japanese success in California, mingled with larger forces of international hostility to Japanese imperial power, triggered a wave of anti-Japanese feeling that pressured the United States government into tightening the quotas on immigration in 1908. This cut into the farm labor force, leaving growers once more in a tight spot by the 1910s.[34]

The growers were willing to try anybody who would work under the increasingly harsh conditions and low wages of California's fields and orchards. Although white racism saturated the discourse on immigration at the time, practical considerations always trumped fear of foreigners among the growers. They tried importing several thousand Punjabis (mistakenly called Hindus) to the Imperial Valley around 1910 on the hope that they would be seen as "white" in the racial categories of the time, but the courts cut off that approach. Filipinos began to be recruited into the fields, now that the islands had been pacified by military might and secured under American colonial control. Some 15,000 or more Filipinos would be present in the fields by the 1920s, primarily in the Delta and Salinas Valley. Then there was a large array of fresh European immigrants ripe for exploitation, above all Italians and Portuguese. The latter were most numerous around Fresno County and the former in the Bay and Delta regions.[35]

At least two dozen nationalities were represented in the labor force by World War I. Anglos would still make up roughly half the harvest labor pool, despite the influx of immigrants, up to World War II. A large number of workers were permanent migrants, or tramps, moving around California with the harvests and even around the whole West. But others were drawn from local townsfolk and out of the pools of casual labor in the cities—Stockton, Oakland, and San Francisco in the north; Fresno, Bakersfield, and Los Angeles in the south. Rural and urban labor pools were by no means entirely separate. No wonder that Stockton had the lowest per capita income among cities in the

United States for decades, and San Francisco had the highest proportion of transient housing in the country. Nor did Japanese and Chinese workers entirely disappear through the 1920s, though they tended to be the most geographically confined, in Delta towns like Walnut Grove and Santa Clara Valley settlements like Alviso.[36]

By World War I, the full-blown harvest of cheap labor was firmly in place, with 150,000–200,000 workers hired over the course of the season. Demand for mass agricultural labor continued to rise through the 1920s, thanks to the never-ending fruit boom and the expansion of cotton and vegetables. In 1924 more than three-fifths of California farms used hired labor, the highest figure in the nation by far. Not until the 1930s, however, would scholars such as Paul

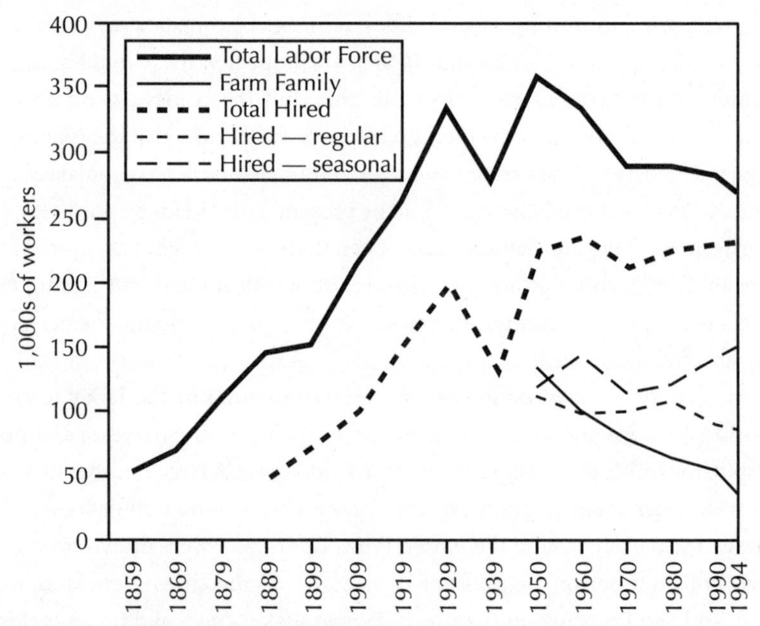

Source: Olmstead & Rhode 1997, 1859–1939 from U.S. Census
Carter & Goldman 1997, 1950–1994 from California Employment Development Department
Hired labor 1889–1939 estimated from various sources.

California's farm labor force, 1859–1994

Taylor and Varden Fuller and journalists such as Carey McWilliams and John Steinbeck finally recognize the system for what it is: a necessary part of the California agrarian landscape and production system.[37]

Southern California and the Imperial Valley had surged to the front ranks of farm production by this time, and there the growers looked longingly across the southern border at the potential pool of Mexican workers—who had been absent from the California scene since the Civil War. Recruitment started up after the war, with the help of the federal government through special provisions in the 1917 Immigration Act and by administrative loopholes. A ready supply of Mexican labor was forthcoming because of the railroads built across the border; several decades of recruitment into the Southwest for mining, construction, and farmwork; and the dislocations of the Mexican Revolution (1910–17). These people went to work in Southern California's burgeoning orange groves, cotton fields, and melon patches. By 1920, Mexican workers already numbered around 70,000, perhaps three-fourths of the regional harvest labor force. In the mid-1920s there were almost 60,000 Mexicans in the San Joaquin Valley, which had the greatest labor demand: 35,000 pickers in cotton alone, four-fifths of them Mexican. The main centers of settlement were around Los Angeles and the Imperial Valley; as Mexicans migrated into the San Joaquin Valley, they ended up in barrios all along the east side.[38]

Conflict and racism once more stanched the flow of workers in the early 1930s, as hundreds of thousands of Mexicans and Filipinos were deported in the midst of the Depression, creating a labor shortage in 1933–34. So a fresh infusion of wage labor was sought among white "Okies" fleeing agrarian collapse in the Dust Bowl (Okies were actually from Arkansas, Texas, and Missouri as often as from Oklahoma). In 1935 they came into California by the tens of thousands, a flow that peaked at over 100,000 in 1937 and had brought around 400,000 into the state by 1940. Forty percent of the new arrivals ended up in the San Joaquin Valley, where they put their stamp on such towns as Oildale, Madera, and Arvin. They were put to work on the burgeoning grape, tomato, and cotton crops. Cotton was by then the most labor-demanding single crop in the state. By 1935 the total harvest labor force stood at 145,000. The Okies' reign in the fields was short, however, as the outbreak of World War II drew them into industrial jobs in the cities. Nonetheless, the demand for field labor rose again during the war, to 225,000 workers.[39]

Growers turned to the government for assistance in recruiting labor in wartime. Convicts, prisoners of war, and interned Japanese Americans were all deployed. But the most important wartime program was the U.S.-Mexico compact known as the *bracero* program, which began in 1942 as a wartime emergency measure and lasted for twenty-three years. This was an agreement signed with the Mexican government to bring men into the United States for agricultural work. It funneled over 200,000 workers into the Southwest during the war, with California growers taking the lion's share (two-thirds during the summer and almost all during the winter months). The first delivery of braceros in 1942 was, fittingly enough, to Stockton, the great wen of California harvest labor. The number of braceros would peak in 1956 at almost 450,000 in the United States. Braceros provided up to one-third of all seasonal labor in California in the mid-1950s and over 90 percent in the Imperial Valley and some southern counties; when Texas cotton was mechanized in the early 1960s, California ended up with three-fifths of the total number. Braceros constituted over half the labor force in key crops such as tomatoes, lettuce, strawberries, sugar beets, and cotton.[40]

Braceros were a kind of indentured labor: contracted by the government, housed in closed camps, bused to the fields, and sent home to Mexico once the season was over. One critic compared the braceros to sprinkler systems that could be turned on and off at will. Bracero hourly wages crept up from $.75 to only $1.25 over twenty years, despite the greatest economic prosperity ever known in the United States *and* in Mexico. Everywhere braceros were employed in large number, average farm wages stagnated or fell. Indeed, it was clear at the time that growers used the program to lower farm wages, as reported by both a Presidential Commission on Migratory Labor in 1951 and a California Legislative Research Service report in 1955.[41]

Braceros were workers of the most compromised kind, an army of cheap labor marching to the government's beat. The program was a dramatic setback for American organized labor, overturning the Anti-Alien Contract Labor Law of 1885, which had prohibited importing contract labor. It was also a reversion of sorts for California's agrarian order to a regime of primitive accumulation—in the midst of vast modernization of the production system on every other front. It is likely that abnormally low wages due to the bracero program helped underwrite the postwar spurt in profits and accumulation, including the rise in land prices and land concentration, and investment in machinery, irrigation, and petrochemicals.

The bracero program was ended in 1964 by a combination of changing national priorities under the Kennedy administration and agitation by the AFL-CIO in support of the fledgling Farmworkers' Organizing Committee of Cesar Chavez. Recommodification of labor returned with a vengeance afterward. Growers reverted to hiring illegal immigrants, who had been coming to California in greater and greater numbers—induced, in part, by the experience of the braceros. The Immigration Reform Act of 1965 is generally credited with opening the door to a revival of mass immigration to the United States, which had ended in 1924. But in the case of Mexicans, it was window dressing, since they were exempt under the quota system in any case. Moreover, the border was notoriously porous and immigration restrictions poorly enforced by the Immigration and Naturalization Service (INS). The INS could hardly do otherwise, since the growers' lobby kept up the pressure for cheap labor supply. With the braceros gone, the INS instituted a program of work permits (green cards) for cross-border "commuters" that filled part of the gap. Many other farmworkers just crossed the border without documents.[42]

In reaction against the new mass immigration, the U.S. government began closing down the border again. The first big step was the Immigration Reform and Control Act (IRCA) of 1986. This law was full of the same contradictions that had hounded the border patrol for years: the conflicting demands of xenophobes and employers over the value of illegal immigrants. On the one hand, it sought to stop undocumented entry; on the other hand, it granted legal status to thousands of people who had been in the country for several years. Naturally, the IRCA contained a loophole for agribusiness called the Special Agricultural Worker provision. Congressman Edward Roybal of California denounced IRCA as a "farm labor bill, a bill that is designed to provide cheap labor for the farmers and growers of this country." The atmosphere in California darkened in 1994 with the passage of Proposition 187, which attempted to cut off services to illegals (it was later nullified by the courts). In the reactionary spirit of the time, the Clinton administration responded with Operation Gatekeeper and built a wall 150 miles along the Mexico-California border. It didn't work, of course, because the horse was long since out of the barn. Moreover, immigration law still provides the H-2A visa program for agricultural workers (under which 10,000 or so Mexicans a year come into California). In 2004, the Bush administration proposed a return to a formal, state-run labor recruitment policy with Mexico, something greatly desired by Mexican president Vicente Fox.[43]

Today around 80 percent of California field-workers are undocumented. The places of origin within Mexico have changed over the years, as well, and about a quarter of the workforce are now Miztecans from Oaxaca. They are joined by thousands of Southeast Asians, from communities of refugees placed in the Central Valley at the end of the Vietnam War. The total numbers of workers in California agriculture remain huge: at least 200,000 to 250,000, and possibly as many as 400,000 per year. The hired agricultural labor force has stayed remarkably constant since 1950, even as the family labor force has diminished—making California (along with the Southwest and Florida) an exception to the general decline of farm labor across the country.[44]

The great paradox of California's agriculture is that it remains fundamentally a cheap labor regime despite the immense growth of farm revenues and productivity. Such prosperity ought to have raised all boats, allowing wages to rise as in most other industries of the postwar period; but it has not. Instead, low-wage labor has been systematically built into labor relations and the reproduction of capital, as Varden Fuller first argued in the 1930s.[45] The ever-new, ever-same condition of the endless recommodification of labor and of labor market oversupply has never allowed any group to secure enough of a foothold to stake a claim to a fairer distribution of the immense value produced. That lack of rights is hardly accidental, since it is systematically maintained by agro-capitalists through labor recruitment, union busting, and lobbying in the halls of government. It is not true that a capital-intensive agriculture *requires* cheap labor, as Mitchell argues; rather, wage rates are set through class struggle that the workers have repeatedly lost.[46]

This is quite contrary to the fate of the land. With land, recommodification means a continuous revaluation upward whenever productivity and crop prices rise. Why is this? Land is a commodity operating in a context of social power. Land is still scarce, because it is still necessary to agriculture, still productive, and disappearing in places under city pavements. Equally important, landowners are in a position to dictate prices because they are members of a powerful capitalist class, either as grower-owners, financial syndicates, estate trustees, or banks.

Workers, on the other hand, are recommodified in a position of weakness. Their numbers are elastic, as new waves of immigrants arrive, and they have never gained a solid footing through unions or other means of asserting their right to a greater share of farm revenues. Indeed, the low wages of farm labor are an important factor in the continuing profitability of California agribusiness, and hence in the rising value of land. In those rare instances where union pressure has driven up wages, chiefly in the 1970s, lease rates on land fell commensurably. Normally, the rising value of land acts as a built-in force, seemingly beyond human control, to pressure individual growers to keep wages low in order to maintain profitability.[47]

3

Enter the Grower

Up to this point, we have established the commercial nature of California agriculture and seen how the process of primitive accumulation of the means of production unfolded. Now we need to turn to the key actors who make a capitalist production system go: those who buy the land and wage labor on the market, command the production process, and sell the crop. These men and women are known in California as *growers.* They, like all modern businesspeople, try to produce their products at a reasonable cost, sell them successfully, and make a profit. This is not always an easy task. One might say that they try to wrest a surplus from labor and nature, without selling too cheaply or incurring too much debt to keep operations going. To do so, they have to be good farm managers, overseeing large areas of land, great numbers of workers, a profusion of plants and animals, and machinery and supplies; planning over long cycles of yearly growth and harvest; and balancing the books at night. In the process, they have to contend with several unruly things—workers, nature, and competitors—and in these contestations lie much of the dynamics of farm production. In order to keep ahead of the markets, growers find themselves searching for new and better crops, ways of lowering costs, and means of increasing the productivity of their land and labor.

These basic functions of the grower ought to be the focus of any study of agribusiness. But too often they have not been, because the discussion has been diverted into a misleading debate over the size of land parcels held by growers or absentee owners. Carey McWilliams set the terms for the treatment of California agribusiness for decades after the publication of *Factories in the Fields.* McWilliams was caught up in the stark vision of the cotton fields of the 1930s and translated this into the absolute sovereignty of the "vast agricultural domains" of the great interior valleys.[1] In this he stood in a line of giants, from Henry George to Paul Gates, who misread the capitalist trajectory in California agriculture as primarily one of large-scale land ownership. As David Vaught has recently argued, the great majority of California farms have not been that big in sheer acreage nor organized on a factory basis.[2] But even as Vaught disagrees with facts on the ground, he nonetheless stands in a long line of scholarly populists, including the admirable Paul Taylor—the man who taught McWilliams, John Steinbeck, Dorothea Lange, Varden Fuller, Walter Goldschmidt, and many more about California agribusiness—who have taken the ideology of family farming and the 160-acre homestead far too literally. Therefore, we need to steer a course between the Scylla of factories in the fields and the Charybdis of the family farm in order to capture the elusive agrarian capitalist in his lair.

ORIGINS AND OUTLOOK

California is one of the few places in the world where capitalist agriculture was established free of serious challenge by any prior mode of production, whether feudal, family farm, sharecropping, or plantation. Almost every other path of modern agrarian development across the globe meant the long emergence of capitalist farming out of deeply entrenched agrarian societies, involving fierce political struggles against peasants, landed elites, sharecroppers, or other rural classes. This contest is at the heart of the agrarian debates from Lenin to Barrington Moore. Here, by contrast, as McWilliams has put it, "the lights came on all at once." What made a purely businesslike approach to farming possible in California was the ability to sweep aside all precedents and start over on a fully capitalist basis.[3]

California was forged in a crucible of absolute capitalism in the years after 1848. Californian agriculture therefore became an oddity even by American commercial standards. It shared with southern cotton, Northern wheat, and Mexican ranching an orientation to the market, to be sure, but in other regards was far advanced at an early stage. The thin substrate of Mexican ranching was swept away in a few years. In any case, the Mexican rancheros, who were already supplying international markets with hide and tallow, lacked two crucial commercial inputs: land was acquired through government grants, not land sales, and labor was mostly indentured or enslaved Indians, not wage workers. California was certainly unlike the Southern states, because it refused to accept slavery into its constitution in 1850 (not that some forty-niners didn't want it). Last, there was the lack of mass settlement of the California countryside with small farmers, in contrast to the Northern states.

This is a crucial distinction, over which much ink has been spilt and a goodly amount of populist fervor has been expended. The differences need careful explication. Farmers are so much a part of American identity and so vital to the whole Jeffersonian myth of the Western frontier that it is hard to see the modern farm operator through the haze of national memory. To begin with, California's total commercialism and capital saturation left the subsistence family farmer nowhere to stand. The California grower has almost never been simply a family farmer in search of sustenance, as were settlers in New England, Pennsylvania, and the Ohio Territory in the making of colonial Anglo-America. California was established at the historical turning point in the mid-19th century when subsistence farming gave way completely to commercial growing throughout the United States; Californians imposed the new order within a single generation.[4]

From the outset, California growers served commercial markets wholeheartedly in the mines and cities. They also sought out land to purchase, and the first ones into the action were often big speculators flush with mining capital. Furthermore, they hired wage labor from the git-go, and this put them squarely in the capitalist camp. To be sure, commercial farmers in the Northern states were also selling most of their crop to urban markets after 1850. But this did not make them into agrarian capitalists. Northeastern family farmers had to buy land, but they did so to homestead—to make a home for their families (an idea enshrined in the free land given away under the Homestead Act of 1862, the last gasp of mass frontier settlement policy). Family farmers employed some hired hands, but these were supplementary to family labor and usually came from other farm families. At the heart of the family farming idea, moreover, was reproduction of the household: staying alive, staying afloat, and setting up the children on farms of their own.[5] Capitalist farming in California differed clearly in its motive: it has always been for profit.

The typical farmer in California has been a money-oriented, businesslike operator. From the time of statehood, California growers have been, almost without exception, looking for opportunities to make a profit. With typical Gold Rush gusto, the early entrants "tended to be part farmer, part entrepreneur, and part speculator," as Rodman Paul put it. Southern California citrus growers, according to historians such as Stephen Stoll and Vincent Moses, were "orchard capitalists" and "revolutionary capitalists" because of the vigor with which they implemented new strategies of cultivation and organization. A writer at the turn of the century observed that the California grower had accepted "the necessity of placing his business upon an industrial basis as firm and attractive as possible." If the farmer did not understand the real nature of California agriculture, they would be culled from the pack. When A.P. Giannini, founder of Bank of Italy (Bank of America), went looking for smaller farmers to lend to in the early twentieth century, he sought out "business-minded farmers who regarded agriculture as a commercial pursuit rather than a means of subsistence." Southern Pacific Railroad declared in its promotion of Imperial Valley that "the irrigated countries are no place for the poor farmer. The man who goes there must use his brains in all his farming. He must be willing to learn. He must work hard, and he must have some capital."[6]

Because there was no deep history of precapitalist farm communities, practices, and ideas to carry forward, in California everyone who took up a shovel or a plow was new to the business. A remarkable thing about California growers is how few had been farmers in an earlier life. The agrarian minions hailed from every kind of background in their home states or countries. They were miners, lawyers, manufacturers, merchants, and even fallen aristocrats, but rarely humble farmers. As one observer noted at the height of the horticultural era, "Most of the fruit growers have come from other states, many of them having succeeded in business or professions other than agriculture. They have, therefore, been favored not only by their special business training but also have not had the handicap of having to forget the general farming methods of the . . . regions from which they came."[7]

Agoston Haraszthy, the father of California wine, was a wealthy landowner who fled Hungary in 1840. The founders of the Riverside Colony in 1870 included a judge, lawyers, doctors, a druggist, and other businessmen. The leading citrus growers in Riverside in the 1890s were a former nurseryman, a banker, a treasurer of the National Biscuit Company, and a graduate of Cornell University. In the Santa Clara Valley, James Reed, a pioneer fruit grower, was a former furniture manufacturer from Illinois; George Hardy, a former New York eye doctor, created the world's largest prune orchard (350 acres by the 1890s); Edward Goodrich, an olive grower, was trained as a lawyer and married to the daughter of a state supreme court judge; and Thomas Montgomery, a prosperous orchardist, was the leading businessman of San Jose. Farther north, "Hop King" D.P. Durst of Marysville was a doctor who jumped into farming along the Yuba River; George W. Pierce Jr., almond grower of Davisville, was the first graduate of the University of California from the Sacramento Valley; George Kellogg of Placer had been a miner before turning to farming; grower Peter Shields of Davisville was a lawyer and judge; and Frederick Ayer of Marysville was a water capitalist who turned to fruit when hydraulic mining was banned. Notably, the majority of students entering the farm school at Davis in 1916 were from the city, not the country.[8]

Entry to farming in California was open to a large number of immigrants with skills and energy that often exceeded those of Anglo-American contemporaries. Chinese were prevented from owning land by racial exclusion laws but still managed to become important as tenant farmers in the Delta and

Sierra foothill regions. Acreage leased by Chinese peaked at 25,000 in the Delta in the early 20th century. The Japanese were even more successful at moving from farm labor to farm operations, after they entered in number after 1900. By 1920, over 5,000 Japanese *Issei* families were cultivating over 350,000 acres (out of over 450,000 acres on their farms) from San Jose to Los Angeles, producing one-eighth of California's farm output by value. Only about one-sixth of this acreage was owned, however; the rest was leased or sharecropped because of racial barriers like the Alien Land Law of 1913 thrown up to impede them. This was followed by the Oriental Exclusion Act of 1924, which halted the expansion of Japanese farming; but Japanese still controlled 5,000 farms in 1940, producing over 40 percent of the state's truck crops.[9] Had the racism been less virulent, one could easily imagine tens of thousands of Asian farmers across the California landscape today, instead of just a few hundred.

Frank Namimatsu on his farm, San Jose, 1945

European immigrants faced fewer barriers but often built themselves up from almost nothing through assiduous labor and good farm management. About one-third of the approximately 90,000 growers in the state in 1910 were foreign born, and they were predominantly to be found working in vegetables and specialty fruit. Italians predominated in wine grapes and olives, Jews in eggs and chickens, and so forth. Many, like the Armenian raisin growers around Dinuba, were wildly successful (an Armenian son, George Deukmejian, become governor in the 1980s). It didn't hurt that Giannini's Bank of Italy targeted small farmers, especially immigrants, for loans to help them thrive. Today, the pattern has been revived by new immigrant Asians (mostly Hmong, with some Vietnamese and Indians) and Latinos (mostly Mexicans), who own about one-eighth of all farms in the state.[10]

Vaught tries to make the case that small growers were fundamentally different from big growers because "horticulture was a way of life *and* a business." But *all* business is a way of life as well as an economic proposition; only a naive economic theorist could think otherwise. No one is purely calculating and businesslike in their dealings, even the prodigiously successful cotton king Jim Boswell. I have no intention of making such an argument, nor losing sight of the tensions between the capitalist dimension of California agribusiness and other dimensions of the social order in the countryside. Certainly many California growers put their families' welfare above profits. Immigrants brought a host of new cultural influences, many of them contrary to American commercial logic. Racism intruded as a blatantly antirational suppression of highly productive Asian workers and farmers. Utopian desires for land and freedom would influence many a rural settler. While all this is true, the businesslike, capitalist nature of California agriculture was always the third rail providing the power that accelerated agrarian development. Despite a longstanding debate over the "land monopolists" and the fate of the "family farmers" in California, the fundamental nature of farming here has never been in doubt. As Vaught himself observes, "farm size alone does not define relations of production," which means—contrary to his own position—that horticulturalists could be just as much capitalist farmers as bonanza wheat kings.[11]

Moreover, the capitalist nature of California farming became more entrenched as time went on. This derives not only from commercial outputs and inputs, wage labor and profit making, but also from the accumulation of capital. Once growers got going, they usually prospered by virtue of the good returns they earned on high-value crops, productive land, and cheap labor. A crude index of this is the relative value of output per farm in California compared to the rest of the country. Net profits would be a truer measure, but there is no general index of profitability. Reports of high returns were commonplace in 19th-century California, and were not all simply boosterism and bluster. Wheat, citrus, cattle, truck crops, hops, and more were all turning nice profits in their banner years.[12] As growers made a lot of money, they turned it back into new capital to build up their operations, planting more vines, acquiring more land, hiring more labor, buying more machinery, and the like.

One sign of this rapid accumulation is how swiftly planting took place in

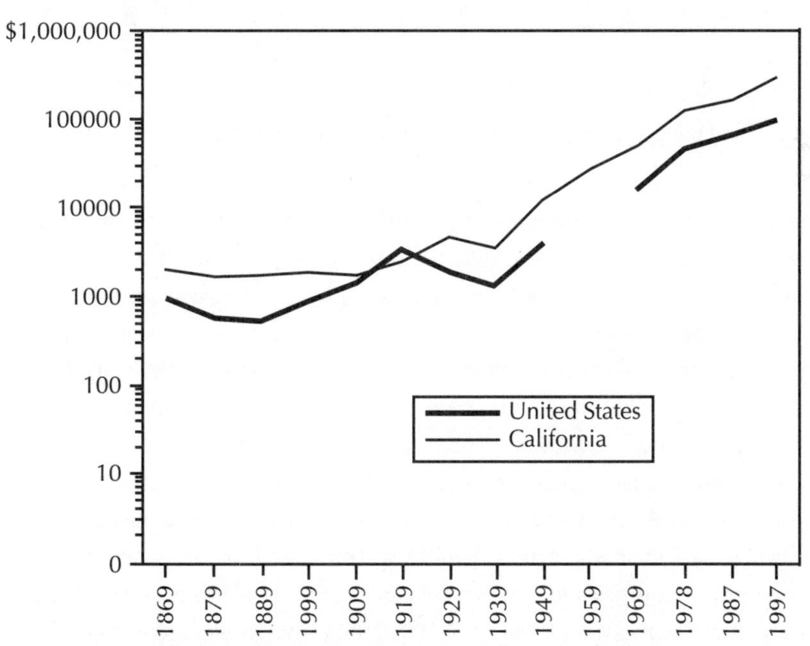

Note: 1959 data missing for the United States.
Source: U.S. Census of Agriculture.

Output per farm: California and United States, 1869–1997

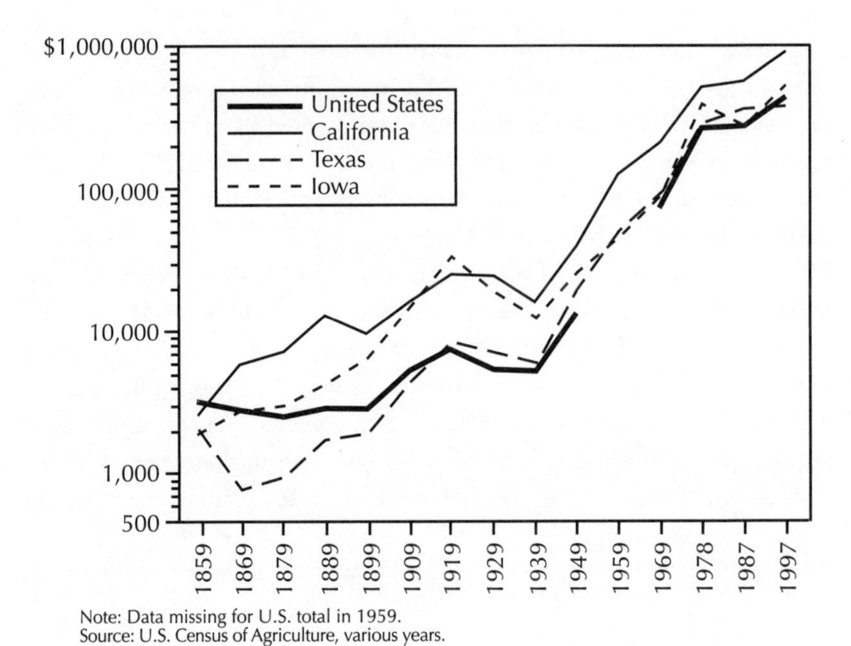

Note: Data missing for U.S. total in 1959.
Source: U.S. Census of Agriculture, various years.

Land and buildings per farm: California, United States, Texas, and Iowa, 1859–1997

profitable crops, like oranges or grapes, with every upswing of the business cycle. Another sign is that California growers acquired substantial capital as they built up their farming operations. This was true regardless of the size of landholding. The capitalization and profit of a large fruit orchard (like Dr. Hardy's prune farm) could easily rival a wheat ranch ten times its size. As Cletus Daniel observes of the new horticulturists of the late 19th century, "when scale was measured in such terms as land value, level of capitalization, degree of irrigation necessary, and labor requirements, it was plain beyond any doubt that what the commercial fruit and vegetable farming of the 1880s lacked in spatial extent it more than made up for in concentration of economic resources and intensity of operation." In the late 1920s, over one-third of all farms grossing over $30,000 in the United States were in California. By 1925, the total value of land and buildings on all California farms came to

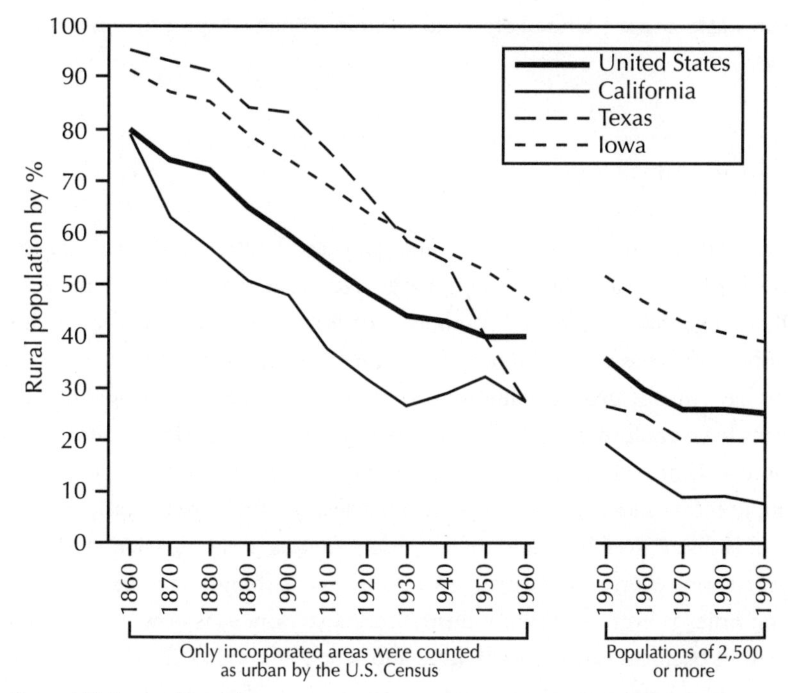

Source: U.S. Census of Population.

Rural population: California, United States, Texas, and Iowa, 1860–1990

the hefty sum of $3.15 billion, exceeded only by Illinois and Iowa—but not for long. By the 1950s, it hit an astronomical $100,000 per farm, and left the others far behind.[13]

A further index of the difference between California and the Northeast is the lack of mass farm settlement in the state. Whereas millions of farm families came to rest from Pennsylvania across to the Dakotas, in California the numbers were always modest. The number of farms did not reach 100,000 until the World War I, and topped out at under 150,000 in World War II. This has always been an urbanized state, unlike states such as Iowa, Indiana, and Kansas. Farm families made up 60 percent of the population in 1870, 45 percent in 1900, and 30 percent in 1940, far below the other leading farm states and the United States as a whole. Today, the proportion is minuscule.[14]

SCALE AND SCOPE

Rather than dividing into two antagonistic classes, California's farmer-growers have walked on two legs. Large and small growers have represented different specialties and land conditions, or slots in the division of labor. Scale in operations has depended heavily on the crop and the location. Large holdings have mostly grown grains or grazed cattle and sheep. Small and medium farms have predominated in orchards, vineyards, truck gardens, dairying, and poultry. That is, farming is organized around economies of scope, not just simple economies of scale. Scope means size appropriate to the task, in which labor skills are well honed to the job, equipment is well utilized, and management is not overstretched trying to do to many things. The logic of scope economies is everywhere in industrial organization, as is now widely recognized in business economics.[15] Scale and scope are not to be confused, however, with ownership and the mode of production.

Even in the early years of the wheat and cattle empires that so exorcised Henry George's contemporaries, the largest-scale operations occupied sites at the far edges of settlement, such as the upper Sacramento and lower San Joaquin valleys and backward Southern California.[16] Along the main axes of commerce around the San Francisco Bay, east along the Delta, and up into the gold country, however, small farms sprang up to serve the mining towns and cities with potatoes, fruit, wine, and vegetables. Then, as the horticultural era unfolded in the last quarter of the 19th century, farm size declined dramatically. The wheat and cattle barons eagerly subdivided prime lands (in both soil and transportation terms) into small farms. Why did they do this? Because they were not *latifundists* of the sort found in Prussia or Argentina, who held land as the principle means of production and social power; and because they could make a killing selling to small holders, who did a better job of converting to horticulture and garden farming. Small holdings multiplied from 23,500 to 72,500 between 1870 and 1900, with the average farm size declining from 482 to 397 acres. This was particularly dramatic in areas outside the initial zone of truck farming, particularly in the San Joaquin Valley and east of Los Angeles. In the most intensive areas, average size was much smaller: in Santa Clara County it had fallen to 50 acres by 1910.[17]

California Farms, Farmland, and Farm Size
1859–1997

Year	Farms (1,000s)	Farmland	Ave. Size (1,000s)	U.S. Ave.	Cropland (1,000s)
1859	19	8,730	459	199	---
1869	24	11,427	476	153	---
1879	36	16,594	461	133	3,321
1889	53	21,427	404	136	5,289
1899	73	28,829	395	146	6,434
1909	88	27,931	317	138	4,924
1919	118	29,366	250	148	5,761
1929	136	30,443	224	156	6,549
1939	133	30,524	230	174	6,534
1949	137	36,613	267	216	7,957
1959	99	36,888	372	303	8,022
1969	78	35,328	454	389	7,649
1978	73	32,727	447	449	8,804
1987	83	30,598	368	462	7,676
1997	74	27,698	374	487	8,563

Farmland = all land in farms and ranches, including grazing land and unused land
Cropland = all harvested land
Average size = mean size of farms based on total farmland

Source: Olmstead & Rhode 1997, Table 1, and U.S. Census of Agriculture.

Relations between large and small growers could be tense, of course. Post–Civil War California was rife with bitter complaints about how big capitalists had elbowed homesteaders out of the way during the crucial years of public land disposal.[18] While land-holding became more dispersed during the horticultural era, the transition was not always smooth. A good example is the fierce struggle between the old order of wheat and cattle and the new orchardists of the Santa Clara Valley from the 1870s to the 1900s, with the former rallying behind the Workingmen's Party (along Irish-Catholic lines) and the later organizing as Progressives. The famous dispute at Mussel Slough on the San Joaquin River that became the basis for Frank Norris's *The Octopus* pitted big and little wheat growers against each other, not just the Southern Pacific. The formation of a raisin cooperative around Fresno in 1890–1910 took fierce jockeying between large and small growers, Anglos and Armenians, in which one side had numbers and the other money and hired nightriders to enforce the collective will.[19]

What this contest between different groups of growers does not represent, however, is a sustained contest between two modes of production, capitalist and family farm. The long national debate over family farms and the future of the republic was always tangential to California. It was a displacement of the Eastern family farm ideology to a place where it never had any real purchase. As Daniel puts it, "California did not represent a departure from the dominant family-farming tradition in America for the simple reason that California was never part of that tradition." As a result, there was never any sustained rebellion against agribusiness from a small farmer class. Compared to every other place in which capitalism has subverted the countryside, from Britain to Texas to Iran, California has spawned no real mass movement of protest and opposition to the capitalist mode of development.[20]

Of course, the Jeffersonian mythos was active here. The vision of California as an agrarian paradise of the small holder seized hold of popular imagination from time to time, particularly at the high tide of horticulture in the early 1900s. The hope lay in the way the number of farms kept rising, hitting 118,000 in 1919 and 136,000 in 1929. California's fruitful farms and irrigated

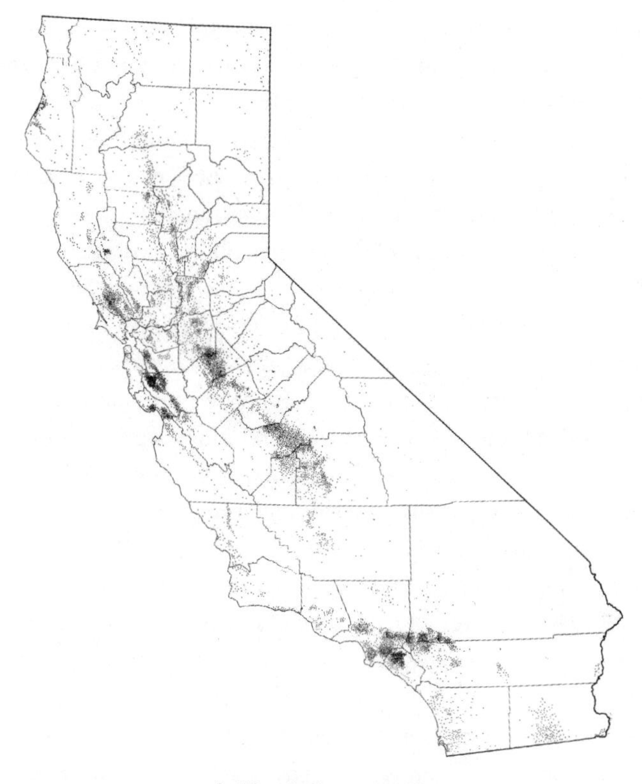

California farms, 1930

acres seemed to offer a utopian promise to the small owner in a fully capitalized society: a perfect marriage of agrarian sensibility to modern business sense, small property to global commerce, individualism to collective order, enlightened operation to high returns, rural living to the urban framework. Yet the idyll failed. There was quite a bubble in irrigation colonies, promoted both by populists like Elwood Mead and capitalist speculators trying to profit from high land prices in good times. But the economic downturn just before World War I undercut most of the irrigation schemes, bankrupting the likes of Borax Smith of Oakland, Will Tevis of San Francisco, and the Kunz brothers of

Pittsburg (Mead got a good government job with the Reclamation Service). Small operators were again badly squeezed by the fruit glut of the late 1920s, followed by the Great Depression of the 1930s.[21]

The first part of the 20th century brought a counterflow to subdivision once more. Big operations came to dominate the Delta islands, the cotton fields of the Tulane basin, and the Imperial Valley, and others grew through diversification, leasing, and buyouts of failed neighbors. Most notorious of these is J.G. Boswell, and there is nothing pretty about the way he, Everette Salyer, and other big producers drove the small Kern County growers and their California Cotton cooperative out of business. Still, after all this, many small farms survived with an infusion of off-farm income, so that in absolute numbers, small farms topped out in 1935 and were still more numerous in 1945 than they had been in 1920. Farms under twenty acres numbered 48,000 out of 136,000 in the state in the 1930s—a farm structure more like Massachusetts or New Jersey than either the Midwest or the South. There is no evidence, therefore, of a transition around 1900 from family farming to large-scale agribusiness, despite what Daniel, Donald Pisani, and other historians believe.[22] If there is a case to be made for such a transition, it would be at the middle of the 20th century. After World War II, there was a renewed concentration of land in the hands of the larger growers as input costs (including land), productivity, and capital intensity increased rapidly. The number of farms in the state fell below 80,000 by the late 1960s, the average size of farms doubled, and concentration of revenues and profits in most crops rose sharply.[23]

Yet even in this context, there were crops and regions where small farmers survived and even thrived. Southern California citrus, dairying, and avocados remained a province of small owners, even if average size of operations grew. Tomatoes were still grown on an average of 32 acres in the 1950s, and while mechanization brought larger tomato operations, the average of 360 acres in 1975 is still not anyone's idea of corporate goliaths. Strawberry farms averaged 34 acres in the 1980s. In tomatoes, rice, and strawberries, small producers survived under a form of capitalist sharecropping, leasing land under share leases and employing very little wage labor. In the strawberry case, these were former workers converted into sharecropper-owners by large operators trying to dodge unionization in the 1970s. With the revival of niche markets in luxury and organic crops in the 1980s, a new bit of space opened up for smaller growers, especially in the coastal margins. The number of farms actually increased in the 1980s, quite the opposite of what the farm crisis of the decade might have indicated, before falling again to under 75,000 by the end of the century. Ironically, the biggest operators on average, ranchers, fared badly over the last fifty years, forced out of the foothills by urbanization and marginalized in the beef production system by feedlots (most ranches only grow calves now).[24]

In wine, small growers did famously in the North Coast, although their success drew in corporate investments by Gallo, Coca-Cola, and others. If anything, the big vineyards suffered the most from the overplanting bubble of the 1970s in the Central Coast. Yet here, too, well-financed and -run operations could grow large (even multinational), like the far-flung, diversified, and integrated empire of Jess Jackson (Kendall-Jackson wines). In organics, as soon as the possibilities of the new high-value crops ripened, large growers—mixing organic and conventional crops—entered the markets in force. Still, new rags-to-riches stories are possible, such as that of Earthbound Farms, which started off as two University of California at Santa Cruz students selling bagged salad mix from their Carmel Valley farm. In the late 1980s they grew rapidly and became "Natural Selection," an umbrella for a conventional line of crops. They later went into partnership with major Salinas growers and handlers, such as Tannemura and Bud Antle, and today dominate the organic vegetable industry with around 8,000 acres in production, the other major growers contracting to them.[25]

The jockeying for class position continues. On the one hand, large companies have some real advantages, and this means that the big bruisers win most rounds against the small challengers. Ample capital and access to financiers buffer bigger operators from the vagaries of product markets and business cycles; the biggest hazard for small business is riding out downturns or product failures. Solid financing also pays off on the upside, in the ability to invest in new land and technologies. Yet large size alone is not enough. Growers still have to produce. The smaller grower has some things going for her, especially paying close attention to the production process from planting to harvest. In agriculture there is still no substitute for hands-on know-how and personal attentiveness to problems as they arise. That is why large, impersonal corporate farming has been stillborn in California. Miller and Lux fell apart after Henry Miller died in 1908. More recently, entries by Tenneco, Heublein, and other outside multinationals had little effect on the trajectory of development, and most have withdrawn from the field or remained on the distribution end. Instead, most California growers (87 percent) are family proprietorships, partnerships, and corporations that grew up organically from smaller farms. Even J.G. Boswell became the nation's largest farmer through personal attention to all facets of his operations.[26]

The result is not so very different from the situation of the vaunted electronics industry of Silicon Valley. The lesson in high tech's history has been that in wide-open industries, the big do *not* always get bigger while the small go on the dustheap of history. This is why Alfred Chandler today speaks in terms of scale *and* scope, rather than an inevitable concentration in multidivisional corporations. Of course, the peculiarities of farming mean more firms than in any manufacturing sector (though fewer than in retailing or real estate). It is not insignificant that there are still 75,000 growers in California agriculture, a number that derives principally from the intimate relation to nature in farming and consequent economies of scope in production. The growers are no less agrarian capitalists for all that.

MODERN FARM MANAGEMENT

At the heart of business farming is the ability to manage a farm operation in an effective, competitive, and profitable way. Growers must be able managers. Such management is not an easy matter in any business, let alone one as multifaceted as farming. There are five fundamental dimensions of farm strategy and management: crop selection, land acquisition, directing the production cycle, keeping accounts, and labor allocation and supervision.[27] We will discuss the first four here and return to the fifth in the next chapter.

The first problem facing the grower is crop selection, or, chiefly, where to specialize. A plus for the smaller operator is the ability to enter new markets, take risks, and deliver high-quality produce, as in the fresh fruit industry a century ago and in exotics, organics, and elite restaurant supply today. Because there is so much hands-on knowledge and experimentation required in certain crops, especially when they are first established, immigrant farmers have often led the way. Japanese, for example, were critical in developing strawberries, snap beans, and cut flowers.

Large companies move more slowly and carefully, because of prior commitments, managerial distance, and risk aversion. On the other hand, large growers in California have been able to beat the small ones at their own game through a strategy of *diversification*. Beginning in the late 19th century, after the pioneer horticulturalists had proved the viability of the fruit markets, bigger landowners began to grow tree, vine, and garden crops on part of their acreage, leaving the rest in cattle, wheat, or other field crops.[28] This was not mixed farming in the sense of the Midwestern hog-corn family farm with a garden plot on the side. Rather, it represented multiple specialties, in the manner of General Motors' line of products from Chevies to Cadillacs. This gave large growers more stability in the face of market ups and downs and, importantly, an advantage in winning contracts with big buyers like supermarket chains, which want a broader product line without dealing with dozens of farmers.

A second strategic issue is land acquisition, meaning buying or leasing land and maintaining a certain area of operations. In the past, this was usually a matter of finding a spread for sale, coming up with the money, and securing title. For those who were cash short or excluded by race barriers, leasing on a cash-or-share basis was the alternative. Leasing was common early on in the Delta, because the job of reclaiming whole islands put the land beyond the means of most anyone but large absentee investors. It was also prevalent in the Imperial Valley because water supply was so insecure before 1940. Putting aside the question of financing for now, the main issue was to buy or not to buy, and plots were contiguous for the vast majority of owner-operators. The decision has become more complex with time, however.

By the middle of the 20th century, many of California's big growers, such as Gallo, Bud Antle, and Bruce Church, were diversifying their holdings *geographically,* acquiring land up and down the state and in Oregon, Arizona, and Mexico (the precedent for this was Miller and Lux's million-acre cattle operations across three states in the 19th century). The justification is twofold: to achieve greater product diversity with lands of different qualities and to be able to harvest crops through a longer season, running from south to north. This meant a wider product array to offer distributors and retailers, as well as more continuity of production and income flow to stabilize company finances in the face of the ups and downs of agricultural fortune. It combines the best of scale and scope economies under one company.

Another telling strategy of larger growers that came into play in the post–World War II era was leasing. Previously, leases came in three forms: cash, share of output, and contract for output. The modern lease is usually a three-year contract for cash. Leasing has several advantages. It allows successful growers to increase the scale of contiguous operations and is an easy way to diversify crops and geographic holdings. Financially, it saves capital otherwise sunk in land—which was only getting more expensive with the productivity increases of the postwar boom—and lowers the risk of moving into new areas or crops, because one can withdraw after a couple years of failure. Postwar leasing reversed the long-standing tradition of smaller farmers leasing from large landowners, as large operators increasingly combined leases from many smaller owners under a single management. Leasing smoothed the way for successful growers to grow larger and uncompetitive ones to exit the field gracefully. Many small growers retired on the rental income, leaving the field to the bigger and more professional operators.[29]

The third aspect of farm management is oversight of the production cycle of planting, weeding, harvesting, and the rest. The grower has to plan every step, especially the timing, based on the condition of the crops, the weather, markets, labor availability, and so on. He has to choose seeds or varieties of vines to plant, and when and how to prepare the ground, and how densely to plant. He has to see to it that pollinators are present in sufficient numbers in certain crops. He has to decide when to fertilize or irrigate, depending on soil moisture, plant vigor, setting of fruit, and the like. He has to thin and weed around the crop, or use herbicides. He has to watch for pest infestations, deciding when and what to spray. With trees he has to plan for pruning and shaping, usually during dormancy, and perhaps spray again. With animals, the production rhythms are different, but they present the same kind of endless cycle of decisions and timing, including when to buy calves or chicks, what and when to feed, how to monitor and treat diseases, when to give shots and hormone supplements, when to slaughter, and the rest.

Every one of these puzzles has to be solved, over and over, calling on a reservoir of experience, shared know-how among neighbors, advice from Agricultural Extension workers or farm magazines, and any and all sorts of knowledge inputs never seen, only realized in the act. Here is where the small grower and the immigrant have often found an angle for competitive advantage. As Masakazu Iwata says of Japanese growers trying to establish themselves in the early 20th century, "the Japanese had taken up much of the worst lands in California and made them fertile." They did it by their skill and persistence "in soil preparation, crop and seed selection, planting, cultivation, irrigation, and spraying. They were . . . able managers."[30]

A fourth dimension of farm management is keeping the books, and keeping one's nose above water financially. This may have started as humble register of accounts, but became increasingly complicated as farms had to juggle longer-term financial flows in and out, and keep out of the red. By the early 20th century, the typical grower would have had a long-term mortgage from the bank, medium-range loans to pay off for equipment, and short-term commercial credit for seeds and stock, in all likelihood from several different banks, merchants, and cooperatives. She would have had to keep up with prices, including selling in forward markets, and might well have had a contract from a processor or buyer with prescribed payment terms.

Another consideration in accounting, by the second half of the 20th century, would be keeping track of multiple lease contracts, as leasing spread to every aspect of farm operation, such as tillage machinery, irrigation pipes, and sprayers (livestock operators can even rent their herds today). As with land, equipment leasing avoids heavy up-front capital costs and inefficient, episodic use. (Leasing became a general trend of American business in the 1980s, but arrived in California agribusiness earlier than most.) To stay on top of all these streams of costs and income, no farm today is without its personal computer and specialized software, and in the most sophisticated corporate operations, they will be able to calculate rates of return on capital, as well.

On today's farms, many productive functions are contracted out. This, too, is a modern business strategy that became the rage in Silicon Valley and beyond over the last generation, but appeared in agribusiness well before that. The modern grower can call on contractors to do just about everything a farm family and its hired hands once did: field leveling, planting, pollination, crop dusting, pruning, irrigating, weeding, and harvesting. In animal husbandry, contracts are made for artificial insemination, hoof clipping, veterinary care, feed preparation, and animal finishing. Specialists have proliferated in every domain touching on farming, for reasons of efficiency or economies of scope, defined as specialization of labor and equipment, combined with fuller utilization of the subcontractors' capacities across many farms over the course of the season.[31]

These contracted functions are called "business services" in the language of business studies, but they are part of the process of farm production. The contract form of business organization is not easily comprehended under the terms "vertical integration" or "commodity chains." Some have called it a *network* type of business. Others have tied it to the formation of *industrial districts* of the kind seen in electronics or moviemaking. Such regions provide a rich mix of inputs and services, a milieu of up-to-date industry practices, and a continuous flow of new technologies. Agribusiness is rife with such industrial districts, from the lettuce zone of Monterey County to the wine country of Napa County. If farming were treated with the same adulation as electronics, scholars from all over the world would be rushing to the Salinas and San Joaquin valleys the way they now scramble to unlock the secrets of Silicon Valley, where such networking is famously advanced. Margaret FitzSimmons was the first to identify this parallel, though she oscillated between the older version of agribusiness as a vertically integrated corporate monopoly and the newer ideas of industrial flexibility.[32]

What this means for farm management is that it is more like being a circus ringmaster than a performer. One observer has gone so far as to declare, "A telephone book and a check book are the only items needed by such a farm operator."[33] This is an exaggeration, but a telling one. In fact, scheduling contractors, tracking crop growth, and balancing the books remain daunting tasks that require considerable skill and that are rife with opportunities for mistakes. Managerial capacity continues to improve through the introduction of computers, specialized software, better feedback from the fields, and much more. In order to keep on top of such a sophisticated and many-pronged labor process, a family member today is likely to have an advanced degree and to focus on farm management. This has parallels in the dawn of the modern corporation, when sons and daughters of industrial magnates started going into the professions and coming back to work in management of the family-founded company.

Contractors have input into today's farm management, as well as being subject to it. They act as consultants, providing sophisticated technical evaluations of plant stress, integrated pest management, or soil improvement. Some growers, particularly the largest, most geographically extensive ones, hire professional managers to coordinate the whole operation. Vineyard managers are ubiquitous and vital in the wine industry today. It is not true for California, however, that farming has been taken over by industrial and merchant capitalists coming from outside, in the manner of the poultry industry of the South. Growers remain remarkably independent, despite constant interaction with and intervention from agro-industrialists.[34]

The extreme version of management bruited today is "precision farming." Its premise is that large field farming is too crude and one needs to move beyond such mass production. While that once offered simple advantages of mechanization and labor saving, it no longer suffices to get the best performance from the crops and minimize total inputs. Because fields, orchards, vineyards, and pastures are not uniform, they need to be micromanaged to account for site-specific differences in soil, water, pest, and plant conditions. Empirical studies show substantial variation across fields in things like soil moisture, root penetration, nutrient availability, pH, organic matter, weed infestation, insect dispersion, and, consequently, plant growth and stress. While this is a fact long recognized by farmers and soil scientists, the technology now exists to deal with it systematically, using ground-based sensors, remote sensing (satellite imagery), computers, and sophisticated software.[35]

Precision farming is information intensive and expensive, but the necessary computers, programs, and data-gathering capabilities are well advanced. The delivery systems—machines, irrigation systems, and controls—are available on the market that allow farmers to vary the delivery of water, fertilizers, and pesticides, as well as the intensity of cultivation. Furthermore, chemical companies, machinery firms, and third-party consultants now offer turnkey systems that combine field monitors, global positioning receivers, geographic information systems, and variable application technologies in a single system. Like most fantasies of total automation, precision farming is still unevenly applied and only partly diffused. Further advances will depend on growers and their advisors' adopting a radically new frame of reference.

FARMERS' KEEPERS

Growers have long been keenly aware of California's distinctive conditions, and their own shortcomings in the face of new circumstances. From the earliest days of commercial farming in the state, they dove into the puzzles of farming with gusto, learning on the job and seeking outside help for improving farm methods. One sign of this is that they were avid readers and self-educators, as can be seen by a glance at the vigorous farm press that began with *California Farmer, California Culturalist,* and *Pacific Rural Press* in the early

1850s. Over the years, farmers in different crop specialties sought advice and counsel from dozens of trade journals, like *Western Feed & Seed, California Cultivator, California Horticultural & Floral Magazine, Dairy & Produce Review,* and *California Citrograph.* Later, the growers would listen to farm broadcasts on the radio. Not surprisingly, the first radio station in the world was begun in San Jose in 1908 for the purpose of reaching Santa Clara Valley farmers. Growers readily adopted practices proclaimed by the trumpeters of progressive agriculture and, later, scientific management.[36]

Farm management does not stop at the farm gate. Over the years, other participants in the agro-industrial system of California have come to play an advisory—and even directive—role in farm operations. Grower cooperatives, at the turn of the last century, were some of the first to intervene by demanding certain quality of fruits, and by bringing in outside advisors to consult on pest control or fertilizer application in the orchards. Irrigation districts, which came into fashion in the late 19th century, helped determine watering schedules for member farmers, and lent advice on soil moisture and water application. Canners contracting for crops such as processing tomatoes were, by World War I, demanding specific planting and harvest times, breeds, and inputs to assure quality, and providing technical assistance. Today the coordination of contract growers is done by computer. Even labor contractors play a role in farm management, and track payrolls, personnel, and production by computer, too. All of these actors will be discussed further in later chapters, but the point here is that their advice is mostly welcome and considered necessary for better farm management.

The University of California has had a prominent place in the collective pursuit of good farm practices. The state was among the pioneers in university aid to farm modernization. The State Agricultural Society was already lobbying for an agricultural college with suitable scientific foundations in the late 1850s. A College of Agriculture was built into the university's Organic Act in 1868 to take advantage of funding from the federal Morrill Land Grant College Act of 1862, and the first building on the Berkeley campus, South Hall (1872), was earmarked for the college (the structure still stands, with its bas-

"Rain for Rent" sign, east of Stockton, 2003

relief sheaves of grain and bunches of grapes intact). The college flourished after it secured the leadership of Eugene W. Hilgard in 1874. Hilgard had connections of great value to the young state. He was a graduate of the University of Michigan, a rising young soil scientist in Mississippi, and someone able to speak to a national audience. He was also a cousin of Henry Villard, the railroad magnate, and of Charles Weber, the founder of the city of Stockton.[37]

The university announced its intention to be an agricultural experiment station as early as 1870, and the legislature began to fund research within a few years. Hilgard set up the pioneering Agricultural Experiment Station in 1877 and began publishing research bulletins in the next decade. The Hatch Act of 1887 made such initiatives national policy, and offered direct aid to research at the land-grant colleges for the first time. But research was only half the charge of the College of Agriculture. Under pressure from the Granger movement, the college staff in 1891 began Farmers' Institutes, a series of traveling lectures to growers around the state. A permanent Agricultural Extension program was installed in 1897 to put farm advisors out in the field, working directly with growers.[38]

Hilgard recruited Edward J. Wickson, a dairy and livestock expert from New York and indefatigable editor of *Pacific Rural Press* (a post he held for over fifty years), to lead the way in farmer education and outreach. The two men made a splendid team, combining basic research with practical counsel on improved farm practices. Wickson wrote a series of guidebooks for growers in every sector, a major popular source of practical knowledge for California farmers from the 1890s to the 1920s. Hilgard and Wickson were "the horticultural leaders of the State for nearly half a century," with Hilgard in the role of chief scientist and Wickson the master of practical advice. Hilgard served as dean of the college until 1905, and Wickson from 1907 to 1913.[39]

Neither the university nor the College of Agriculture were very well funded in the 19th century, and Hilgard complained repeatedly of the poverty of his facilities. The college did get some injections of money from its agribusiness constituency: Hilgard's first outside research contract came from the wine growers in the 1880s; and poultry, dairy, and other trade associations would weigh in with funds in the early 20th century. The Citrus Association even taxed itself to support the university's Citrus Research Station in Riverside. The Southern Pacific Railroad induced the state to pay for a UC Experiment Station in the Imperial Valley in 1908. Phoebe Hearst, widow of mining billionaire George Hearst, solved the university's financial problems by taking it under her wing at the turn of the century, transforming it into the academic flagship of the Pacific Coast.

The Agricultural College was able to grow more rapidly once it had continuous injections of state and federal money after the turn of the century. Funds were ratcheted up significantly with the state farm bill of 1905 and doubled from the federal government by the Adams Act of 1906. An enlarged staff of researchers, teachers, and farm advisors promoted the college's vision of scientific and rational agriculture for the state. They weighed in with scores of bulletins and books on everything from mealy bugs to labor relations.[40] Meanwhile, the education and outreach function of the university was augmented in 1906 by the founding of a university farm at Davis, in Yolo County, allowing for hands-on training of a kind the college had not been able to do before.

The period around World War I ushered in a new era for the College of Agriculture. A new dean, Thomas Forsyth Hunt, arrived at Berkeley in 1913 (the university wanted Liberty Hyde Bailey of Cornell, but got one of his colleagues instead). Funding rose again under Progressive governor Hiram Johnson and through the federal Smith-Lever Act of 1914. The latter mandated a new outreach effort to be called Cooperative Extension, which combined university extension and state agents. The scope of teaching and inquiry at Berkeley and Davis (with an overall faculty of nearly one hundred) broadened to include such fields as agricultural economics, home economics, and nutritional science.[41] Appointed by Dean Hunt, B.H. Crocheron organized the new Cooperative Extension service in 1915 and directed it for the next thirty years. Increased research linked to vastly extended outreach could move ideas of good farm management from the campuses out to the farmers as quickly as possible.[42]

The college grew so enormously that it became a force of its own within the university for the next fifty years. This shift was clear by 1921, when the newly organized Agricultural Legislative Committee, a creature of the growers' cooperatives, forced a thorough overhaul of agricultural education through the legislature, which created a kind of master plan for the college. The university farm was closed, the broadest definition of research and training for agribusiness (defined as "rural progress") was confirmed, and Davis was promoted to be the northern branch of an enlarged College of Agriculture (it would not become a full UC campus until the 1950s). Claude Hutchinson was brought from Cornell to run the northern branch, and became dean of the whole college from 1930 to 1952. Hutchinson, like his predecessors, was a national figure who had worked for the Rockefeller Foundation abroad. Similarly, Davis's leadership passed in 1937 to Knowles Ryerson, former head of the USDA's Bureau of Plant Industry. In 1938, the northern branch became simply the College of Agriculture at Davis, signaling its equality with Berkeley, at last. Budgets would grow with increases in state spending and federal aid from the Purnell (1925) and Bankhead-Jones (1935) acts.[43]

One of the most important developments in the college was the creation of a specific field of investigation called general farm management. Its leading figure was Richard L. Adams, a member of the Berkeley faculty, whose path-breaking textbook on the subject, *Farm Management,* appeared in 1921 and was reissued into the 1950s. Adams's goal was to rationalize farm organization, bookkeeping and accounting, marketing, land assessment, and labor use, among other things, and to make it possible to assess farm costs and profitability accurately. He issued a string of pamphlets on all these topics over three decades, along with his textbooks. For Adams, farming was like any other business, "for farming is under a broad construction the manufacturing of food products." It is instructive to look back on Adams's work with contemporary eyes and see how much he was a transitional figure from Country Life to Modern Management. He still tended to personalize (and racialize) the virtues of the good farmer, to advocate for diversified farms (while noting that Western farms tended toward specialization), and to err in his calculations of things like the rate of profit.[44]

Adams represents an important shift toward economics in the College of Agriculture in the 1920s, in which production, marketing, and the rest of farming would have to be evaluated in terms of sophisticated accounts and economic frameworks. This project was enhanced by a bequest from the Theo Kearney estate and the founding of the Giannini Foundation in 1928. The Giannini Foundation has for decades issued "Agricultural Outlook" bulletins to support farmers' self-improvement. Extension agents in the 1920s began doing efficiency assessments for growers to show them whether their operations were up to standards. California was far ahead of the field in agricultural economics and farm management for years.[45]

Adams's legacy continues in new, high-tech guises. University researchers have to proffer advice on the widest scope of farm technology and economics. The capacity to farm with increasing precision alters significantly the kind of scientific advice and management philosophy needed from the university, farm consultants, and professional managers. This can be seen in the way that new fields such as GIS and optics are being applied. It also demands a more integrative approach to agricultural science than ever before, blending knowledge of soils, genetics, chemistry, entomology, mechanics, and the rest. Finally, an essential consideration in modern farm management has to be the environmental effects of such things as fertilizer runoff, pesticide dispersion, and soil maintenance. Farm management and advising have come a long way in this age of science and information, but it still serves the essential profit-making purposes of the agrarian capitalist and the on-farm production of plant and animal commodities, to which we now turn.[46]

4

Down on the Farm

Farming is the heart of agriculture, and the pivot of all food provision. Yet farms are mostly foreign places to those of us who work in offices, factories, and cities today, something seen from afar on the highway to Los Angeles or Yosemite. So we need to step out into the countryside and take a look at what goes on in the fields and orchards of California agribusiness. We don't want to be so innocent of farm practices that we believe, as a youthful friend of mine once proclaimed, that "cows eat meat" (and she was headed for medical school, a truly frightening thought). If growers were that benighted about the ways of the natural world, we would all soon starve.

Farming is a business in which nature plays a fundamental role. Growers cannot turn the trick of making money without making nature move to their command. In agriculture, however, nature can be both a fertile partner to human ambition and a recalcitrant helpmate with a will of its own. One needs good land and soil, favorable rainfall and sunshine, and the right sorts of plants and animals. Given the right conditions, the crops will sprout, flower, bear fruit, and seed, and the animals will grow, fatten, and calve. Given adverse circumstances, plants will wither, animals sicken, soil turn to mud, or pest infestations devour the harvest. This much is obvious, and makes agriculture both a wondrous thing, in which nature quietly does much of the work for us, and a gamble, in which the results depend heavily on providential events.

In modern times, of course, human ingenuity has brought nature's abundance to heel to a remarkable degree. This has been especially true of California's countryside, which has yielded an astonishing abundance. So a good part of our story in this chapter will be how that profusion of crops was brought into being and nurtured to high levels of output. Yet the kind of mastery of the production process that one finds in car assembly or plastic moldings still escapes the farm operator, and there is much that can go wrong. This is particularly true of pest depredations, on which so much attention has been lavished over the years. California growers' run-ins with natural obstacles must be acknowledged, and the damage they do on the process of reengineering the landscape as well.[1]

Farming in a Mediterranean climate zone has given California agriculture a history at odds with the rest of the United States. This has been a real advantage for growers in the state because of the variety of things that can be grown, the long growing season, fertile soils, and much else. If biology has made agriculture everywhere recalcitrant to capitalist processes of homogenization of production, that has been even more true in the diverse environments of California. But that very diversity, and difference from Europe and the eastern United States, has presented a distinctive set of barriers for growers to overcome.

We will begin, in section one, with the introduction of plants and animals, and the process of discovering what would or would not prosper in this new environment. The first item on the agenda of California farm operators in the 19th century was to assemble the living things set loose on the land, and then improve them. The second section will move on to the soil, and the process of discovery and mastery of the geological facts of the land. This engagement has its own distinctive historical geography and record of progress. That discussion will be followed, in section three, by early irrigation development and the opening rounds of combat against depredations of insects and other natural competitors of humans for the bounty of the land.

In the last section of this chapter, we will turn to that other nature-given force of production: human labor. Farm work is, of course, a social thing, with human purpose, imagination, and sweat. Yet the labor process is deeply ensnared with the natural processes of plant and animal growth. It must move to the rhythms of the seasons, through a series of biologically and geographically defined tasks, and it must attend to the specific needs of different living things. Cross-cutting this interaction of nature and nurture are social relations of production beset with tension and confrontation. Farmworkers are people with their own needs and desires, out to earn a living. But once they muster at the farm gate, they have to march to the drumbeat of the grower, whose purpose is to produce profits as well as crops. That imperative for profit has led growers to impose rigorous work regimes; strict divisions of labor; and tools and machinery that dictate the parameters of farm work, and often render it hazardous to the natural bodies of the workers.

In this chapter, I am most concerned with how modern agriculture was established in a new place. California's growers first had to get the lay of the land, introducing new plants and animals, opening up new acreage to farming, dealing with new kinds of pests, and putting a wage-labor force to work on the fields. Over time, however, the story evolves, with the patterns of cropping and farm practices morphing into different configurations as growers introduced new varieties and breeds, soil enhancements, and greater pest controls, and as the labor process adapted to changing conditions. We will touch on this evolution here, as the production regime shifts from discovery to the age of hybrids to petro-farming, particularly as this rests on improvements on the farms.

In so much of the agrarian studies literature, the natural *limits* on farming get the greatest emphasis. Agriculture seems to pale by comparison to the achievements of factory-based industry in the modern era, and this ought to pose a serious barrier to the development of capitalist production on the land.[2] In one sense this is clearly true: capital cannot just force nature to dance to its command, ringing up profits at will. Yet in the case of California agribusiness, one should not exaggerate the travails of the agro-capitalist, any more than any other business owner beset by problems of the market, labor, and technology. This is, after all, one of the world's great agrarian success stories. California growers have clambered over the natural limits on farming by such strategies as growing high-value crops, adopting new breeds, adding water and fertilizer to the land, and employing cheap labor. And, as we shall see in chapter 5, they had considerable help from outside, in the provision of off-farm inputs in every facet of the production process.

FIRST THE STOCK

Seeds and breeds are the most basic inputs to farming. Field and vegetable crops must be planted each year, trees and vines set in place, and animals birthed. All living organisms have to be reproduced in this fashion. Of course, some plants (and now animals) are clones that are not grown from seed or egg, but the problem of reproduction and establishment remains. Throughout the long history of human agriculture, farmers have had to save seed for the next year's crop, grow vines from clippings, breed their stock, or otherwise renew the crops over the years. Hence Jack Kloppenberg's phrase, *First the Seed.*[3] Most discussions of seeds and breeds in modern agriculture begin from the assumption that capitalism enters from outside the farm, in the form of seed companies or breeders selling to farmers. But in California the introduction of commercial farming came all at once: everything had to be introduced. It was as much a stocking up of the country as it was a seeding.

The mass importation of plants and animals to California began in 1850 and was frenzied for the next one hundred years. It was marked by a continuous experimentation with what would and would not grow in this Mediterranean land of many microregions. The supremely confident growers, with their new mode of production, were able to make the land flourish in a thousand ways. From there it continued into more sophisticated realms of breeding and scientific genetics, but the impulse has been constant: the unleashing of ingenuity and pragmatic experimentation in search of workable, profitable crops. It drew ultimately on progress in biological science and genetics, but it represents much more than "science" or systematic knowledge applied to pro-

California Farm Inputs, 1997
(in dollars)

Category	Dollars
Total	16,817,253
Seed and Stock	
Livestock	759,223
Seeds and Nursery	526,323
Feed and Fertilizer	
Animal Feed	2,588,982
Plant Fertilizer	746,325
Energy	
Fuel	488,226
Electricity	526,592
Machine hire	595,872
Chemicals (pesticides)	957,006
Labor	
Hired	3,392,577
Contract	1,386,159
Maintenance	777,097
Financial Payments	
Interest on debt	958,431
Rent of land	554,077
Property taxes	360,348
All Other	2,200,014

Source: California Agricultural Statistics, 1998, California Department of Food and Agriculture.

duction: it bubbled up from everyday practice on farms, nurseries, and dairies; took advantage of formal science; and cycled through the rural press like a nutrient flux in a complex ecosystem.[4]

Livestock were the first thing to feel the influence of agrarian modernization in California. Hundreds of thousands of cattle, sheep, horses, and mules were driven west during the 1850s, bringing in new breeds from Europe, the East Coast, and Texas (Shorthorns, Devons, and Aylshires among cattle; Leicesters, Cheviots, and Merinos among sheep). The American ranchers used them to replace the lean and feral longhorns of the Mexican rancheros, then began their own calving and breeding operations. Improved stock was important in driving out the rancheros by competitive means (in terms of adaptation to introduction forage and consumer taste for softer meat). Dairy cows also had to be introduced en masse from Eastern breeding grounds.[5]

Horticulture developed willy-nilly through the importation of diverse fruits that might prosper in the microclimes and soils of California. Mission orchards and vineyards established by the Spanish friars provided valuable cuttings, even before the Gold Rush, in places like San Jose, Los Angeles, and the San Gabriel Valley.[6] But roots, seeds, and shoots began immediately to be brought from New England, Oregon, and Europe. G.C. Briggs brought peach, apple, and pear trees from New York to Yuba City in 1851. The three brothers Seth, Henderson, and John Lewelling moved down from Oregon, bringing East Coast cherries, apricots, pears, and other fruit trees to Sacramento and San Jose. Louis and Pierre Pellier brought the French prune and wine grapes to the Santa Clara Valley in 1856. Agoston Haraszthy tried out European winegrape stock in the mid-1850s in San Francisco, San Mateo, and Sonoma. Charles Lefrance brought cuttings in to start Almaden Winery in 1857, and was the biggest vineyardist in the state with 350 acres in 1875.[7]

Hundreds of thousands of trees and vines were planted before the Civil War. Experimentation was rife. "During the 1850s pioneer orchardists, nurserymen, and viticulturists made heroic and very well publicized efforts to develop their specialties," writes Rodman Paul. An early assessment showed almost 1,200 varieties of fruit introduced to the state, of which over 500 passed muster. California was not exceptional in its fascination with plant introductions, and California benefited from outside developments. The horticultural age of discovery was quite general in Europe and the United States at the time. Kew Gardens and other botanical centers were gathering plants from around the world and aiding their transfers to new realms where they might support the great colonial enterprise (a practice that Lucile Brockway calls "economic botany"). As Edward Wickson observed later, "California received immediately all that European and East African collectors had secured during several previous decades of ransacking the five continents."[8] Californians put this wealth of the world to good use. Wickson added that "California was very fortunate in numbering among the early settlers so many men with horticultural tastes, skill and experience." Paul Gates echoes this sentiment: "Rancheros, small farmers, nurserymen, fruit growers, and agricultural journalists were showing minds open to new ideas, willingness to experiment, . . . receptivity to innovations in the introduction of new crops . . . , [and] concern about the quality of their products." Taste and skill were aided, one should add, by gold in their pockets to buy and plant millions of seeds and cuttings.[9]

There was, of course, considerable misfortune within the generally upbeat tale of new introductions. In 1860, pomologist Wilson Flint despaired that "of all the fruit trees sold in California, not more than one-third of them ever survive to a bearing age; and it is safe to affirm that not more than a half of this third will ever produce any considerable amount of good fruit." The most notorious fiascoes came with mass introductions of mulberries and silkworms, tobacco and coffee, all supported by state "bounties" in the 1860s. Agoston Haraszthy was sent on a fool's errand to Europe by a governor's commission to develop the wine industry; he collected 100,000 grape cuttings but was never paid on his return because he was on the wrong side of Civil War party politics. Disease, drought, fog, and other problems selected out many more varieties and regions. Some 800 varieties of strawberries had been winnowed down to about 200 by the end of the century, for example.[10]

But the post–Civil War era brought another wave of success. Nurserymen continued to import new plant stock, special envoys were sent by growers to search out new varieties, U.S. consular reports on foreign agriculture were perused for ideas, and the new United States Department of Agriculture helped out. The late 1860s saw the introduction of sustainable varieties of French and Chilean walnuts in the North and South, respectively. In Southern California the arrival of Brazilian (Bahian) navel orange "sports" in 1873 (thanks to the USDA) and summer-ripening Valencias in 1876 provided the key to the subsequent citrus boom. Frederic Pohndorff imported the successful Manzanillo olive in 1874 and an amateur, the Reverend Charles Loop, the Picholine and many others in the 1880s. New introductions from abroad continued well into the 20th century. Acala cotton arrived in the 1910s, with the aid of the USDA. The USDA sponsored more than fifty excursions to the far corners of the world between 1900 and 1930 to gather plants, and it introduced most of the commercial avocado varieties to California. Yet experimentation with new breeds meant a lot of failures; for example, two hundred kinds of olives were brought into the state, but only three were widely planted.[11]

This vast array of specialty crops would not have come to California without the aid of a host of human immigrants, beginning with Frenchmen and Scots, and renewed with every wave of new arrivals. Croatians brought in their own stone fruits; Italians artichokes, olives, and white figs; Japanese plums and ornamentals. The role of immigrants in plant introductions is crucial because they brought with them a grounded knowledge of many crops and cultural awareness of their benefits.[12] It points up the limited experience, taste, and know-how of many Anglo growers, who were in it for the money, not out of devotion to farming or foodstuffs.

Merely introducing new plant and animal stock could only take California's growers so far. Almost from the beginning, they started manipulating that stock through systematic cross-breeding, and keeping an eye out for chance crosses that appeared in their orchards and nurseries. Biological manipulation is as old as agriculture itself, of course. But it accelerated in modern times with the energetic animal husbandry and horticulture of 17th-century France and

18th-century Britain, where agrarian capitalism first got established. In the United States modern breeding began in earnest around 1800, and it was in full flood by the time of California's entry into the Union.[13]

Building on this tradition, local plant breeders soon gave California several of its most successful varieties of commercial fruits. A.T. Hatch discovered the importance of cross-fertilization in almond trees and propagated several new higher-yielding varieties. G.W. Thissell found the Muir and Lovell freestone peaches among chance sports, and these turned out to be ideal for canning. In the 1880s, N.A. Griffeth converted growers to the Gravenstein apple, which seems to have been a local volunteer hybrid. J.P. Onstaott of Yuba City found the table grape sport he began selling under the name Thompson Seedless. James Waters of Pajaro Valley Nursery introduced the highly successful Melinda variety of strawberry in 1890; Charlie Loftus stumbled across the Banner variety in Shasta County around the same time. Working in isolation in Humboldt County, Albert Etter contributed mightily to the development of better strawberries by systematic cross-breeding with California native varieties.[14]

California added more modestly to the development of animal breeds. Colonel Younger of San Jose was the best-known shorthorn breeder in the mining era, and John Patterson made a name as a breeder of Merino sheep. Yolo County had a number of prizewinning breeders in the early years, led by Jerome Davis. Large cattle ranchers like Miller and Lux and Hiram Merritt invariably became important stock breeders. The most important quality animal breeders sought was uniformity for the market. This was not yet the age of pumped-up bulk, higher fat content, and rapid maturity.[15]

California's most famous plantsman was Luther Burbank. Burbank's reputation as a man of inventive genius rivaled that of his contemporary Thomas Edison on the national scene. He personified the craft of the individual plant breeder, before so much of practical science was moved inside corporate and university labs. Young Luther moved from Massachusetts, the hub of American horticulture, to Santa Rosa in 1875, just in time to catch the fruit planting wave in California. He responded with over 800 introductions and varieties for everything from walnuts to peas to blackberries, not to mention flowers. Most famously, he bred plums (e.g., the Santa Rosa), as befitted the prune yards going in all over Sonoma County, and we have him to thank for the Shasta daisy and the Russet Burbank potato—mother of all fast-food french fries today. He also helped increase the supply of trees by showing California growers the technique of "June budding." His obsession with breeding spineless cactus to make desert agriculture bloom is much less memorable.[16]

Burbank was embraced wholeheartedly by the California scientific estab-

Luther Burbank and Atlee Burpee, Santa Rosa, early 1900s

lishment at Berkeley and Stanford. But he had little of the formal scientist about him, keeping few records and having little truck with theory; he was, instead, an artist with nature's palette. Burbank was akin to Linnaeus a century earlier: gathering every plant he could and praying it would grow in the home soil to fulfill the national destiny. Plant and animal breeding in the 19th century were still very much practical arts, quite hit-and-miss in results. But they were not ineffective, for all that. The general principles of variation and characteristic improvement through selection of offspring were understood before Mendelian genetics addressed them scientifically.[17]

Long before the university scientists made their interventions in agriculture, there were nurserymen, farmers, and gardeners all pushing the envelope of plant and animal variation. Of course, as California agriculture filled up its spaces and became more businesslike, the approach to plant and animal improvements would develop along narrower paths. Scientifically trained breeders started to overtake the artisanal ones in the early 20th century with the confirmation of Mendel's experiments, and men like Burbank would fade into obscurity, looking more like cranks than geniuses to later generations. But to tell the story without Burbank and his ilk is like telling a history of electronics without Edison.

Scientific input to breeding, especially from the University of California's College of Agriculture, did not come into play immediately. In the 19th century, the northwest quadrant of the Berkeley campus was set aside for fields, barns, and gardens, and the college proceeded to meet its charge to gather seeds and cuttings from around the world—very much in keeping with the time. E.W. Hilgard grappled with every aspect of agricultural science, even though he was a soil specialist by training. He looked into the cultivation of wheat, olives, and citrus, and took a particular interest in grapes and wine at his campus farm and experiment station. Thanks to the viticultural industry's ability to secure state grants for his work, the first off-campus experiment station, set up in Cupertino in 1884, was for the study of vines. The citrus industry got a Pomona field station established a decade later, which helped test new varieties of plants for Southern California, including strawberries.[18]

Yet, for all their effort, the college men had, so far, surprisingly little of scientific merit to offer to the state's breeding and seeding process. Few of the college's early plant introductions were adopted, and after considerable experimentation Hilgard had to admit that the Mission olive, Bartlett pear, and Blenheim apricot were already the best fit to California. Before 1920, the Davis university farm's only major contribution to stock improvement was an award-winning livestock breeding program based on importing fine specimens and carefully monitoring lineages, not on laboratory science. Meanwhile, plant introductions continued to issue forth from growers and nurseries well into 20th century in such lines as table grapes, avocados, citrus, and walnuts. For example, several varieties of canning peaches were developed in California up to World War II without the help of the university; a Pasadena mailman, Rudolph Hass, patented the avocado that bears his name in 1935.[19] It will not do, therefore, to treat university science as the heart and soul of biological modernization, ignoring the vast resources of commonplace agrarian practice and artisanal breeding efforts that came before—and continued after—the Mendelian revolution.

What were seekers and breeders of new varieties after? A whole range of qualities in plants and animals could be useful, and not only to the growers but in the context of the whole commodity chain. Simplest was the search for varieties that thrived and produced good fruit, meat, or wool, and with the natural diversity of California, different varieties had to be found to suit a wide range of locales up and down the state. Breeders sought to enhance valued qualities, by developing sheep with more wool or sugar beets with higher sucrose content. But that is not all. Good growth habit, as in dwarf citrus, could make picking and harvesting easier. Strawberries were bred to be self-fertile and to bear through a longer season. Reduction in pest damage and better resistance to diseases like mildew or root rot were often critical. Equally important were qualities that served processors, such as easy pitting of stone fruits, high fat content in milk used for cheese, and easy shelling of nuts; or those that aided retailers, such as reduced spoilage and extended shelf life. The godsend of the fresh lettuce industry in California was, without question, the "iceberg" vari-

Brussels sprouts, Watsonville, 2003

ety developed in 1883 by Atlee Burpee (in Philadelphia). One of the most important ways of advancing market share was to find or develop fruits that ripened earlier or later than elsewhere, or develop products that would attract the consumer's eye, like redder tomatoes and whiter eggs. Breeding should also be understood as motivated by ideology as much as sheer functionality. For example, plant and animal breeds were long selected for their uniformity, which, while functional in some ways, became its own fetish of standardization that did not necessarily make for better taste or lower costs.[20]

The history of introductions and breeding provides a countertendency to monoculture. A certain efflorescence and variety of cropping occurred through transplanting and crossing plants and animals, and California became a sort of gigantic horticultural garden of the Victorian age. As the new introductions fell into place and favored varieties won out, however, specialization got the upper hand, and the successful breeds, such as those best suited to canning, were planted in vast numbers. Later, as modern science came to play a bigger role in the manipulation of genetic stock and the control of nature, monoculture reached levels never anticipated by the most utopian of 19th-century agrarian capitalists. (We shall take up the advancing front of farming biotechnology further in the next chapter.) At the same time, considerable damage was done to native ecosystems by setting loose invasive species with little forethought. California's unique landscape is plagued by blue thistle, pampas grass, and garden snails, among thousands of other dubious introductions.

GROUNDWORKS

If the seed or the sapling is first, the soil cannot be far behind. Here, too, there was a long period of trial and error in the geographic expansion of the frontier of cultivation, combining practical reason, reckless adventure, heavy investment, and scientific insight. This had five facets: recognition, treatment, reclamation, preparation, and fertilization.

The first problem facing farmers was to find crops appropriate to the character of different soils. In California, the land presented its own puzzles, different from the eastern United States. Practical experience was again the leading source of knowledge, with a great deal of trial and error. For example, an observer could still complain after the orchard planting boom of the 1920s that "there are very large areas where, on account of soil limitations, fruit culture must always be precarious or unprofitable. Unfortunately, some of these areas have been planted with fruits and the growers must inevitably fail."[21]

Five main types of land are available to growers and ranchers: peat in the flooded Delta and Tulare lake beds, fine clay and alkali in the valley bottoms, deep alluvium on the valley floors and lower slopes, clay hardpan on valley margin terraces, and shallow soil uplands. Recognizing soil types and fitting crops to them took time. In the early days, grazing animals could be set loose on the extensive grasses and forage of all areas, but with time other crops pushed them back to the terraces and uplands, or to forage on the fine clays

Vineyards in winter, Napa Valley, 2002

of wet bottoms and floodplains. Corn, tomatoes, and asparagus did well on Delta organic peats, but waterlogged Sacramento river floodplain and Tulare lake lands remained largely untouched until the arrival of rice after 1900. Wheat and barley grew well on relatively level alluvium, later replaced by row crops like beans and sugar beets. Trees and vines preferred gravelly river deposits, especially natural levees and fans. With time, orchards moved up onto terraces around the major valleys, where growers sometimes had to blast holes through clay hardpan in order to plant. Terraces left to grazing in the cattle era were often converted to orchards later, without the intervening step of wheat cultivation. Grapes and melons were found to do well on sandy alluvium soils, while berries became a standard on lowland clays in the early 20th century.[22]

Within the five major categories, there are hundreds of specific sites and soil types, well suited to certain crops and prohibitive to others. Even today as one travels through the valleys, sharp breaks occur in cropping patterns depending on soil profiles. For example, calcareous soils of the Metz series were discovered to support vigorous vegetable cropping in the Salinas Valley. The most perplexing soil problems facing California's newly arrived farmers, affecting millions of acres in the San Joaquin and Imperial valleys especially, were alkalinity and clay hardpan. These are characteristic of the arid lands of the western United States, and they presented a set of unprecedented challenges both to farmers and to the nascent field of soil science in the 19th century; debates long raged over their proper treatment.

As soon as he arrived, Professor Hilgard leapt into the breach with a program of soil sampling, and he began to make headway in understanding the chemistry of arid soils. Hilgard recognized the fundamental fertility of arid soils (which have not been leached of their nutrients by eons of heavy rainfall). Hilgard's work on alkali soils, in particular, "catapulted him into world fame" in soil science, as Hans Jenny has recounted. Hilgard made his greatest mark promoting soil treatment that used gypsum to neutralize alkali and render hard clays friable, and he opposed the popular Eastern practice of liming (suitable only for acidic, not basic, soils). He also taught farmers to use deep drainage to leach water "perched" on subsurface clay hardpan and eliminate excess salts from the root zone. This advice opened up the southern San Joaquin to wheat cultivation in 1880s and 1890s. UC advice on soils, cultivation, fertilization, and irrigation practices would do the same for the Imperial Valley a generation later. Outside the university, practical plantsmen also contributed to improving soils, as when Felix Gillet recommended iron sulfate amendments for chlorosis, a common affliction of plants in the state.[23]

Despite Hilgard's efforts California lay outside the prevailing systems of soil classification until well into the 20th century. Because the legislature provided little support, Hilgard had to produce the first crude soil survey of the state in 1880 with federal funds. Detailed soil surveying did not begin until after 1900, and it continued for the next forty years. University researchers G.W. Shaw and Earl Storie put the classification on a more systematic basis, and the Storie Index became the practical guide to soil productivity used by lenders between the world wars. Further advances were made by university scientists on the crystal structure of clays, on the use of sulfur and alum on black alkalis, and on deficiencies and excesses of micronutrients such as boron and zinc.[24] All of these had very specific applications to locales with those problems.

Reclamation was a third means to secure good agricultural lands. The best lands in terms of flat terrain, fertility, and access were often those on floodplains or marshlands. The half-million acres of the Sacramento–San Joaquin Delta are the foremost example of wetland farming. Given the Delta's pivotal position in the geography of early California, it was the first target of large-scale reclamation, beginning in the 1870s. Delta wetlands had to be leveed, pumped dry, and cleared of tules, backbreaking work done mostly by Chinese laborers and tenant farmers. By 1920 almost nothing was left of the natural vegetation, and the Delta was growing asparagus, corn, beans, tomatoes, and pears.[25]

Tule Lake bed, drained for agriculture, 1992

Clay bottomlands took somewhat longer to master, and along the upper Sacramento flooded rice cultivation—a kind of partial reclamation of paddies and controlled flooding and drying—proved to be the answer to decades of frustration. It was Japanese farmers, who took up lands no one else wanted and introduced rice growing around 1910, who salvaged the Sacramento Valley from its long post-wheat depression. The Tulare and Buena Vista lake beds of the southern San Joaquin Valley suffered a different fate. Long used as wet pasture for cattle, they were dried out and turned into cotton fields in the first half of the 20th century. The reengineering of these large segments of California's land and waterscape has been profound, with immense environmental consequences. Vast stretches of migratory bird habitat were eliminated, rodent populations exploded, and anadromous fish suffered. Once dried out, the Delta's organic soils began to subside, creating an endless drama of levee building and flood protection.[26]

Vegetation removal—not usually considered reclamation—has also been a regular feature of opening up new territory to planting. Terrace and hillside clearance for orchards and vineyards frequently requires the cutting of dense chaparral or oak savanna, as on the western slopes of the Santa Clara Valley in its prime horticultural days. Once again, it took Hilgard to convince farmers that such lands were suitable and the labor of Chinese workers to turn them to good effect. That was in the late 19th century, yet things that seemed completed decades ago recur with new rounds of agro-investment. In our time, land clearance has taken off again with a vengeance for premium wine grapes, especially in the North Bay. Thousands of acres of oak woodlands, chaparral, and redwood groves have been cleared to make way for vineyard.[27]

A fourth process, field preparation, is a repeated necessity of good farming. Californians were quick to apply machinery to such matters, coming up with all kinds of special plows for various soils (wheel plows, gang plows, rotary spades, rotary harrows, etc.). Deep tillage was all the rage for a time, on the erroneous theory that this would bring more water to the surface of arid lands. When this proved wrong, plowing practices changed. As this shows, the use of mechanical power was secondary to discovering the best preparation and planting methods for different soils and climatic zones: bed planting, furrow planting, flat planting, etc. The Chinese and Japanese played an important role in solving problems of farming damp clay soils, and they were able to turn unused lowlands, like those around Alviso and Agnew at the foot of San Francisco Bay, into successful truck farming zones.[28]

Equally important was leveling, and Californians came up with the Fresno scraper to remove high spots. Enterprising Japanese farmers in the Santa Clara Valley improved on this machine so that it would also fill in depressions (which tended to collect water in the lowlands, where they were growing berries and vegetables). In the age of irrigation, leveling became even more important to achieve an even distribution of water down the rows and across the fields. Given the immense flatness of the Central Valley, a level of smoothing could be achieved that few farmers in Iowa or Ohio could aspire to. In recent times, field preparation has become a fine art, as laser leveling was introduced to create the ultra-rationalized field.

Plowed field, San Joaquin Valley, 1985

One might expect that California agro-industry would have taken a leading role in the production and use of fertilizers to enhance soil productivity but this was not true until quite late. The prime reason is that California soils are relatively fertile and had not been depleted by centuries of farming. For example, wheat yields from virgin soils were fantastic, on the order of forty to fifty bushels per acre, compared with fifteen to twenty bushels in the prairies. Nitrogen levels were robust, though could be depleted. Potassium and phosphorus are naturally abundant in California soils, with some exceptions, such as the potassium-deficient east side of the Sacramento Valley and west side of the Santa Clara Valley, and the phosphorus-short soils of the Salinas Valley. Wheat growers did not fertilize, but they did fallow from spring to fall (sometimes they double-cropped with sorghum for stock feed). Nor did orchardists and vineyardists fertilize during the great horticultural boom. North Coast vineyards, for example, neither fertilized nor irrigated until World War II. Excessive nitrogen was thought to reduce glucose content in sugar beets. Hilgard's sanguine observation was "The need for fertilization may be expected to arise within from five to twenty years, according to the nature of the land and the kind and intensity of production thereon." As late as 1912, California farms were utilizing relatively little fertilizer compared to the northeastern and southeastern United States.[29]

Nitrogen is, of course, something almost all soils and plants can benefit from, and California is no exception. Traditionally, nitrogen came from animal manures or legumes (green manures). California was not an intensive mixed-farming region, however, and most manure came from work animals such as horses and mules (recall that the horsepower of California farms was high). Sheep were often grazed on stubble after harvest before summer pasturage in the mountains. Manure from penned livestock in dairies or chicken farms could be spread on fields, but this had a sharply limited geography. Heavy applications of manure to push up nitrogen levels was almost always prohibited by the danger that salts would burn the plants. Imports of Chilean nitrates were similarly limited because sodium concentrations turned California clay to hardpan. So rotations of alfalfa and other legumes were a common way of maintaining nitrogen levels in the soil up to World War II. For example, in the 1930s nitrogen depletion in the newly opened rice lands of the Sacramento basin was addressed through the introduction of winter legumes.

The humble work of land preparation occupied much of the growers' attention in the 19th century and even well into the 20th. This grounding of agrarian capitalism was essential to the proliferation of crops and the overall expansion of the California agriculture in its first century. Intensive application of industrial fertilizers would come to dominate in the petro-farming era after World War II, and output would shoot up as soils were exploited on a continuous cycle of double, triple, and quadruple cropping. This subject will be taken up again in the next chapter.

WATERBUGS

A great many observers have argued that irrigation is the key to the prosperity of California agriculture. The most extreme version of this position is found in Donald Worster's *Rivers of Empire,* in which he goes head to head with Carey McWilliams. What, he asks, has been the central fact of agribusiness: control of labor or control of water? For Worster, the conquest of the rivers is the defining element of California's agrarian development and the capstone in the arch of social power across the arid West. In his version of agrarian history, a hydraulic empire built by the state holds sway over landowners and workers alike. It makes for terrific historiography—the most ambitious theoretical work on California's countryside ever undertaken—but it is, in the end, magnificently wrong.[30]

Attempts to put irrigation first among causes for California's agrarian development are doomed. Irrigation was not the vanguard of agricultural advance across the state, but a laggard. Empirically, it followed the spread of agriculture by half a century. Theoretically, this ought to be expected. Water is, of course, a crucial element of the natural production cycle, and it must be present in the soil for plants (including the grass and fodder supporting animal grazing) to grow. Yet the soil and the plants have to be in place and ready to receive water. Moreover, water is given by nature without any human effort, as rainfall, infiltration, and ground flow—if episodically and unevenly. Throughout most of the history of world agriculture, it has been received in this manner. In arid or semiarid climate zones, irrigation has been developed to supplement rainfall, as would be the case in California. But even here, most such efforts were local and on-farm in the first instance.

At the dawn of modern agriculture in California, the wheat-barley-cattle era, irrigation water was hardly used at all. Wheat and barley were sown in winter to benefit from rainfall and harvested in early summer; it was typically a one-crop season, but yields were so high that no one complained. Since this

Furrow irrigation, San Joaquin Valley, 1984

era is commonly regarded as not part of the era of intensive agriculture, it is written out of water history as an aberration. Yet the horticultural revolution that followed did not depend on irrigation in the first instance, either. This was true because of the abundant groundwater and high water tables in the alluvial bottoms and fans most favored by orchardists and vineyardists. Once established, trees and vines could grow quite well without watering, providing a harvest in most years. Where supplemental water was needed to establish trees or green up animal fodder, it normally was found on the farm, either as small stream diversions, artesian wells, or windmill-driven pump wells.

Since commercial agriculture in California started mostly in the north, where rainfall was more abundant, irrigation was less of a problem. Pioneer horticultural areas like Yolo County, the Santa Clara Valley, and the Napa Valley had abundant artesian water. This was true even in large parts of Southern California. Riverside colony, out in what might be thought a semidesert, was plagued at first by a high water table that drowned the roots of citrus trees; colonists discovered in the 1870s that the uplands provided not only a frost-free climate, but a deeper acquifer. Irrigation would, of course, become vital as water tables were drawn down. Even in Imperial Valley, water development could not proceed until proper soil surveys were taken, and once water arrived in quantity, it meant an enormous new expense of deep-tile drainage of the waterlogged, alkali soils.

What is clear from the aggregate data is how far irrigation lagged behind improved acreage (basic land preparation) and harvested cropland. Irrigated acreage was a paltry 60,000 acres around the state in 1870. It had risen to a more robust 1 million acres by 1889, yet this only represented 8.2 percent of improved acreage. By 1899 irrigated acreage had only increased to 1.5 million acres. The real jump came after the turn of the century, up to 2.67 million acres in 1909 and 4.2 million acres in 1919. The first figure comes to about a quarter of improved acreage, the latter to one-third. This reflects, above all, the massive expansion of modern agriculture across Southern California, which had 75 percent of the state's irrigated acreage in 1900.[31]

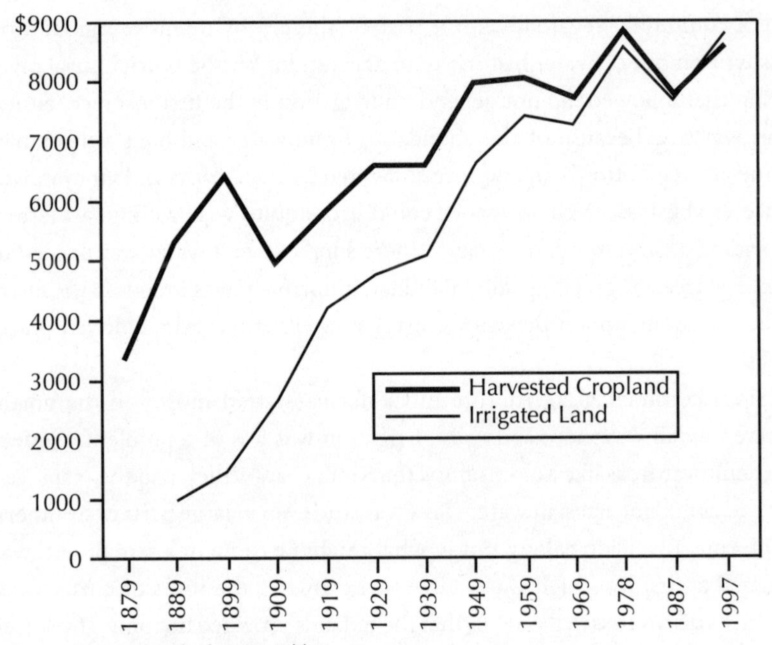

Source: Olmstead & Rhode 1997, Table 1.

California irrigated acreage, 1879–1997

It would not be until the mid-20th century that irrigation would become completely generalized across the state. It did so as part of the petro-farming, multicropping cycle, where big inputs of water made perfect sense to make the seasons endless, to pump up the size of fruit, to deliver fertilizers, and so on. What those studying postwar agribusiness may have done is to read history backward: because irrigation appeared so fundamental in that period, they mistakenly thought it must have always been the principal cause behind California's verdant fields and orchards.

Pest problems are as old as settled agriculture, but accelerated in the United States with the advent of modern commercial farming in the 19th century.

This had much to do with specialized monoculture in crops like wheat and cotton and with extensive land use practices peculiar to this country (American farms were always large and continuous relative to European ones and never employed hedgerows). A host of new pests, like the Colorado potato beetle, swept across the country in the second half of the 19th century. Farmers started to look for relief from the infant science of entomology (insect studies), pioneered by Thaddeus Harris at mid-century. The Hatch Act of 1887 led to the almost universal appointment of official entomologists by the states. Yet by the 1880s it was modern chemistry that intervened to provide farmers with easy solutions to their pest infestations—or so they hoped. Paris Green (cupric arsenate) was the first widely popular formula, Bordeaux (copper sulfate and lime) the next. Then, in the first half of the 20th century, lead arsenate became the poison of choice, mixed with oil for adherence; it was broadcast over orchards and fields with abandon, leaving heavy residues on foods (the danger of which farmers, chemical companies, and their research affiliates steadfastly denied for decades).[32]

California agriculture enjoyed a long honeymoon with nature in the 19th century as far as pests were concerned. Its geographic isolation served as a barrier to many common pests (insects, viruses, mollusks, fungi, etc.) found elsewhere across the country. Grain and livestock suffered no major epidemics, and orchards were remarkably free from maladies up to the 1870s. The first big losses were Southern California grape vines destroyed by Pierce's disease in the 1860s. With importation of new crops, intensified plantings, and monoculture, however, problems began to multiply. One of the first major infestations was the cottony-cushion scale, which ravaged Southern California citrus in the 1870s. The new Federal Bureau of Entomology came to the rescue by finding a natural enemy of the scale (the vedalia beetle) in Australia and bringing it back, where it brought the epidemic rapidly under control. The ironic result of this success—the first scientific instance of biological pest control in the United States—was a long-standing faith among Southern California growers in biocontrol long after it had been abandoned by other farmers in favor of chemical controls.[33]

The next large-scale attack came from *phylloxera,* a root louse that ruins wine grape vines, in the late 1870s. A principal reason for establishing the Board of Viticulture in 1880 was to combat this infestation, while the state Board of Horticulture, founded in 1883, had the citrus scale in mind. The university's College of Agriculture met the demands of farmers to intercede in pest outbreaks. E.W. Hilgard did pioneering work on *phylloxera* and the use of Bordeaux mix on grapevines. Hilgard saw that the cure lay in grafting European *Vinus vinafera* shoots onto American root stock, since the offending bug was an American native, but he erred in thinking that *V. californicum* would provide rootstock that could resist the louse—and lost his own Mission San Jose vineyard in the 1890s (Dr. Thomas Munson later figured out the best rootstocks were those adapted to calcareous soils from his native Texas, which saved the wine districts of the world).[34]

By the turn of the century, the honeymoon was over. Trying to maintain its physical isolation, California imposed the first plant quarantine law in the United States in 1899. Pest outbreaks led Dean Hilgard to appoint C.W. Woodworth to head up a program of plant pathology and entomology in the College of Agriculture (both subjects would soon become full departments). In 1903 Hilgard and his staff (including the newly hired entomologist Ralph Smith) worked successfully on asparagus rust at the request of farmers in the Delta. Another off-campus experiment station, at Petaluma, was set up in 1904 at the behest of chicken farmers. The next year the legislature funded research on walnut blight and pear disease. Calls for help from citrus growers led to the establishment of the Southern California Pathological Laboratory at Whittier in 1905, which evolved into the Riverside Citrus Experiment Station (and eventually UC Riverside). Citrus growers also called on the USDA for assistance in the person of plant pathologist G. Harold Powell, who helped control the mealy bug and subsequently became the leading figure in the industry as the director of Sunkist Cooperative.[35]

Some of the crop introductions in the 20th century had pest honeymoons, too. Cotton was pest-free when introduced in 1908 and enjoyed a huge advantage as a result over the devastated fields of Mexico and the American South.

Indeed, up to 1945, surprisingly little pesticide was applied in California: Bordeaux mix for grape fungus; Paris Green and lead arsenate for codling moths in apples, pears, and walnuts; lime sulfur and organics such as kerosene, soap, and oils for insect control. A handful of botanicals such as tobacco (nicotine sulfate), pyrethrum, and ryania were also used. California was the one place in the United States where pyrethrum was widely adopted, and in the 1930s there was a brief fling with fluoride compounds. All of these were not without their dangers or critics. For example, the Federal Bureau of Chemistry tried to limit the amount of sulfur in dried fruit in 1907, but California growers beat back the regulation. A pair of state chemists had detected high levels of arsenic in wine by 1905; the FDA put a limit on lead arsenate residues in food in 1927; and the USDA was telling Western fruit growers, especially pear and apple orchardists, not to overspray. But such warnings had little impact on growers trying to bring nature to heel.[36]

In short, pest control long ran counter to the tendency to introduce manufactured inputs in California's industrialized agriculture. This came as a surprise to me, as it may to many others brought up in the postwar pesticide age. What this tells us, along with the delayed history of irrigation, is that the capitalist logic of the control of nature and the industrialization of agrarian production are not immutable. It will not be followed in lockstep if these things don't make sense in terms of need, cost, and profitability of farming. Here, as elsewhere, the logic of industrial agriculture would move onto another plane entirely in the mid-20th century with the introduction of the electric pump and high dams for water supply and petrochemicals and fumigants for extreme forms of soil preparation pest control. We shall continue that story in the next chapter.

LABOR PAINS

Employment on the farm has a special quality, thanks to peculiarities of the labor process. Workers must move over extensive farmlands; work outdoors in sun and rain, dust and mud; and attend to the changing needs of plants and animals. Being close to nature in the labor process is something farmworkers feel all too acutely, and with little of the romance often attributed to "Country Life." Every crop has its specific needs and skills. Above all, labor time must mesh with the rhythms of the seasons through a series of biologically defined tasks. This makes farmwork notoriously episodic, ruled by the rhythms of natural plant and animal growth. As a result, labor is difficult to allocate smoothly over the course of the year, and has to be assembled at key points in the growing season to undertake time-specific tasks such as planting, weeding, and harvesting, or to attend to calving, roundups, or shearing.

Managing regular labor is by no means simple, given the complexities of natural rhythms, market demands, and the human nature of workers—who have wills and problems of their own. Managing seasonal labor means, most of all, working rapidly (but carefully) to avoid spoilage. As one observer of California agriculture puts it, "Harvests were times of crisis, anticipation, and short tempers." The harvest can be done in a few days, as with apricots, or stretch out for months, as with cotton. Because harvests are so labor intensive in fruits and vegetables, harvest labor constitutes from 30 to 60 percent of total production costs.[37]

The nature of the labor process affects the way workers are mobilized and the shape of the labor market. One way to handle the seasonal flux is to divide California's farm workforce into two groups, regular and harvest labor. Those who can be employed full time on a year-round sequence of tasks are "regular labor," and those applied en masse to seasonal tasks are considered as "seasonal labor" or "harvest labor." Family labor, which was considerable in the past but is less important today, falls mostly under the first heading.[38] Regular labor undertakes a round of jobs such as field preparation, machine maintenance, pruning, removal of diseased trees, pollination, irrigation, and fertilization. Seasonal labor is used mostly for planting, weeding and thinning, and harvesting, but occasionally for other jobs.

Hiring seasonal labor poses a tremendous problem: how to bring huge numbers of people to the farm site when planting or weeding must be done

Use of Hired Farm Labor
California and United States
1870–1990
(Wages as Share of Farm Product*)

Year	California	U.S.
1870	20.8	12.7
1880	--	--
1890	--	--
1900	19.6	7.6
1910	22.2	7.7
1920	16.4	6.3
1930	21.4	9.9
1940	25.3	11.7
1950	21.8	11.0
1960	17.7	8.5
1970	16.2	7.4
1980	14.7	6.4
1990	17.1	8.0

*Gross value of farm produce, 1870–1920, value of farm products sold, 1930–1990.

Source: Olmstead & Rhode 1997, Table 3.

or, worst of all, the harvest is due. Hiring can be quite hit and miss, with growers hoping that word of mouth in local communities or along the migrant routes will bring workers to the farm gate. In the past, they often recruited harvest workers using handbills that promise good wages, good work, and good housing—all of which rarely existed once the workers showed up. Indeed, the point of recruiting drives has been to attract a labor surplus to keep wages in check. The maintenance of harvest labor has rarely concerned the growers at all. Farmworker housing and sanitation have been notoriously bad in California's fields. Compared to the effort put into health insurance packages, workers' compensation cases, and the like by urban employers, there's hardly any management at all for harvest labor, other than getting the job done and paying wages—then hoping the workers will go away until next season.[39]

On the farm, workers march to the drumbeat of the farm manager and gang boss, whose imperative is to produce profits as well as a crops. California growers have always driven a hard bargain for the wages they pay. Given their historic weakness, farmworkers have been all too exposed to the force of the employers to extract long and arduous hours of labor for minimum pay. Farmwork is thus notoriously subject to long hours under miserable conditions. The usual means of payment has been piece rate, and child labor has been commonplace.

The favored method by which labor has been employed on California farms is in crews or gangs. Gang labor is easier to manage, recruit, and pay, because it is normally mediated by labor contractors, or gang bosses. Such contractors are usually former workers who are deeply embedded in the migrant world, mustering workers from extended-family, national, and regional social networks—whether Chinese, Japanese, Mexican, or Miztecan. The gang system means, in essence, the self-organization of the working class. It offers growers tremendous advantages in mediating a chaotic labor market and managing an extensive production process in the fields, while leaving the details of payment and the workers' livelihood to the contractors. This arms-length relation delegates enormous power and responsibility to the contractors, who can be either the workers' benefactors or another layer of exploiters. In the past, such labor contracting was a system commonly used in steel, longshoring, and other industries, one that was normally transcended as capitalist labor relations became more institutionalized and regulated, and skill requirements went up. But not in California farming, where it remains central to this day.[40]

From the point of view of agriculture as a whole, managing a cyclic and geographically shifting labor process presents a particularly thorny problem.

Because of the range of crops grown in California and the varied timing of the seasons north and south, coast and inland, there is a year-round flux of agricultural work up and down the state. Geographic mobility is fundamental to the farm labor process *as a whole,* as Don Mitchell has argued so brilliantly. This mobility demands a migratory labor force that follows the crops—what John Steinbeck called "the harvest gypsies." The individual grower may not care where these people come from or how they survive, as long as they show up at the farm gate at the right time. Yet agribusiness has a collective interest in managing the overflow flux of labor around the state. This has been a vexing matter, which has often called for grower organization at the highest levels, aided by the state.[41]

The early bonanza wheat growers initiated the gang system to handle the harvests over their extensive domains. The overall paucity of field labor, however, meant the use of large machinery in grain operations in the 19th century, making for a curious juxtaposition of a high level of mechanization with the beginnings of the gang labor system. Cattle and sheep ranching had a very different labor process that demanded a higher degree of skill and independence among the many vaqueros, working over vast ranges and drives across long distances. Miller and Lux were the first operators to worry about the statewide circuit of migrant labor, no doubt because their ranching system was geographically so far-flung; they established a system of feeding tramp labor called "the dirty plate route," which kept the supply of workers alive and on favorable terms with the company.[42]

The spread of fruit and garden crops in the last quarter of the 19th century brought many more small farmers into play and a different circuit of farm labor. In horticulture, the tasks of caring for trees and vines involve more off-season labor, especially planting and pruning. Horticulture could therefore be undertaken to some degree by a permanent labor force of the farmer, family members, and a hired hand or two. Italian, Chinese, Japanese, and Mexican workers were commonly employed for their skill in tending new trees and vines. Fruit harvests, on the other hand, made extraordinary demands on farmers and workers. They usually require mass labor, applied rapidly, but feature the additional problems of uneven ripening and easy damage by poor handling (a quality shared with truck crops like tomatoes and berries). But because of the cheap harvest labor market, California farms rarely undertook diversification for the purpose of keeping family labor occupied, as operations did in the Midwest.[43]

Chinese workers were preferred in the late 19th century for all manner of farm labor not simply because they were paid less but because they could be hired through Chinese labor contractors and worked in gangs. The Japanese who followed were equally desirable in this regard because they were largely self-organizing through ethnic contractors.[44] But with the diminution of the Japanese labor supply after 1910, the growers had to turn to a collage of nationalities, only some of whom were recruitable through contractors.

Asparagus cutting, El Centro, Imperial Valley, circa 1910

As the harvest labor force became larger and more diverse in origin, growers began haphazardly to provide labor camps, installing a pathetic geography of primitive settlements around California. They had no incentive to build good housing for a temporary labor force, in terms of either the capital outlay or the possibility of giving the workers a sense of permanence. These camps were so wretched as to provoke labor unrest, so the state got involved in monitoring the condition of migrant labor after 1912. This took place under the California Commission on Immigration and Housing (CCIH), created by the Progressive administration of Hiram Johnson and led by Simon Lubin, a Sacramento merchant and philanthropist. Through the 1920s, the CCIH investigated housing and sanitation conditions on farms, and in several cases forced growers to improve things. The CCIH reported that more growers were turning to contractors in the early 1920s because the courts had exempted growers from responsibility for meeting housing and sanitary standards if contractors mediated the employment relation.[45]

A crucial problem facing the farm manager was how to apply such a polyglot mass of labor to a wide variety of tasks. University intellects offered their advice on this aspect of managing the labor process, especially the ubiquitous Richard L. Adams, author of *Farm Management*. According to Adams, different races should be parceled out to their appropriate slots in the division of labor according to their inbred dispositions. In a devastating reading of Adams, George Henderson shows how the distinctive California farm labor market that mixed together a host of nationalities was made to seem an advantage. In this, "The body, once the subject of the farmer's own hard work, now becomes the object of a class obsession—the natural limits of labor power. In Adams's text, this general truth about labor power and its constraints takes on the distorted form of an absurdly retentive, racialized bestiary." Ironically, Adams steered clear of the more rabid racial exclusionism so popular at the time precisely because of the "great variety of men needed to meet all farm needs." He saw a good use for every race. Adams was in synch with the growers, whose racial theories were conveniently posed so as to assign inferior peoples their proper place in the sun—in 110-degree weather, with no shade, no water, and pittance wages—but always remain flexible enough not to interfere with practical economic calculation.[46]

The arrival of a large Mexican labor force in the 1920s corresponded to the great expansion of melons and vegetables in the Imperial Valley and cotton in the San Joaquin Valley. Cotton is a crop notorious for its association with slavery, and many of the growers who had come in from the American South brought with them a fierce ideology of class and racial supremacy. The early cotton growers of Imperial brought in African American and Punjabi workers expressly to pick cotton. Mexicans could be slotted into preexisting prejudices without too much trouble. Mexican men worked the fields, Mexican women the canneries—but never in the dairies or with machinery. Mexicans, like most of their predecessors, were worked in field gangs. Crew chiefs were commonly men of importance at home; initially drawing from family, friends, and village, they sometimes grew into major labor contractors, or merchants of labor. But they, like their Asian counterparts, might show solidarity with their workers in times of strife.[47]

A notable step taken in the 1920s was state involvement in collective labor recruitment and wage setting. A series of labor bureaus or labor exchanges were established under the auspices of the county farm bureaus, ag extension, and local chambers of commerce. The largest was the Agricultural Labor Bureau of the San Joaquin Valley, established in 1926 by the powerful cotton interests: ginners, large growers, and financiers. Each spring, the labor bureaus would establish wage rates for the coming season, then help recruit workers and enforce the prevailing wage by keeping both growers and workers in line. Former CCIH worker Frank Palomares served as the middleman for the San Joaquin bureau. By establishing an industry-wide labor market and wage rate, however, the labor bureau set the stage for industry-wide union organizing and strikes in the 1930s.

In the bracero era, labor recruitment was the business of the federal government, and labor contractors receded in importance. The chief dilemma was in choosing between braceros and "wetbacks." The racial logic of production was simple: Mexicans were seen as amply fit for "stoop labor." But the now-upstanding braceros were contrasted favorably to the new outcasts, the wetbacks (a racially tinged term for undocumented Mexicans used as freely in government reports as in everyday slang). Wetbacks were seen as less desirable because they were more mobile, and hence able to flee to another job in another place if wages and working conditions were too outrageous. Not that they fared well, by any means, since being illegal meant being unprotected by American law and custom, and subject to nonpayment and INS raids. But they had the one freedom—movement—that the braceros lacked. And labor mobility, as Mitchell has shown, is something to be tightly constrained.[48]

After World War II, California entered the age of petro-farming, and farm productivity shot up. The new production regime did not eliminate labor, however. Unlike the case in midwestern grain or southern cotton farms, total labor hours did not fall off with the rapid rise in productivity. Some hand labor was reduced by harvest mechanization, as well as herbicide application, aircraft spraying of pesticide, and addition of fertilizer to irrigation water. Yet overall labor demand included a host of new tasks. Multiple cropping meant a greater amount of work on field preparation, planting, weeding, and harvesting cycles. Irrigation meant laying in pipe over the prepared fields or maintaining sprinklers in orchards. Mixing and spraying pesticides required more labor in the end. Most of all, with California's large proportion of fruits and vegetables, greater output meant more harvest labor.

The ongoing importance of hand labor is crucial. Field mechanization became a subject of intense debate and scrutiny in the 1970s after the United Farm Workers had succeeded in organizing many of California's fields and vineyards. Workers and their allies felt that machinery was developed and used explicitly to foil the UFW.[49] A widely held view at the time was that technology was going to end the days of mass harvest labor in California. That turned out to be false. Fruits and vegetables have proven notoriously difficult to mechanize, and the reversion to greater fresh produce in the commodity mix has only emphasized this. Hand labor using simple cutting tools is still de rigueur for fresh fruit such as apricots, peaches, and grapes, and tender garden fruits and vegetables such as berries, melons, and lettuce. Hand labor has remained popular for another reason: it is cheap. What occurred in California was not mechanization but *Mexicanization* of the labor process, as Juan Vicente Palerm has called it.[50]

Where this kind of cheap wage labor is available, the management of the labor process remains much as it was before. Labor contractors mediate the relations between growers and workers on most large farms. Contractors were often replaced by union hiring halls in the heyday of the UFW in the 1960s and 1970s, but there are few union contracts today. The number of labor contractors doubled in the 1980s, and they continue to be popular to this day.[51] Working and housing conditions also remain scandalous. Workers were found living in caves near Salinas in 1991, for example. Federal OSHA regulations on portable toilets and ample fresh water supplies in the fields arrived only in the 1970s, with much resistance from growers.

Despite the emphasis on harvest labor in most discussions of California agriculture, only half the labor force is seasonal. There is a regular workforce that tends to the less episodic tasks of preparation, planting, and care throughout the year. With multiple cropping and larger diverse grower operations, this part of the labor force ought to have grown proportionately larger, but it has not. The official labor force count might be even larger if not for the employment of third-party contractors to do much of this work.[52] This development in farm operation is driven, in part, by the specific skills associated with the tasks and the need to use capital equipment efficiently—i.e., economies of scope. That is, it makes more economic sense to hire a pesticide sprayer for a day here and there than to have the equipment and the skills lying fallow around the farm most of the year. But one could say that subcontracting rests on the deplorable state of class relations, too. Because farmworkers have been kept in such a state of poverty and transience, they rarely have been able to improve on their skills or get credit for them; this is the antithesis of, say, the traditional German manufacturing labor process and its skill upgrading. Instead, the grower turns to subcontractors to handle the skilled jobs, and the contractors can, in turn, hire unskilled immigrant labor as their work teams. Where skilled labor is employed full time by growers, it stand outs, as in the high-quality wine-grade vineyards of the North Coast, where some Mexican workers have had the same jobs caring for and harvesting the vines for thirty years.[53]

In a sense, the present labor regime is the gang boss–seasonal labor model raised to a fine art. Almost no one is permanent, and everyone may be an independent contractor. Capitalism has seemingly turned the whole agrarian labor process—not just the migratory or multiethnic labor force—into a thing of replaceable parts. In this it is but a perfection of the mode of production by commodities, discussed in chapter 2, and extension of the farm management system treated in chapter 3. It is, furthermore, but a microcosm of the larger input supply system that has evolved as an extension of the agrarian division of labor. This development will be taken up in the next chapter.

Now we have California's agricultural production system in place. This has been a necessary foundation, just as the establishment of the basic commodity system and the capitalist farmer-grower was before. Yet this chapter only begins to tell the story of how agrarian development unfolded in the state over the last 150 years. The farming system just described is a lively one, and dynamic up to a point, but it is only a precursor of what was to come. California agriculture quickly became famous for its exceptionally high levels of productivity, and it has never looked back. Many students of California agribusiness, such as Carey McWilliams, Paul Taylor, and Don Mitchell, have been so caught up in the exploitation of harvest labor that they have downplayed the productivity dynamic, but we cannot make the same mistake.

California agricultural productivity shows up in crop after crop as early as the wheat era. While some tales of fantastic yields and pumpkins the size of stagecoaches are apocryphal, there are many well-documented cases of sustained productivity above that found anywhere else, from rice and cotton to milch cows and strawberries. Some examples: wheat production was up to forty bushels per acre in California, versus a maximum of twenty in the Northeast in the 19th century; average yield of tomatoes per acre in 1955 was

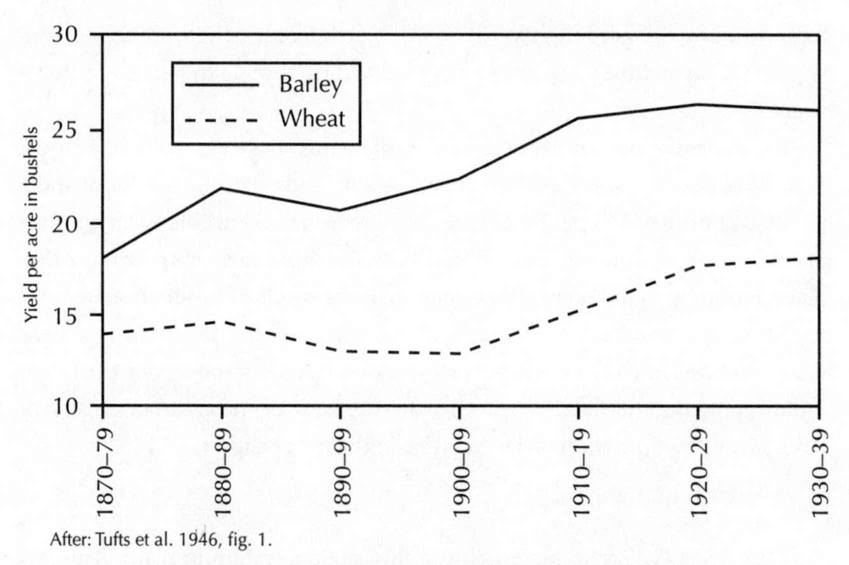

After: Tufts et al. 1946, fig. 1.

Productivity in California barley and wheat, 1870–1939

over twice that of the rest of the United States; cotton output per acre was twice the national average from the 1920s onward; milk per cow was running ahead of all other states in the same period; and strawberry yields hit an astounding five times the national average in the 1980s. Overall, yields per acre in 1980 were the highest in the country across almost the whole spectrum of crops.[54]

In part this productivity gap has been due to California's geographical advantages of growing higher-value crops and opening fertile lands to production. But even with these advantages, California growers continued to chalk up extraordinary productivity gains year after year. There are no long-term aggregate productivity statistics available, but figures have been calculated for some sectors. For example, tree crop output per acre nearly doubled between 1889 and 1919, and in strawberries, productivity by weight grew by seven times between 1945 and 1988.[55] The only long-term productivity statistics available are in field crops: cotton, rice, alfalfa, and grains.

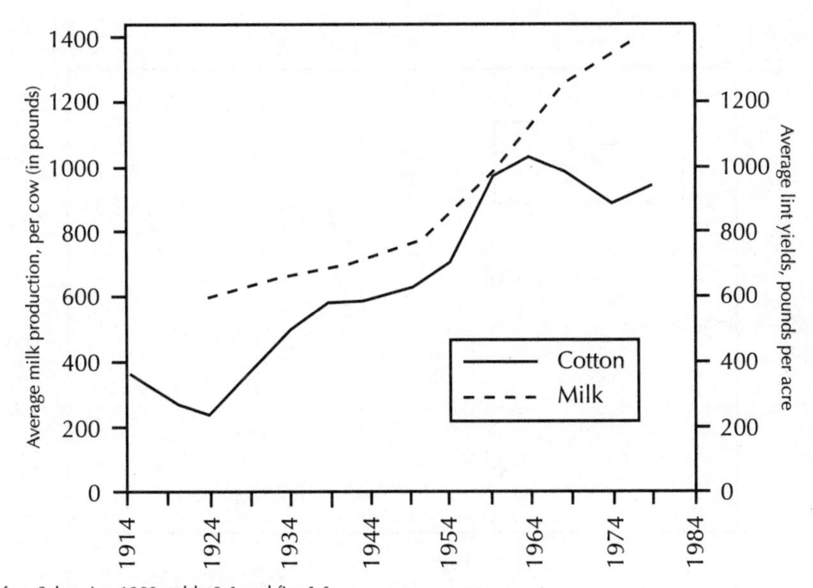

After: Scheuring 1983, table 9.6 and fig. 6.6
Sources: Data from California Crop and Livestock Reporting Service.

Productivity in California cotton and milk, 1914–1984

The contrast with national experience is instructive. It is well known that aggregate farm productivity in the United States was flat from 1870 to 1940, when it turned dramatically upward thanks to the Green Revolution package of grain hybrids, petrochemical fertilizers and pesticides, and tractors. Average yields increased 2.1 percent per annum in the years 1940 to 1960. Agricultural historians thus focus on what Willard Cochrane refers to as "the great transformation" in his superb history of American farming.[56] The postwar regime had a similar impact in California, when productivity also soared. Nonetheless, in California the contrast between the two eras is less marked, because there is no prior era of stagnation. This shows up across all crops for which there are long-term statistics. The principal grains, wheat and barley, slowed down in the glut of the late 19th century, but then show brisk advances after 1900. Cotton and rice production both moved briskly upward in the first

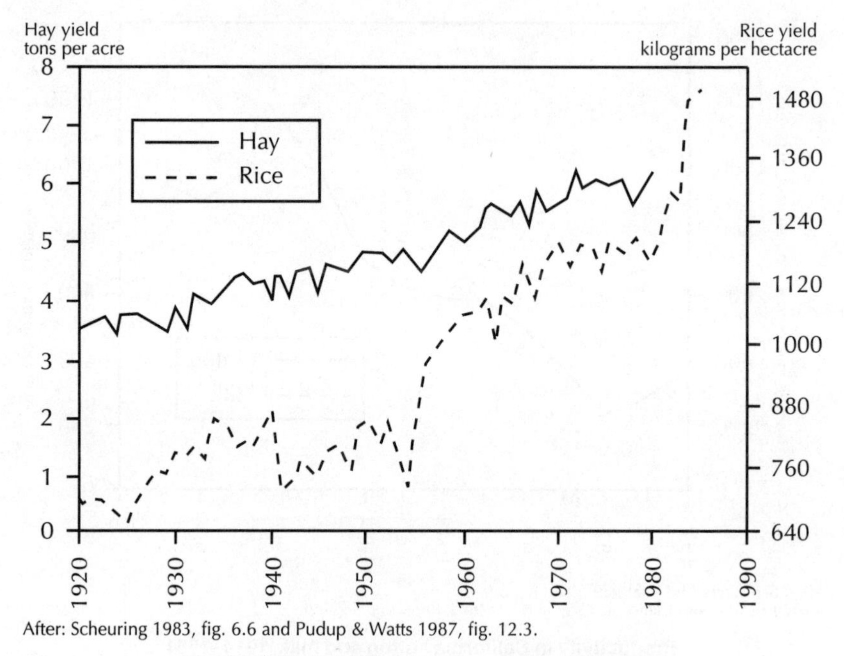

After: Scheuring 1983, fig. 6.6 and Pudup & Watts 1987, fig. 12.3.

Productivity in California hay and rice, 1920–1990

half of the 20th century, only halting during the rapid planting of the two world wars. As Olmstead and Rhode observe, "California's history does not conform to the standard paradigm that treats biological productivity changes as primarily a post-1930 phenomenon in American agriculture."[57] Indeed, California conforms to very few standard paradigms about modern agriculture. But it does follow one line of development quite clearly—a growing reliance on industrial inputs and science, and a consequent expansion of the division of labor and rapid technological advance across many fields of importance to growers. To this story of agro-industrialization we now turn.

Productivity of Eight Leading Crops California vs. United States 1900–2000

(output per acre)

Commodity	Units	U.S.	California
	1900		
Corn	bushels	28.1	27.4
Wheat	bushels	12.5	13.6
Rice	pounds	807.5	NA
Cotton	pounds	194	NA
Sugar beets	tons	7.2	8.6
Tomatoes	bushels	72.8	83.2
Lettuce	bushels	404	289
Strawberries	quarts	1701	3181
	1920		
Corn	bushels	26.7	29.5
Wheat	bushels	12.9	15.5
Rice	bushels	39.1	53.1
Cotton	running bales	0.34	0.53
Sugar beets	tons	9.42	7.56
Tomatoes	1920 dollars	122.2	113.9
Lettuce	1920 dollars	396.2	276.4
Strawberries	quarts	1481	2172
	1940		
Corn	bushels	29.9	32.1
Wheat	bushels	14	16
Rice	bushels	51.5	75.6
Cotton	running bales	0.5	1.38
Sugar beets	tons	11.9	15.8
Tomatoes	1940 dollars	80.2	107.3
Lettuce	1940 dollars	101.2	96.5
Strawberries	quarts	1547	3247

(continued)

Commodity	Units	U.S.	California
	1960		
Corn	bushels	52.8	68.2
Wheat	bushels	21.3	24.2
Rice	bushels	74.8	97.3
Cotton	running bales	0.95	2.18
Sugar beets	tons	18.4	23.3
Tomatoes	NA	NA	NA
Lettuce	NA	NA	NA
Strawberries	quarts	3276	8669
	1980		
Corn	bushels	91	135
Wheat	bushels	33.5	74.3
Rice	pounds	4413	6440
Cotton	pounds	404	969
Sugar beets	tons	19.8	25.7
Tomatoes	tons	24	25
Lettuce	hundredweight	249	278
Strawberries	pounds	15500	42980
	2000		
Corn	bushels	136.9	170
Wheat	bushels	42	76
Rice	pounds	6281	7940
Cotton	pounds	632	1342
Sugar beets	tons	23.7	34
Tomatoes	tons	37.49	37.96
Lettuce	hundredweight	390	377
Strawberries	hundredweight	399	570

5

Industrial Agriculture

Agrarian capitalism could not do the work of farming without the necessary tools for the job. Inputs have flowed from factories, ditch companies, powerplants, and other sources to be incorporated into on-farm production. What are the key agro-industrial inputs? Again, we look to the basic elements of biology, geography, and the labor process. Farming begins with plants and animals, and needs a constant resupply of seeds and stock from nurseries and breeders. Second, farms use a host of tools and machines coming from the machining sector. A third key input in dry summer regions is irrigation water, a curious product with its own specific forms of supply through dams and ditches. A fourth type of input is fertilizers, which eventually became an output of the chemical industry—as did the fifth, pesticides to control outbreaks of insects, weeds, and diseases. A sixth and final group is animal feed, which has several possible origins.

In California, an array of specialized suppliers arose to meet the demand for farm inputs. Nurserymen and stock breeders were some of the first to get established, usually in the leading farm areas like the Santa Clara Valley. Agricultural implement makers popped up almost immediately among San Francisco and Oakland's pioneer machine shops. Irrigation water came from a host of ditch companies, then irrigation districts, complemented by groundwater raised by manufactured pumps, and finally from gigantic state irrigation schemes. Fertilizers were dug out of the ground, imported, and later manufactured from petroleum. Pesticides were, at first, other organisms, then heavy metals and sulfur, and finally organic chemicals from petroleum. Feeds came from other farms, fisheries, and specialized merchants. This chapter will provide a brief overview of each facet of agro-inputs.

This development of suppliers has meant a growing division of labor across a larger domain of agribusiness. It has also meant the commercialization of a whole host of farm inputs—in other words, an extension of the process of commodification begun with crops, land, and labor. Such an unfolding of the whole agro-production complex is an essential facet of modern agriculture; Goodman, Sorj, and Wilkinson call the process "substitution and appropriation," but we can refer to it simply as agro-industrialization. That this is a long-term process of capitalist modernization in agriculture is readily apparent from the development of American farming generally, and is not unique to California. The long-term expansion of output and productivity has required the participation of other sectors of the economy and the technological know-how of the wider industrial system.[1]

Agriculture is not alone in the provision of key elements and technologies of production by outside suppliers or scientists. Every industry benefits from upstream and downstream divisions of labor, technical complementarities and innovations, and external economies of scale and scope. Such a process is fundamental to capitalist development in general (try to imagine Ford's assembly line without electric motors or conveyor belts). Yet many students of agrarian development have held that something peculiar is going on in agro-industrialization. For Goodman, Sorj, and Wilkinson, an evolving division of labor between farms and industry has meant the disassembly of farming as an organic whole. In Richard Lewontin's view, industrial capitalists have sought to wrest control of more and more farm production for themselves, reducing farmers to semi-proletarians under the command of an externally based agribusiness. These views flow from two key assumptions: that farmers are petty commodity producers outside the capitalist mode of production and that farm production is held back by the biological basis of agriculture. Certainly, such disassembly and domination of farming by external capital has happened in some places, like the hog farms of the contemporary South, where a robust on-farm capitalist class never emerged (outside, perhaps, of tobacco). But this was not the case in California.[2]

In California's version of agrarian development, the growers do not offer such an abject lesson in corporate conquest. There are three probable reasons for this. One is that nature does not present such an insuperable barrier to capitalist agriculture as these authors think, a point driven home in chapter 4; hence prosperity and productivity in California farms have not lagged industry so clearly as they did elsewhere. Second, California farmers have been capitalists themselves, for the most part, and able to seize nature by the root on their own behalf. Moreover, their class standing has given them countervailing power in confronting outside capitals. This does not mean that they have not lost battles, or position, but the outcomes of such struggles are not what one finds in the rural South. Indeed, for much of the 20th century, growers ruled the roost in California politics (a point taken up further in chapter 7). Third, the relations among capitalists across broad divisions of labor and extensive production systems are not necessarily antagonistic. Goodman et al., for example, seem to have a rather idealized, Fordist notion of systems integration in factory and corporation, toward which modern history is marching inexorably. But not many companies are like DuPont in its heyday. Small and medium-sized firms are capable of building coordinate relations that hold the structure of production together while allowing for competitive struggle, pricing battles, and innovation to take place.[3]

The robust development of California agriculture has been built on more than increases in external supplies of materials and machinery. It has relied on technological advance across a full spectrum of inputs, as economic historians Alan Olmstead and Paul Rhode have seen so clearly.[4] Agrarian capitalism has embraced a full range of technologies—biological, pedological, hydrological, chemical, and mechanical—and farm technology and the competence of input suppliers have evolved together. This has been true of American agriculture as a whole, of course, given the commercial cast of most family farmers and the supportive linkages to manufacturing and urban markets. Yet one can still make the claim that California has been in the vanguard of agro-technological advance. Equally intriguing is the way its agro-technology has taken unique turns because local conditions are distinctive.

The examples are many, as we shall see. California plant and animal breeders came up with a host of new and more vigorous hybrids, some of which became generalized across the country, and the state led the way into scientific hybridization and, more recently, agro-biotechnology. California machinists came up with a wide variety of new farm implements, from plows to harvesters, some of which were peculiarly adapted to our terrain and crops, and some of which were world-shaking breakthroughs, like the caterpillar tractor. California developed the world's largest plumbing system and the first concrete high dams, to water its farms. California had peculiar needs in fertilization and pest control, and hopped on the petrochemical bandwagon very quickly. Lastly, California revolutionized animal feeding and raising, creating the first confined chicken hutches and cattle feedlots.

California's farm technology and extended supply system has evolved continuously over a century and a half. In the late 19th century, the main introduced inputs were seeds and breeds, plus machines for grain harvesting. Other industrial inputs could not be effective before the on-farm production system was firmly in place and ready to be pushed along to higher levels. This does not mean, however, that California agriculture's marriage to industrialism had to wait for a 20th-century breakthrough into "industrial agriculture," as Lawrence Jelinek puts it; it was industrial from the very beginning of the American era.[5] Nonetheless, irrigation, fertilization, and feeding made great advances in the first half of the 20th century, and modern science began to have a more dramatic effect on agricultural practices. Those, in turn, were overshadowed by productivity gains after World War II, when the petrochemical-

Aggregate Productivity Growth in Agriculture
California and the United States
1949–1990

(Composite Index, 1949=100)

Year	California	U.S.
1949	100	100
1950	100	98
1955	118	112
1960	118	121
1965	128	128
1970	148	143
1975	175	169
1980	191	180
1985	209	214
1990	203	218

Source: Alston & Zilberman 1997, Table 3, based on data provided by B.J. Craig and G.T. Pardey.

hybrid complex took hold, bringing new plants, fertilizers, pesticides, and irrigation schemes into play. Petrofarming was not one thing—machinery, water, or petrochemicals—but a whole complex of farming practices, technologies, and inputs.

Why did agro-technology progress so swiftly in California? Why this forcing of agriculture to perform greater and greater miracles? Favorable climate and soil have played their part, to be sure, but natural conditions cannot explain the social origins of such productivism, and why it has only increased with time. Olmstead and Rhode rightly put rapid technological change at the core of California agricultural development. They attribute this agrarian dynamism eclectically to markets, institutions, tinkering, and learning—all common themes in economic history generally, and all true. What is missing is any reference to capitalism as an economic system and the capitalist imperative that set in train the industrial revolutions of the last two centuries across half the world. Why shouldn't one see the same revolutionary process at work in an agrarian order as in manufacturing, especially one so thoroughly commercial, businesslike, and well capitalized as California agribusiness? While the latter cannot be distinguished absolutely from Midwestern family farmers (who, after all, were embedded in a capitalist commercial nexus as thorough as any in the world), it makes their histories diverge significantly. The divergence from Southern sharecroppers and Brazilian latifundists is greater yet.[6]

Because it had an agrarian capitalist order from the outset, California farming could be pulled along rapidly by the booming market demand of the state and exports to the rest of the country, and it could be pushed along by remarkable innovations coming from irrigation engineers, machinists, and plant scientists. Much ink has been spilt over the fruitless debate in agricultural economics between demand-pull and supply-push theories of induced technological change. The secret lies not in the horse in front or the oxen in back, but in the way the farm wagons could be propelled so easily because capitalist behavior of the growers has been to keep the wheels of progress well greased. In short, technology under agrarian capitalism is propelled from many directions, but has always been eagerly sought after and taken up by agribusiness itself.[7]

ROOT AND BRANCH

The first thing California's growers needed to do their work was a supply of plants and animals. This remains true today. This function could not be handled by the growers alone, and became the specialized province of commercial orchardists, seedsmen, and breeders, with the backup of the university's College of Agriculture. As soon as horticulture began in California, the nursery business took off. The first nursery was begun in 1850 by L.L. Warren, a former nurseryman in Massachusetts who came with the Gold Rush to Sacramento and led the way in serving the farms of the gold country in the early years. Some forty more sprang up in the following decade, mostly around the Bay Area. By 1883 there were 160 across the state. As the century wore on, the expansion of agribusiness into the San Joaquin Valley and Southern California, along with the rapid influx of people seeking garden plants, brought into prominence those regions' nurseries and horticultural leaders, such as John Armstrong of Ontario, Timothy Carroll of Anaheim, and Kate Sessions of San Diego.[8]

Nurserymen were responsible for many, if not most, of the plant introductions of the 19th century. G.G. Briggs of Marysville was a pioneer in peaches, Felix Gillet of Nevada City in walnuts and strawberries, John Rock and Louis Pellier of San Jose in prunes, George Roeding of Fresno in figs. They freely exchanged cuttings and grafts among themselves, partnered, and started up new businesses, spreading the trade. Nurserymen were key reservoirs of knowledge about plants and their performance in California, and often had to teach growers their business. Warren started the first agrarian magazine, *California Farmer.* Wilson Flint of Alameda and John Rock were heralded as the pomological leaders of the state in the 19th century, while Felix Gillet became one of the foremost plantsmen on the Pacific Coast. Nurserymen were also great organizers. Warren agitated for the creation of the State Agricultural Society in 1854, and Gillet served on the first Agricultural Commission.[9]

The biggest cluster of nurseries in the 19th century formed around San Jose. E.L. Beard, a pioneer farmer, established a nursery near Mission San Jose in 1852, sending an agent east for seeds and scions; Beard and his stepson, Henry Ellsworth, developed a large horticultural garden, in which many of the first plants imported from Japan and Australia were to be found. Louis Pellier, a forty-niner, returned to his former trade as nurseryman and settled in west San Jose in 1850, using pear starts from the Peralta adobe next door; brother Pierre came with grape cuttings, then returned to France in 1856 to gather a host of starts ordered up by Louis, including the *petit d'agen,* or French prune. Robert Stockton ordered various fruit tree cuttings from Massachusetts in 1852 to start a nursery of his own. Along with the order came a professional nurseryman, Bernard Fox, who opened Santa Clara Valley Nurseries and Botanical Gardens in 1852 on Milpitas Road. Fox had over 300 pear varieties for sale in his San Jose nursery by 1860. Suyetaro Araki worked with Fox for many years before starting his own orchard and selling stock after 1900. John Rock popularized prunes commercially after 1868. Rock and Fox's nephew established California Nursery Company at Niles, in 1884, and it became the largest on the West Coast well into the next century. J.K. Kennedy, who moved to Los Gatos, pioneered orchards in the western foothills, while fellow nurseryman George Tarleton is credited with the first graftings of French prune onto Damson plum stock.[10]

Luther Burbank of Santa Rosa was a commercial nurseryman as much as a plant breeder. He had to make a profit on sales to growers, including new introductions. In this regard he was a true American son, a notorious salesman and self-promoter, putting out florid catalogues to sell his many wares (though his catalogues were probably no more exaggerated than other sales bromides of the time). He was manufacturing new plants to feed the insatiable appetite of commercial agriculture. In the rush to meet demand, he sometimes falsely presented new breeds as his own—some of which turned out to be renamed imports rather than new varieties. Burbank did his best work before 1900, after which he became a victim of his own reputation and creature of the enthusiasms and humbug whipped up in the popular press of the day, including *Sunset* magazine and the Hearst newspapers. He is largely forgotten today.

Nurserymen, even more than growers, were commonly immigrants and often businessmen in their previously lives. John Rock was a German (born Johann Fels), as was George Roeding's father, Frederick, who started the nursery. Bernard Fox was from Ireland, John Armstrong from Scotland, the Pellier brothers and Felix Gillet from France. Suyetaro Araki and other Japanese entered the nursery industry in the 20th century. One such was Toichi Domoto, one of premier nurserymen of the state, operating out of Hayward. His parents owned Domoto Brothers Nursery in East Oakland, from the 1880s to the 1930s. He was influential in importing camellias, among others garden plants.

Seed companies soon dotted the landscape, as well, as California became the country's premier seed production district. John Horner of Alvarado was the first, in the 1850s, followed by D.L. Perkins of Alameda and A.P. Smith in Sacramento. But seed was not grown extensively until the 1870s (given the self-sowing qualities of wheat), and even then it was mostly grown for the California market. Then C.C. Morse got started in the Santa Clara Valley; he began growing vegetable seeds for D.M. Ferry of Michigan. (Ferry and Morse would later merge.) Bodger Seeds of Gardena and El Monte, Germain Seed in Los Angeles, Henry Fish of Santa Barbara, Walter Rohnert of Gilroy, and Sunset Seed in San Francisco were others of note by the turn of the century. Flower seeds even had their own "Woman Burbank" in Theodosia Sheppard of Ventura in the 1880s and '90s. Santa Clara County was so rich in seed farms that they generated $1 million in revenues out of a total of $8.3 million for all farm products in 1908. In that year, Atlee Burpee began to shift most of his seed growing west to California. World War I provided a huge boost to the seed industry in America, thanks to interruptions in deliveries from England. California became a leader in a wide variety of vegetable and flower seeds, providing 90 percent of all flower seed grown in the United States in the mid-20th century. The heart of the seed industry was Lompoc, on the Central Coast. Burpee and Ferry-Morse were the best-known names in seeds for three generations.[11]

The Mendelian revolution at the turn of the last century ushered in a new era of breeds and seeds. Mendel was rediscovered by Hugo de Vries of Holland and others, and the principles of inheritance transmitted across the country by groups like the American Breeders Association, founded in 1903 as a joint convocation of land grant college scientists, USDA researchers, and commercial breeders. With acceptance of the new genetics, out went such false notions as the blending of parental traits or the inheritance of acquired characteristics, which had dogged breeding and evolutionary theory alike. Of particular importance was the move to what was called "systematic" breeding: away from promiscuous cross-pollination to in-breeding for pure strains, then crossing for vigorous hybrids, and finally back-crossing (and double back-crossing) to arrive at consistency in new hybrids.[12]

The scientists at the university's College of Agriculture at Berkeley, and its new branches at Davis and Riverside, would now become an important source of plant and animal hybrids. This paralleled a national shift, which was greatly assisted by a rise in funding from the federal government. In 1904 a Department of Agronomy was established at Berkeley with the help of funds from the wheat industry. Almost immediately, college scientists, in concert with USDA's Bureau of Plant Industry, began a wheat hybridization program that raised yields within a decade and led eventually to Green Revolution varieties. In 1908 research on lima bean selection was begun, financed by the Lima Bean Growers' Association. In 1912 experimental work on cotton varieties began in Imperial County. In 1922, the strawberry growers got the legislature to fund UC research on xanthosis and root weevils, which was quite successful. In the early 20th century, Davis faculty were active in stock breeding and fruit studies on the university farm and Wolfskill Experiment Orchard, as well.[13]

University agricultural research budgets soared with the help of the federal largesse and state matching funds in the interwar period. The results began to pour forth by the 1920s and became widely known by the 1940s. After lagging in research behind Berkeley, the Davis campus became a center of scientific hybridization after Ben Madson took the reins of the Agronomy Division in 1921. Davis agronomists initiated the California Approved Seed Plan in 1934, in cooperation with the state Department of Agriculture and the Farm Bureau Federation, to guarantee the quality and general distribution of "pure" seed to growers. In the postwar era, research budgets increased again, as California poured more money into its Colleges of Agriculture and Extension Service than any other state, and UC Davis flourished. Tellingly, the new dean of the College of Agriculture in the 1950s was Fred Briggs, a former back-cross hybridizer working on plant diseases. Many more departments were moved to Davis from Berkeley in a general reorganization after the war. In this period, University of California ag scientists would generate a slew of new plants and animal lines, and would become famous for their efforts—or notorious, depending on your perspective. The single most famous case of college-based hybridization for the benefit of agribusiness must be UC Davis's postwar development of the hard tomato for machine harvesting.[14]

Paradoxically, the growing importance of research science and university breeding did not eliminate the private providers of seeds and breeds; indeed, it gave them new life. Kloppenberg is wrong, therefore, when he declares, "prior to the development of hybrid corn, the private seed industry was little more than a merchandiser for the 'college bred' varieties developed in public agricultural research institutions . . . and there was little incentive for private research."[15] This is simply not true for California. The two halves of the input chain to the growers worked side by side, and continue to do so to this day.

Seed farms and nurseries were not often a high-profit business because of the continual leakage of innovation from suppliers to farmers, who could just gather seed one year for the next year's planting or take cuttings of trees and vines. With hybrid plants, however, a new level of commodification of living organisms was reached. Hybrids often produce seed that does not grow true to the parent, or that is sterile. If so, new seeds must be purchased on an annual basis by the growers. This was already happening with the Green Revolution hybrids coming on the market in the mid-20th century. Many conventional breeders thought this was a good thing because it rewarded hybridizers and seed companies for their investment. The particular pathway taken by the Green Revolution breeders was, as Kloppenberg points out, a conscious choice of certain ag scientists not to pursue a strategy of "recurrent selection" that mixed in-breeding and cross-breeding over time to avoid seed sterility.[16]

University scientists almost never generated new plants and animals by themselves, however. There has been an ongoing partnership of growers, private horticulturalists, and university researchers. Take the case of tomatoes. The College of Agriculture was only one player in a long drama of plant improvement. The standard California canning tomato before 1920 was the Trophy variety, which had been introduced by an Oakland seedsman in 1874 and later improved by Berkeley agronomists in 1908. The USDA developed a wilt-resistant canning tomato, the Marglobe, for release in 1925 (its parent was a mutant found in a *fusarium*-infested field in Louisiana). More tomato varieties were developed in the 1920s by Ferry-Morse, UC, and Campbell's Soup Company. Between the wars, seeds and plants were supplied to growers by the canners; after the war, by specialized companies. In the 1940s, an agricultural extension scientist at UC Davis, G.C. Jack Hanna, went to work on breeding a tomato for machine harvesting. His VF145 tomato was released in 1961 and California's share of processing tomatoes mounted rapidly from one-third to 90 percent of the national total.[17]

Another example of collaboration is the wine grape sector, which was rebuilt after Prohibition with the help of the Viticulture Department at Davis. All through the dry years, Davis professor Albert Winkler and Berkeley students including Harold Olmo and Maynard Amerine studied grape varieties and the effects of climate, coming up with the modern classification system of climatic zones. In 1951 Olmo, now a professor, began a program to import cuttings of high-quality wine grape varieties, recapitulating the mass introductions of a century before. UC's researchers were on the lookout for better varieties, particularly better grafting stock, rather than developing hybrids, since the old wine grapes are still the favorites. In 1960 they began promoting the AXR rootstock for *phylloxera* resistance with high yield (although later sad experience would show the lack of full immunity of the AXR stock). A large nursery sector grew up with the expanding demand for grafted vine stock, allowing the huge plantings of the 1960s, the 1970s, and thereafter.[18]

A final example of plant supply and breeding is the strawberry sector. California got a major boost in productivity and marketability with the generation of new hybrids such as Sierra, Donner, and Lassen, introduced by UC scientist Harold Thomas at the end of World War II. These were followed in the 1960s by Tioga and Tufts and in the 1980s by Douglas, Chandler, Pajaro, and Selva. Breeders, both private and university, have manipulated such characteristics as yield, size, plant resiliency, pest resistance, growth rate, and ripening time—the last to smooth out annual production cycles. Strawberries used to be left in the ground for years, but are now raised on an annual basis by a separate nursery industry in the colder climes of Shasta County before being transplanted to fields on the Central and Southern coasts.[19]

Biotechnology is the new frontier in plant development, one that takes the manipulation of living organisms beyond anything previously possible. It is the means by which genetic material (DNA, but also RNA and mitrochondria) can be transferred from one organism to another in the laboratory, without recourse to sexual reproduction. Unlike all previous breeding practice, biotechnology makes it possible to transmit genetic sequences across distant species boundaries—mutations rather than crosses—creating what are called *transgenic* organisms (also called "genetically modified organisms," or GMOs). This means the introduction of exotic characteristics from animals to plants, bacteria to animals, viruses to cells, and so forth. With this ability, the new laboratory breeds can do things previously unimaginable, like carry a fish gene that adds more vitamin A to rice. In the 1980s, as biotechnology took off, a revolution in agriculture was predicted, with much hyperbole from its advocates and many fears by opponents of an untried and poorly regulated technology. The reality has fallen well below the hype—so far.[20]

In one sense, the new biotech is not unlike the Victorian search for plant material around the world. But it functions over a vastly wider range because of the ability to take advantage of microorganism genes (by far the greatest reservoir of genetic material on earth) and at a much faster pace. With the new techniques, researchers can go to a pond in the tropics; extract a tube of unknown insects, plankton, and bacteria; throw it in the digestor; and have a source of DNA ready to aim at the germ cells of a conventionally bred plant species. The altered cells will be grown into mature plants to see what happens. With laboratory advances developed in the 1990s, this kind of mass mutation can be done at an extraordinarily rapid rate. But it is still "a shotgun approach to breeding," as Larry Busch puts it, with imperfect controls on where the new genetic string ends up in the chromosome and what changes result in the genetic code and resultant organism. Most are stillborn. Others grow but express unexpected characteristics as a consequence of the new gene string (it is never a single gene, but a desired one harnessed to a string of presumably inert DNA). And, as with all new technology, some of the best results are happenstance, like finding a bacteria in the waste stream of a chemical factory that has developed resistance to a popular weed killer.[21]

California played a significant role in the emergence of biotechnology. Gene splicing was first discovered by Paul Berg at Stanford, and gene cutting by Stanley Cohen of Stanford and Herbert Boyer of the University of California's San Francisco medical campus. It was taken commercial for the first time by the latter pair as the company Genentech in 1980, with the help of venture capitalists. This was quickly followed by Chiron and Amgen in medical biotech, then by Calgene, Agracetus, and Mycogen in agro-biotech. Tens of billions of venture capital dollars would flow into biotech companies over the next decade. California, the world center of venture capital, became home to the world's largest concentration of biotech companies in the process. The first agricultural transgenic product to be tested outside the laboratory was the Ice-Minus bacteria, which was meant to put a protective coating around potatoes to save them from frosts; it was created by a UC Davis researcher and his private company, Agro-Genetic Science, in 1987 and tested in Berkeley. Another pioneer agro-biotech life-form, the Flavr Savr tomato, came out of Calgene, another spinoff from UC Davis, in the 1980s; it was the first genetically engineered food product when introduced in 1994.[22]

Then, in 1996, four new genetically modified (GM) field crops were introduced, and they swept across the United States and Canada. These were transgenic corn, soybeans, canola (rapeseed), and cotton. By 1999, almost half of the corn, soybean, and cotton crops in the United States were GMOs, covering some 40–50 million acres. The curious thing is how little impact biotech has had on California agriculture as yet, since neither corn nor soybeans are big crops here. The biggest GMO grown in the state is Roundup Ready cotton, amounting to about one-third of the state's crop. As for potatoes, the Ice-Minus bacteria failed to work well in the fields, and the Flavr Savr tomato was abandoned (by Campbell's, which had bought the rights) when it did not hold up as well as hoped in transit from the fields. Most of all, California's specialty crops—fruits and vegetables—are individually too limited a market to justify the investment on R&D to develop GM varieties, at least from the point of view of the big corporations leading the biotechnology charge today. There is some research going forward, but the only GM fruit or vegetable grown is a virus-resistant crooked-neck squash.[23]

We may well be standing on the verge of another Green Revolution owing to the spiral of genetic engineering. How this will unfold, however, is as yet unclear. Biotech appears ready to make agriculture more intensive, more synthetic, and less subject to the rhythms of nature. It also promises to vest even more influence on the input side of agribusiness, and reduce the traditional independence of farmers even further. The rapid entry of multinational chemical and pharmaceutical giants into agribusiness in the 1990s evoked considerable disquiet about their monopoly power over global agriculture. Monsanto purchased several seed companies and formed a new agro-biotech division, then merged with Pharmacia; Dupont bought the largest seed merchant, Pioneer Hybrids; Zeneca seeds merged with drugmaker Astra to create AstraZeneca, which was then bought out by Advanta; Novartis created a new arm, Syngenta, around Northrup King seeds; and so on. Almost all the agro-biotech start-ups of the 1980s were absorbed by agro-chemical companies by 1995, and several of the big companies invested in huge new research facilities or consortia with the universities. Of particular significance for California were Novartis's gleaming new laboratory near San Diego and its research collaboration with the Berkeley campus's College of Natural Resources (successor of the College of Agriculture). More generally, the shift toward private investment in biotech (coming on top of a generation of agro-chemical research) brought to an end, by the 1980s, the century of dominance of public agrarian research by university-based scientists.[24]

Following the strategic lead of Monsanto, outside corporations were induced to enter the field by the seemingly vast horizons of biotechnology as it exploded on the scene in the 1980s, when it suddenly seemed that laboratory science was going to become the propulsive force in the whole food chain.

They went after seed companies because it appeared that new transgenic seeds would be the key to everything. The corporations were also seduced by the apparent synergies in basic genetic science between the fields of drugs, chemicals, and agriculture (which they began incorporating into "the life sciences"). Finally, they were encouraged by the new legal climate after 1980 in which biological organisms could be patented and copyrighted as "intellectual property" (legitimated in the Supreme Court's crucial *Diamond v. Chakrabarty* decision and in the federal Bayh-Dole Act). Previously, seed companies had to live with farmer skimming, and indeed, many plant genomes were considered common property, in the public domain. Now, with biotech, a company could splice in a bit of genetic material to act as a tracer of its patent rights or, more drastically, introduce a "terminator" gene to render all seed sterile.

Nonetheless, the promise of biotech has so far exceeded its payoff. Monsanto's profits and stock plunged after 1998, and several of its competitors withdrew from the field (e.g., Novartis spun off Syngenta to pursue agrobiotech, leaving it few resources).[25] The companies' misplaced enthusiasm may have been a repeat of the mistake made by resource corporations like Tenneco that invested in California land in the 1970s. And the fears of critics about monopoly power in the food chain may similarly be exaggerated, as fears of corporate land takeovers were then. As I have previously suggested, monopoly is not the best lens through which to look at the developmental dynamics—and sins—of agribusiness. Moreover, since data on seed monopolies and other such concentration are almost always drawn from figures on corn, soybeans, and other crops secondary to California, they do not give us a good picture of what is taking place here.

There are several reasons for the setbacks felt by the biotech/agrotech giants. Two of them are integral to the technology and how it has been applied. For one, biotechnology has been less able to generate useful varieties than one might think. There are too many genes and dimensions of plant performance for simple gene splicing to work well. In the end one is still working with specific organisms that are not infinitely malleable, that must succeed on several counts all along the production chain, and that must be accepted as food. As a consequence, conventional plant breeders may still have more success than GM companies in bringing new crops to market: the Bakersfield company Sun World, whose DiVine Ripe tomato was developed over a twelve-year period, is one example.[26] Sun World is no traditional company (it was founded by a former executive of Tenneco), but it reminds one of how Silicon Valley start-ups repeatedly confound dire predictions of monopoly power in electronics. Another reason for GM's lack of success is that so far agro-biotech has been aimed at the production chain and has had little payoff for the consumer. That may change as a new generation of more sophisticated varieties are introduced that provide better taste or nutrition.

Finally, the political furor over GM is having an effect on agribusiness strategies, making some crops unpalatable and others unprofitable. Genetic modification got a great deal of bad publicity worldwide as a consequence of Monsanto's efforts to foist terminator seeds on the world's grain farmers, an activity they have subsequently been forced to abandon in the face of global outrage. Because GM's full effects on human health and global ecology are highly controversial, despite assurances from their promoters, there have been peremptory bans imposed in Europe, Brazil, and elsewhere.[27] The United States has been more welcoming, but by the year 2000 things were not going so well. BT potatoes have recently been rejected by McDonald's Corporation because of consumer doubts. California's rice growers tried to get GM rice banned, fearing adverse impacts on their export markets. The spread of GM is also spurring sales of organic foods, and consumer interest in things like heirloom tomatoes and apples has been piqued, sending California farming along a different trajectory more akin to its classic reliance on high-value specialties.

Biotechnology is radically new, in certain respects, but very much established in the long traditions of the imbrication of science with the practical—and often nefarious—designs of those seeking power and profits. After all, scientists in California have been deeply compromised by aiding and abetting Western conquest, militarism, and mining adventures, among other things. Why should we expect something different in the realm of agribusiness? Additionally, we have to acknowledge that intentions are not the same as results, many of which are unintended. Science in the service of mammon has led to many profound breakthroughs in human knowledge, from the farming of alkaline soils to the creation of lasers, and biotechnology is sure to have the same ambivalent legacy as all the rest.[28]

THE MACHINE IN THE GARDEN

The role of mechanization in California agriculture has aroused the passions of many critics of agribusiness, but should not be overplayed. The adoption of machines has proceeded quite unevenly. Machines have been employed with enthusiasm where appropriate and left aside when there is no call for them. Mechanical inventions of great wit have been brought to market by machine makers around the state, but little progress has been made in broad swaths of crops. The main reason for this reluctance has been the difficulty of mechanizing horticulture and fresh vegetable growing, California's great strengths. In sum, mechanization has not been the defining feature of California agriculture, even though that is how it is sometimes portrayed, and human labor remains an enduring fact of production.

One would not have guessed this outcome from the early years of grain farming, which manifested a high degree of mechanization. "Few if any parts of the world carried to greater extremes the process of mechanizing wheat raising and harvesting." The mechanization of California farms in the early years of wheat and barley was so striking that "nineteenth century observers watched in awe."[29] California immediately jumped up to a level of capital per worker comparable to the northeastern United States. Several things combined to stimulate the rapid adoption of farm machinery. The major crops in the early years were grains, wheat, barley, and sorghum, all susceptible to machine harvesting, and the land of the valley floors was flat. Economically, the scale and financial wherewithal of growers made purchasing machinery feasible, while the cost and scarcity of labor made it imperative. And to supply the machines, there was a host of local machinists already working on mining equipment and protected by high transport costs from eastern competition.

California machinery of the 19th century was larger and more sophisticated than that adopted in eastern states. Gangplows were built wide, wheeled, and with up to eight plowshares; broadcast, endgate seeders were used instead of seed drills; and headers (which cut off only the heads, not the straw) were employed instead of reapers for harvest. Other unique machines of the time were the straw-burning portable thresher and the reversible side-hill plow (for use in orchards). Most impressive was the general use of combine harvesters by 1880, a generation earlier than elsewhere.[30] All this big equipment impelled California growers to shift from oxen to horses and mules (the number of horses vastly exceeded any other state in the 19th century, and California had three times the number of draft animals per farm in 1870 as the U.S. average). California also moved to tractor power very early. The first steam tractor was constructed in California in 1868, and the number of tractors ran far ahead of the nation (behind only the Dakotas). A most notable

Caterpillar tractor in the Delta marshes, circa 1910

adaptation to California conditions was the revolutionary continuous-tread device, developed in 1904 for use on soft Delta peat soils; it became known as the "caterpillar" tractor. All of these machines were developed by local manufacturers such as Ben Holt and H.C. Shaw of Stockton, Dan Best of San Leandro, and Joseph Enright of San Jose.

Nonetheless, machinery was not the overwhelming feature of California farming. One indication of this is that animals were the major capital stock of agrarian California in the 19th century, and they still outweighed machinery and equipment in 1912, $175 million to $39 million. Tractor adoption may have been ahead of other states in the 20th century, but only 10 percent of California farms had tractors in 1910 (compared to 3.6 percent for the United States as a whole), and barely 20 percent by 1925. Indeed, a closer look at the figures shows that there was a slump in equipment used per worker, not felt in other states, from 1880 to the 1920s. California fell behind the Northeast and Midwest (north-central) states in level of mechanization (Olmstead and Rhode's comparisons with national averages that include the South are misleading).[31]

Why was this? The key is that mechanization was not feasible in orchards and vineyards, where the action had shifted during the golden age of horticulture. Fruits have been the most difficult crops to mechanize because of the nature of the organisms. Trees and vines have tender fruit that can be bruised by mechanical handling, mixed-size fruit, an unevenly distributed fruit, and irregularly ripening fruit that makes one-time harvesting impossible. Fruits thus demand careful and recurrent hand picking. As one textbook puts it, "There are relatively few crops for which selective mechanical harvesting is economically feasible."[32] To this must be added the maintenance of orchards and vineyards, which must be trimmed and weeded in situ and cannot be cleared, plowed, and disked de novo. Vegetables and truck crops, especially tomatoes, which became so important in the 20th century, pose similar problems, although the plants are mostly annuals—with some notable exceptions, such as berries and asparagus. The only invention in this domain in the 19th century seems to have been Emil Horst's hop-picking machine and natural air hop drier.

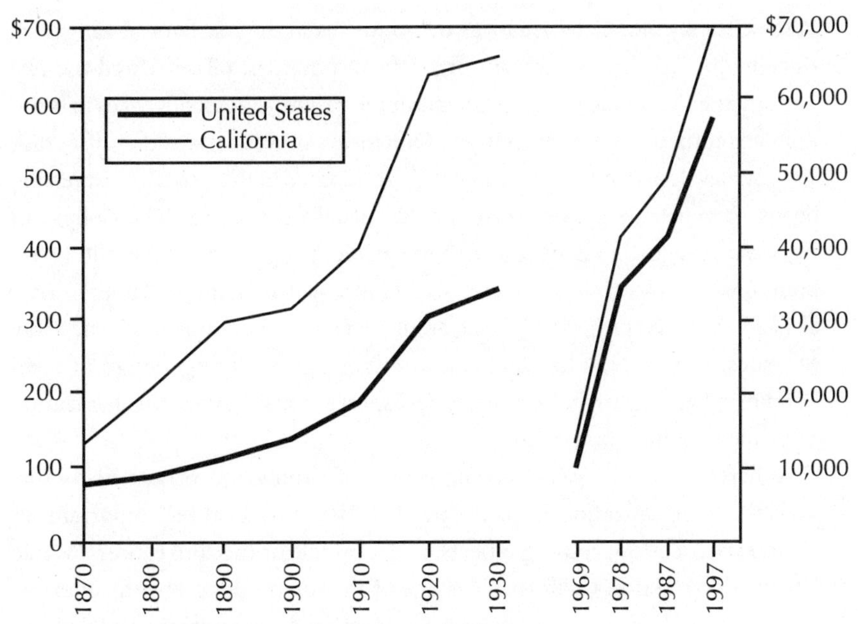

Source: Olmstead & Rhode 1988, fig. 3, for 1870–1930. Figures for 1969–97 from U.S. Census of Agriculture.

Value of machinery and equipment per farm, 1870–1997

Given the recalcitrance of fruits and vegetables to mechanization, the most important farm machine adopted in the first half of the 20th century was the irrigation pump. From 1910 to 1940, California used over two-thirds of all farm pumps in the United States, and had the highest levels of electrification to go with them (50 percent of farms in 1929 versus 10 percent for the nation).[33] An important modification of the pump was John Bean's invention, the modern pump sprayer. Another orchard adaptation was Adolph Greeninger's goose-neck spray truck, which could maneuver through narrow spaces and take sharp turns. These were entirely local developments, because in addition to all the other difficulties, the scale of orchards and vineyards was quite small by national standards, and offered no great market to enterprising agricultural machinery makers in the eastern states.

Mechanization proceeded apace in California field crops, as it did elsewhere in the country. But California was by no means in the mechanical vanguard, since major field crops such as corn, wheat, potatoes, oilseeds, and the like were by then secondary in the state's cropping system. Most advances in those domains took place in the Midwest. One local innovation was the offset disk harrow, developed in 1925. Combines were adopted in the 1920s for harvesting beans—a vegetable grown in large fields, usually for drying. The College of Agriculture lagged in getting a farm mechanics program started in the 1920s, but it finally took off with the arrival of Harry Walker and Roy Bainer in 1928 to head the Davis program. The ag engineers got to work on an eclectic array of issues, such as bulk handling and transport; grain drying; design of farm buildings; lighting; orchard heaters; pest sprayers; and harvest mechanization of prunes, walnuts, and cotton.[34]

Where California was a leading center of production of any field crop, however, mechanization leapt ahead. This was true in three important instances, rice, cotton, and sugar beets, in the middle of the 20th century. While late in adopting rice, California was one of the top producers by the interwar period. By the late 1930s, rice fields were being seeded by airplane and rice harvesters were in general use. The fully integrated combine was universal in rice fields by the 1950s. California rice growing was particularly striking in its level of productivity; as MaryBeth Pudup and Michael Watts observe, "Despite the fact that California rice growers are relative neophytes, they have been in the forefront of farm mechanization, technical innovation, and varietal improvement."[35]

The same could be said of cotton production, which was late in coming to California but took off like a shot. Growing here quickly became more specialized, electrified, tractor driven, and productive than cotton growing anywhere else in the country. Not surprisingly, mechanical cotton harvesting was begun here in the early 1940s and generalized by the early 1950s, a generation earlier than in the South. The fact that almost all California cotton was a single variety, Acala, made the task easier.[36]

Sugar beets have long been a major California field crop, but mechanical harvesting only came along in the 1940s—when the industry appeared to be on its last legs. The breakthrough came from a ten-year project under Roy Bainer and John Powers in the Engineering Department of the College of Agriculture, undertaken at the behest of the U.S. Sugar Beet Association in 1938. A harvester was perfected that could dig, clean, and top precisely each beet. Its success was linked to the development of decorticated seed (producing single beets rather than bunches), machine planting, and thinning. Together these advances reduced hand labor by two-thirds and saved the sugar beet sector.[37]

In the age of petrofarming that followed World War II, however, advances in machinery were less revolutionary in raising output and productivity than those in hybrid seeds, fertilizers, and pesticides. Tractors and trucks were universal by this time, and aircraft became a common implement for spreading seed, fertilizer, and pesticides. Field choppers for hay making were an aid to the supply of alfalfa and other silage to California's burgeoning animal-feeding operations. One notorious success in mechanizing a tender crop was the pro-

Cotton harvesting by machine, Los Banos, 1998

cessing of tomatoes, thanks to another ten-year project by botanist Jack Hanna and agricultural engineer Coby Lorenzen at Davis. Introduced in 1961, the mechanical harvester rapidly swept the field—but mostly in California, where Hanna's rock-solid processing tomatoes were grown, not in other states, where different tomato varieties predominated.

While this was a striking case of substitution of machinery for hand labor, and took place in connection with the end of the bracero program, it was never as symptomatic of California agrarian technology as Bill Friedland, Jim Hightower, and other observers took it to be. Nor is it clear that the tomato harvester was introduced because of labor shortages at the end of the bracero program, as is sometimes claimed. The project was launched in the midst of the bracero era, when labor was plentiful, as were the cotton and sugar beet harvester projects a decade before. While adoption rates of machinery are surely influenced by labor costs, control, and supply, it is a mistake to fall for the old idea of "induced innovation"—whether in a neoclassical guise or a Marxist one—because the pressure for technical change is quite general under agrarian capitalism.[38]

Despite higher yields almost across the board, California was only marginally ahead of other parts of the country in mechanization during the postwar era.[39] The reason was again the continuing difficulty of mechanizing the harvest of fruits and vegetables. High-pressure sprayers were undoubtedly the most useful item introduced for orchards and vineyards at the time. But machine harvesting almost always takes too great a toll on the crop, especially where freshness and quality are imperative. Mechanical tree shakers for nuts arrived in the 1940s to great fanfare, but were later found to damage the root systems of many trees; they worked acceptably with prunes, processing peaches, and canning cherries, but they have not been very effective with apples, pears, apricots, citrus, fresh peaches, or olives. Grape harvesters were developed in the 1960s, but are never used for valuable wine grapes. Strawberries, melons, cucumbers, and fresh tomatoes do not submit to machine picking, while celery, broccoli, cabbage, and asparagus do. Millions were spent by UC and other researchers on a lettuce picker and wrapper that has been a bust.[40]

The most significant arena of mechanization at present is precision farming—the managing of large farms with greater attention to differences in the landscape. Variable-rate application technologies, which began to appear in the 1980s, made it possible to tend fields and orchards with a radically new level of care. Electronic sensors became available to judge site-specific conditions of soil moisture, nitrates, topsoil depth, organic matter, weed presence, and crop growth. Improved machinery included fertilizer spreaders, seed drills, and herbicide and insecticide applicators, as well as regulated irrigation systems. In some cases, global positioning and remote sensing are used to map and manage the whole farm. Costs became reasonable and adoption spread in the 1990s. Given precision farming's multidimensional character, it is ironic that mechanics are in the leading position. This is quite a reversal from petro-farming. Yet precision farming compensates for decades of homogenization of field practices and the suppression of natural diversity under agriculture's version of Fordism. Whether it offers an out from the many defects of this system remains to be seen.[41]

PUMPING UP PRODUCTION

The external mobilization of water for agriculture has an astonishing history, and one that has now been well and often told. Irrigation has featured more prominently in discussions of California agribusiness than any other subject, and a shelf of impressive books can now be found on the water history of the state, from *Cadillac Desert* to *The Great Thirst*.[42] These tell the heroic and nefarious tales of canals, dams, and irrigation districts, and, most of all, of the massive state and federal water projects of the middle of the 20th century. They show how the state came to have the greatest hydraulic system for storing and moving water on earth. It has made for great drama and has changed the face of California forever. Only now are we trying to undo the environmental damage of such unmodulated engineering of rivers, lakes, and estuaries.

Yet, as we've seen, irrigation has not so much created California agribusiness, as many have claimed, as it has pumped up the level of agrarian production. Irrigation should, therefore, be treated as part of the package of external inputs to farm production, along with machines, pesticides, and fertilizers. Irrigation is an add-on to the supply of water already existing in the soil and falling as rain. Irrigation history shows that it came into use first as a supplement to natural supplies.[43] Only later did it become an integral part of almost all farming across the state. A brief look at the highlights of irrigation development should make the case.

Irrigated pasture has been the single greatest user of agricultural water in California since the Civil War. The first large-scale water-transfer project in the state was the 120-mile San Joaquin and Kings River canal, built in the 1870s by a syndicate of Miller and Lux, Friedlander, Chapman, and William Ralston, the kingpins of California capitalism at the time. Ralston and Friedlander both died, leaving Miller and Lux with the canal. How did they use it? To irrigate pasture to stabilize cattle feeding through the summer, and as insurance against a repeat of the disastrous drought of the 1860s. Similarly, the famous *Lux v. Haggin* legal squabble of the 1880s was over water used for irrigated pasture in the Kern River delta (the Buena Vista lake basin). The legal case settled nothing, but Miller and Lux and Haggin and Tevis (the Kern County Land Company) stopped fighting and agreed to build a dam on the Kern River that would provide water for both their pasturelands. Storage and diversion became the preferred solution to water disputes in California from then on.[44]

At the turn of the last century, the leading counties for irrigation development lay up and down the San Joaquin Valley, where pasture for beef cattle and dairy cows was doubtless the leading use of water. In the northern counties and east of the Sierra, pasturelands were being irrigated during the 19th century, as well. In the mid-20th century, the revolutions in dry-lot dairying and cattle feedlots promulgated in Southern California and the San Joaquin Valley depended on feed from irrigated pasture. What is striking about this whole sequence of pasture development is that irrigation was not the cause but the effect of cattle feeding, which was in turn deeply linked to the English breeds and the Anglicized taste for soft and marbled beef (Argentine beef, by contrast, was free range, never dependent on irrigated pasture nor on feedlots). An extreme case of the same thing occurs today, when California alfalfa is being exported to Japan to feed Kobe beef, the national meat delicacy that fetches quadruple ordinary prices. This supports my contention that irrigation is a follower, not a leader, in most of California's agrarian history.[45]

The institutional history of irrigation begins with the invention of the irrigation district in California by the Wright Act of 1877. Irrigation districts were a way of bypassing ineffective private ditch companies and assuring compliance on a regional scale. Their arrival is clearly linked to the coming of horticulture, and they took off in the 1880s before foundering in the recession of the early 1890s. The district idea languished for many years until it was rescued by new legislation in 1911 that met the standards of San Francisco and Los Angeles financiers. Once satisfied, they began to pump capital into irrigation projects, and the number of districts multiplied until they covered most of the cropland in the San Joaquin Valley and Southern California. These irrigation districts were quick to put up small dams on most of the rivers flowing out of the Sierra, diverting water to their growers by networks of canals and ditches.[46]

By the turn of the century, the biggest political battle was the passage of the Federal Reclamation Act of 1902. The frenzy leading up to the passage of the Reclamation Act came mostly out of California, the home of the National Irrigation Congress. The Reclamation Service's officer in California, J.B. Lippencott, was useful in helping Los Angeles spirit away the water rights of the Owens Valley, so it could ship the Owens River from the east side of the Sierra to the south coast (a project completed in 1912). But that was a city project, not a federal one. Furthermore, most of the Owens's aqueduct water did not go to the citizens of Los Angeles, but to the San Fernando Valley to grow oranges. The citrus boom was in full force at the time, and citrus are particularly thirsty trees, which benefited from irrigation as water tables dropped from overpumping. Later in the 1920s, the aqueduct would begin to irrigate suburban lawns as the city grew. Lippencott got cut in nicely for his trouble by the Los Angeles Suburban Home Company, a real estate investment syndicate that made a killing on rising land values.[47]

Lippencott surveyed the future sites of Boulder Dam on the Colorado and

Shasta Dam on the Sacramento, but nothing was built in either place for many years. California's growers did not want the Reclamation Service in charge of their water, because it came with a stipulation that no farm could be larger than 160 acres. Thus, the Reclamation Service did all its work outside California for the first thirty years and, as a result, less than 1 percent of the state's irrigated lands were supplied by the federal government as of 1930. When the renamed Bureau of Reclamation finally came to California, it did so at the invitation of the state's agribusiness, not the other way around.[48]

The explosion in irrigation from 1900 to 1930 can be attributed less to law,

Irrigation well and pump, San Joaquin Valley, 2003

politics, or institutions than to water pumps. Wells were supplying only 10 percent of irrigation acreage in 1900 (most of this in Southern California and the Bay Area), but an astounding 50 percent by 1950. The number of pumps shot up from around 10,000 in 1910, to 50,000 in 1930, to 75,000 in 1950. The key innovation was the deep-well turbine pump. The first one was designed and built by Byron Jackson of San Francisco in 1901 for Pabst Brewing of Milwaukee, but the first agricultural device came from Layne and Bowler of Chicago and arrived in Chino in 1907. The secret to these pumps' effectiveness was putting the impellers on a vertical shaft that could be lowered easily into a well. Electrification made pumping even easier, and it spread like wildfire from 1910 to 1930.[49]

The Imperial Valley has a particularly lurid history of water diversion, and it is the clearest case of the mythology of "making the desert bloom" in California. Now one of the state's premier farming regions, it was at the outset a particularly recalcitrant piece of ground, given the intense heat and lack of water. The solution was to divert the Colorado River westward on a route going through Mexico. But the original plan of George Chaffey—a brilliant engineer and promoter who helped launch irrigated agriculture in Riverside and in Australia—came to grief in the flood of 1905–07. An ill-conceived cut in the bank allowed the rising river to change course and flow unchecked into the Salton Sink for two years, creating the Salton Sea. Even with Southern Pacific's financial might, it took that long to turn the great river back into its bed. After that debacle, Imperial Valley farming finally took off. It would be anchored institutionally and politically by the Imperial Irrigation District (IID), formed in 1911—the largest such district in California.[50]

Environmental historian Donald Worster, in his majestic *Rivers of Empire,* makes the Imperial Valley the centerpiece of the argument that state power is the crux of California's hydraulic society. But the truth is exactly the opposite. The real imperialists of the southern desert were capitalists out of urban Southern California who sent rivers of investment to open up the Imperial Valley during the first twenty years of the 20th century. The agrarian order was securely in place there long before the Reclamation Service got involved in the 1920s. Indeed, the growers and IID made the government build them a better dam to hold back the Colorado floods. Worster also takes Carey McWilliams to task for overplaying the battle of labor and capital in the San Joaquin Valley in *Factories in the Fields,* and for overlooking the irrigation ditches. But Worster errs again. The secret of the cotton fields on which McWilliams focused was not large, centralized river projects (which don't serve the main cotton lands) but small dams and levees built by irrigation and water districts, as well as the little factory in every field by the 1930s, the pumping shed. *Rivers of Empire* makes a better title, no doubt, than *Pumps of Plenty.*[51]

And what a dam the Imperial Valley got! The world's first concrete high dam, erected at Boulder Canyon, was finished in 1932. This marvel of California engineering prowess, carried out under the direction of Henry Kaiser and the Six Companies, is the model for river control around the world to this day. California took the idea to its logical end over the next fifty years, setting in place the world's most massive water storage and transfer system. The anchors of the system are Boulder (Hoover) Dam on the Colorado and Shasta Dam on the Sacramento, completed in 1947, although they are complemented by several other dams on each river and its tributaries. The dams are part of the Colorado River Storage Project and the Central Valley Project, each of which provides several million acre-feet of water per year, funneled along hundreds of miles of canals. The Central Valley project includes the Delta-Mendota Canal (from the Delta to the upper San Joaquin River) and the Friant-Kern canal (from the San Joaquin to Bakersfield); the Colorado project utilizes the Colorado aqueduct (to the Los Angeles basin) and the All-American Canal (to the Imperial and Coachella valleys). A third big piece of the puzzle is the State Water Project, completed in 1972, with a massive earth dam on the Feather River at Oroville and the California aqueduct, running from the Delta to San Diego.

With these mammoth water projects, irrigation finally became universal on all improved acreage in California. The Bureau of Reclamation and the

State Department of Water Resources could deliver millions of acre-feet per year to irrigators (and cities), at a price far below marginal (let alone average) costs, because the projects were completely subsidized by power revenues from the dams, urban users, and the federal and state treasuries. Suddenly, it seemed like water was ubiquitous and cheap as the warm sunshine of California summers. But the economics were askew—so much so that the Central Valley Project could not even pay its *operating* bills, let alone its capital debt, by the 1970s. Imperial Irrigation District hogged the Colorado River so completely that it could deliver water on a regular schedule to its growers at $2.50 per acre-foot whether they needed it or not. As a result, the Salton Sea actually rose for fifty years as the Colorado continued to flow unchecked through the Imperial Irrigation District's pockets.[52]

The dramatic opening of the gates from the big reservoirs understandably led many observers by the 1970s to attribute California's irrigation age to big dams and big government, in the manner of Worster. An even more extreme environmentalist thesis was propounded by Marc Reisner in his widely read *Cadillac Desert* (made into a superb documentary film by John Else). As Reisner puts it, "No other state has done as much to fructify its deserts, make over its flora and fauna, and rearrange the hydrology God gave it. No other place has put as many people where they probably have no business being. There is no place like it anywhere on earth." For Reisner, the problem is Anglo-American hubris in believing that the Western deserts were meant for civilization at all. But he is less on the mark than Worster, even though imperial Euro-American arrogance is certainly at work in the California landscape.[53] California is not a desert, nor is its agriculture unsustainable; but it and the hydraulic system have been vastly overextended in irrational ways.

A striking thing about the three megaprojects is that none of them serves the state's prime agricultural lands, which had already been watered in the first great wave of irrigation expansion from 1900 to 1930, i.e., the east side of the San Joaquin Valley, the Los Angeles plain, the Delta and southern Sacramento Valley, and the Salinas Valley. As Reisner himself states, the Central Valley in the 1920s was already the largest stretch of irrigated farmland in the world. In-

Aerial view of the California aqueduct and the Delta-Mendota canal, 1980

stead, most of the water from the big projects goes to the state's most marginal farmlands on the west side and south end of the San Joaquin Valley and the interiors of San Bernardino, Imperial, and Riverside counties, which receive fewer than five inches of rain per year and can be maintained for intensive production only through massive water imports. These are, truly, the desert lands of California.[54]

But should they have been irrigated at all? Phil Le Veen, an agricultural economist at Berkeley, once calculated that farming in the Westlands Irrigation District—the huge district on the west side of Fresno County, along the California aqueduct—is a losing proposition: the true costs of all inputs (including water without the subsidies) exceeded the revenues from the crops by a fair margin.[55] Of course, one could argue some benefits of the projects: Shasta Dam stabilized the flow of the Sacramento, making rice farming easier along the upper valley; Imperial Valley produces melons, berries, and vegetables in multiple crops throughout the year; the southern San Joaquin kept citrus and other fruit crops viable in California when the best lands were paved over along the south coast (or sometimes simply displaced production from more northerly farmlands, as in olives and almonds). But all in all, the big water projects represent an age of excess, not efficiency, as we now know.

In the 1970s, the mega-irrigation schemes came under intense criticism from several quarters. One was groups that grew out of the farmworker struggles, particularly National Land for the People in Fresno, led by George Ballis, and the California Institute of Rural Studies in Davis, directed by Don Villarejo. They hammered on the refusal of the federal government to enforce the Reclamation Act's 160-acre provision and others on large California landowners. They also made public the way the State Water Project, which has no such acreage restrictions, delivered most of its water to a handful of enormous corporate landowners in Kern County. How did the federal government respond? By ending the acreage limit.

Soon thereafter, environmentalists entered the fray. They pointed out the detrimental effects of the reduced outflow of the Sacramento River on the Delta and Bay estuarine system, which included the collapse of the striped bass fishery. They also became enraged over the discharge of toxic drainage water from Westlands Water District into the Kesterson sink, where it was killing waterfowl. The problem was the same, at either end of the pipe. The Delta is damaged by excessive water withdrawals, and then the water is used to flush the soil of salts, leaching heavy metals into the drainage water. A climactic battle unfolded over the Peripheral Canal, which would have allowed the state to divert more of the Sacramento River around the Delta to the south (and permitted a series of add-ons to the State Water Project). After the scheme was approved by the legislature and Governor Jerry Brown, it went to a referendum and was voted down soundly by the public in 1982. The defeat of the Peripheral Canal was a body blow that staggered the agricultural interests for several years, but they recovered in the 1990s to slowly push back the environmentalists under the CALFED program—a state-federal effort to find some compromise that would improve the fisheries while increasing water deliveries. After years of talk and study, the curtain of illusion on CALFED has been drawn back enough to see the pumps within.[56]

The market has done its own critique on the inefficiencies of the irrigators. With water prices rising and two sharp droughts (in 1976–77 and 1988–91) coming on top of the farm crisis of the early 1980s, some marginal lands have gone out of irrigated production, especially in the southern San Joaquin Valley. Harvested farmland shrank for the first time in a century, although in some areas, high crop values and drip irrigation have sent vineyards and orchards up hillsides into new territory. Growers have also had to attend to the economics of irrigation with more care than before, aided by their advisors in Ag Extension, the university, and private consultancies. As a result, water management has improved markedly. This involved investment in more sprinkler and drip systems to replace open-furrow or flood irrigation, along with better monitoring of soil, atmospheric moisture, and crop stress. A radically new kind of weather-monitoring network was introduced in 1985, under the auspices of the State Department of Water Resources, the irrigation districts, and the university: the California Irrigation Management Information System (CIMIS). This provides current data via radio and Internet on rainfall, evaporation, vapor pressure, wind, soil temperature, solar radiation, and so on, with which growers can make daily decisions about irrigation.

Still, the middle of the pipe is pretty leaky, with a great deal of water wasted on low-value crops and inefficient delivery systems. Irrigators at the extremities of California today are embroiled in fierce, highly publicized struggles with other water users and are holding on by their fingernails. The Klamath River, saved as a Wild and Scenic River, is dewatered by upstream diversions for irrigation—mostly in Oregon—and Klamath Lake, a major wildfowl breeding ground, is reduced to a rump marsh. Massive salmon and steelhead die-offs in 2002 raised the ire of the downstream Indians, who have fishing rights, but Washington came to the rescue of irrigators. Only the Bush administration could defend such egregious ruin of a river system. At the other end of the state, the Colorado's overtaxed flows are monopolized by the Imperial Valley, where growers pay almost nothing for the water, while San Diego and

Tijuana are desperate for increased supplies. IID has been recently forced to sell off some of its water contracts to the coastal cities, after a howl of protest went up when they turned down the offer.[57] No one should be deluded into thinking the urban developers are benign, but agribusiness's days of absolute control over four-fifths of the water resources of California are past.

PETROFARMING AND FERTILIZER

A central thread in modern agricultural history is the growing application of fertilizers to supplement the soil. The ultimate goal now appears to be to transcend soil altogether as a nutrient source and planting medium. Curiously, California agriculture lagged behind most of the advanced capitalist regions in the application of industrial fertilizers. This may appear to contradict my thesis about the developmental logic of capitalist farming, but in fact it turns out to confirm the dialectic of capital and nature that lies at the heart of agro-industrialism.

Until the industrial revolution, farms were largely closed systems on the input side—even in Britain, the most commercialized country. That began to change in the early 19th century, when farmers opened up to commercial inputs of fertilizer, in what F.M.L. Thompson calls "the second agricultural revolution." The most important additives before 1840 were ground bones (bonemeal), wood ashes, and oilseed cakes (later mixed with corn) that were fed to animals. These not only nourished the livestock, but supercharged their manure, which could be spread over fields and gardens. The use of manufactured oilseed cake indicates the close relation between animal feed and fertilization in agricultural modernization.[58]

In the 1840s, Justus von Liebig of Germany put agricultural chemistry on a new footing, when the key role of supplemental nutrients for plant growth was recognized. From there industrial methods could come increasingly into play. For nitrogen, guano was imported in large quantities until cheap sulfate of ammonia could be derived from gaswork wastes in the 1860s; direct synthesis of nitrogen and hydrogen in natural gas came after 1913. For phosphorus after 1850, bones and coprolites were poured into sulfuric acid vats to produce superphosphates; treble superphosphate followed in the early 20th century. Potash imports, for potassium, were vastly increased with the opening of the (industrial) mines in Prussia in 1861. Britain was far ahead in industrial fertilizer use and production until the latter part of the 19th century, by which time the German chemical industry had overtaken the rest of Europe.[59]

The United States ran behind Europe, largely because of its supply of frontier lands, but the new science of agriculture was catching on by 1850, and the practical interest of farmers was rising quickly in the well-worn regions of the East Coast. A commercial fertilizer industry developed for guano, superphosphates, and mineral supplements such as lime and gypsum. The discovery of large phosphate rock deposits in the Southeast spurred the emergence of a superphosphate industry in the last quarter of the 19th century (most U.S. phosphorus today still comes from Florida). Potash still had to be imported from Germany until the 1930 opening of mines in New Mexico, where 90 percent of U.S. potash still derives. The use of commercial nitrogen fertilizer doubled between 1910 and 1920, then increased by 50 percent from 1920 to 1930, with the derivation of cheap ammonia from natural gas.[60]

In California, commercial fertilization hit first in a big way in the southern citrus belt. This was because citrus trees are such big nitrogen feeders and because of the extremely progressive stance of citrus growers (Hilgard had alerted growers early on that citrus would need nitrogen supplements, and subsequent extension staff helped promote fertilization). Nitrogen was expensive, coming from Chilean guano or from coke-oven by-products. Then, in 1931, Shell Oil Company built a plant to make anhydrous ammonia from natural gas at Martinez, and prices dropped; naturally, the first commercial application was made to a citrus grove in Tustin, Orange County.[61]

California agriculture was distinctive in that it took its nitrogen in the form of ammonium sulfate (rather than ammonium nitrate), because local oil refineries were replete with sulfurous petroleum, sulfur can be used like gypsum to break up clay hardpans, and many California soils are sulfur deficient (especially in the far north and on the east side of the San Joaquin Valley). California growers also took a liking to calcium nitrates imported at great expense from Norway, on the theory that they improved fruit and vegetable quality.[62]

The cities seem to have been the major source of bonemeal and superphosphates, which were derived from slaughterhouse and industrial by-products.[63] But some superphosphates were imported all the way from Chicago, and phosphoric acids from South America. California later opened up its own deposits of rock phosphate, though most of its supplies come from Idaho (processed by Best Company, Lathrop, and a co-op near Five Points, or imported by Wilson and George Meyer). In 1916 California developed the first significant alternative to German potash in North America, muriate of potash from Searles Lake in the Mojave Desert, packaged under the Corona brand. This was replaced circa 1930 by better potash coming from deposits in New Mexico and Saskatchewan. Another popular source of all three nutrients was fish offal (called "fish emulsion"), derived mostly from sardine reduction at Monterey.[64]

California fertilizer production prior to 1940 was climbing, but it was still low compared to what was to come. Soils were good and fertilizer prices high, and high-response hybrids had not been introduced for most specialty crops. Moreover, nitrogen application makes little sense without adequate water, and thus usually developed in tandem with irrigation. Indeed, water and fertilizers were often joined at the hip; for example, the Pritzer brothers invented an applicator that introduced soluble fertilizer into irrigation water for citrus in 1928. Fertilizer use took off with the shift to petrofarming around World War II. In this California was, as one might expect, a step ahead of the rest of the country: nationwide use roughly doubled every decade from 1940 to 1970; in California it tripled in the 1940s and shot up twentyfold from 1940 to 1980. California built eight ammonia factories in this era, as many as in the rest of the United States combined, and fertilizer mixing and blending plants dotted the landscape by the score.[65]

Petrofarming triumphed in California, as everywhere else, as a whole new on-farm production regime set in, vastly more productive than anything that had preceded it. Crop yields took off. Chemical fertilization stimulated plant growth and was combined with irrigation to allow double and triple cropping each year. Fallow and crop rotation were abandoned on most crops. This represented a further perfection of California's pattern of hyper-specialization and monocropping. As Guthman puts it, "Land could be made to work harder and faster."[66] Gasoline, pesticides, and fertilizers were all derived from petroleum and natural gas.

Petrofarming's shift toward greater external inputs changed the face of agribusiness across the country. The chemical industry was suddenly the dominant player in agro-industrialism, with farmers put in a much more dependent position to secure fertilizer and pesticides. The proportion of value added

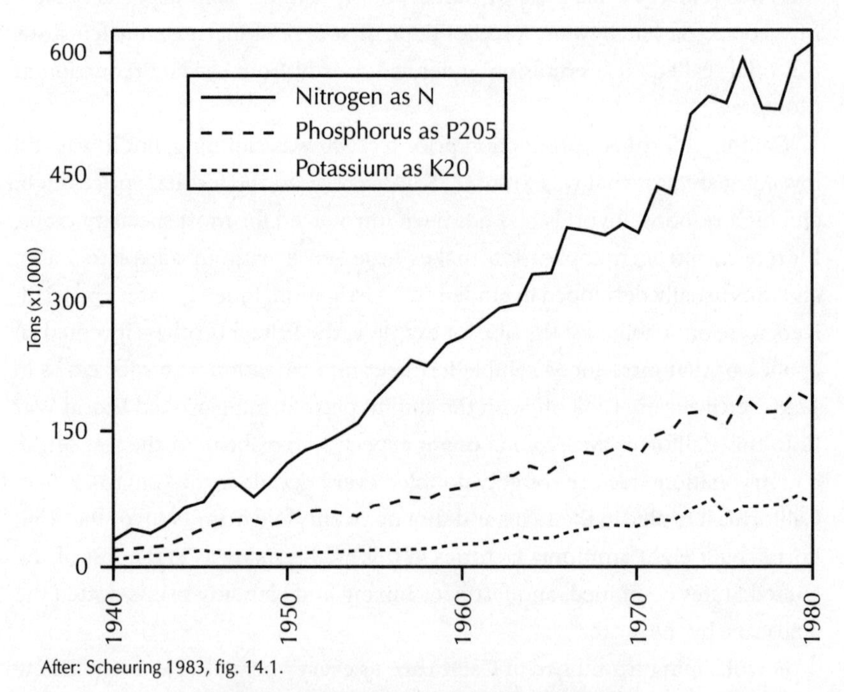

After: Scheuring 1983, fig. 14.1.

Fertilizer used by California farms, 1940–1980

off the farm in the food chain grew disproportionately (though it should be remembered that while inputs are more than double the share of value added by farming, they are still less than half the value added in output processing and merchandising). Furthermore, along with this growing commodity input part of the chain, the large industrial companies—chemicals particularly—became more a part of farm management, giving advice, selling solutions, and, to a considerable degree, forcing their vision of petrofarming on the farmer. In this, California became more like Midwestern agribusiness, not less. Today there is some talk of genetically engineering biofertilizers, by giving bacteria or non-leguminous plants the capability of fixing nitrogen. Progress has been made, but no commercial application of the technology has yet been introduced. If it were, the implications for cropping patterns, alfalfa haying, and soils would be enormous.

AWAY WITH ALL PESTS

Pesticides became a basic part of the petrofarming package at the middle of the 20th century. They key was the development of the organo-chlorine, organo-phosphate, and carbamate families of insecticides, made possible by advances in petroleum refining and organic chemistry, and hastened by the logic of warfare. These products were descendants of World War I nerve gases, and only later discovered to have insecticidal properties. In World War II, DDT proved valuable for killing mosquitoes in tropical climes and saving the military from epidemics of typhus and malaria. DDT was the vanguard product, given a hero's welcome at the end of the war as "the atomic bomb of the insect world." But there were dozens of others, many of them more lethal. And there were other families of fumigants, the most deadly of all.[67]

Within only a couple of years, DDT and its cousins had virtually replaced arsenicals and other metal-based pesticides in American agriculture. California adopted petrochemical pesticides with particular gusto, and applications shot up after 1940. California applied more pesticides than anywhere else on earth over the next quarter century, by all accounts. Over 100 million pounds per year were being used by the late 1970s, costing the state's growers about $500 million in purchases and spraying. One estimate is that California was using one-quarter of all pesticides in the entire United States at the time, another that it was 5 percent of the *world* total.[68]

The use of so much pesticide was made necessary, first of all, by the intensification of production. Well-fertilized and -watered crops were all the more juicy for aphids, caterpillars, and all the rest of the feeders. The end of crop rotation was even worse, for it rendered the fields an almost pure monoculture year-round. As Michael Pollan has put it, "Monoculture is at the root of virtually every problem that bedevils the modern farmer." Insect infestations increased because of the elimination of habitat for their natural enemies (including, importantly, mixed-rotation crops); they were zapped with DDT, aldrin, dieldrin, parathion, and more across fields and orchards. Weeds grew more rambunctious with the addition of so much fertilizer, particularly nitrogen, and the weed killers 2,4,5-T and 2,4-D became ubiquitous in rice fields and elsewhere. Nematodes and fungi grew more insidious with irrigation and maltreatment of soils, making fumigation more popular, particularly for Central Valley grapes.[69]

Pesticide use mounted further as farmers sought to beat back new outbreaks and tougher breeds of their enemies. The new insecticides, in particular, were wide-spectrum biocides that killed with indifference both the harmful bugs and their natural enemies, including insects, spiders, and birds. Chemical companies responded with hundreds of new varieties of chemical killers. But the harmful bugs developed resistance as fast as the new chemicals could be introduced. As a consequence, farmers got onto a pesticide treadmill they could not step down from, and their pesticide costs mounted as fast as crop losses would have. Cotton, California's number-one crop at the time, also became its number-one user of pesticides—a far cry from the halcyon days of pest-free Acala in the 1920s.

The petrochemical age had greatly increased the input of big chemical and oil companies in agribusiness across the country, and especially in California.

Dow, Chevron, Monsanto, DuPont, and many more entered the fray with enthusiasm. Pesticides had become big business, and the logic of sales and profits drove the suppliers to promote their use as much as possible. As Robert van den Bosch put it, in his characteristically acerbic way, "[P]esticides are an ideal product: like heroin, they promise paradise and deliver addiction."[70] The companies advertised, lobbied, and sent their salesmen out to the growers. This was hardly benign, for the kind of farming these fellows were selling was petrofarming, not organics. The chemical age was eagerly promoted by university researchers and cooperative extension. In a particularly egregious act of state aid to agribusiness, pesticide salesmen were given official status as "licensed pest control advisors," as if they were knowledgeable about anything more than the killing qualities of their own line of poisons. This was a sign of growing private influence and diminishing independence of government extension agents.

The negative effects of the rain of pesticides were several. The nerve toxins were extremely hazardous to anyone entering the fields before they could degrade. Widespread farmworker poisonings were reported, and many more went unreported. California also manufactured many pesticides, with some disastrous health effects in the supply line, as in the case of DBCP poisoning of workers in Lathrop (near Stockton) and its spread into much of the groundwater in the Central Valley. Residues on food were suspected of long-term health effects on consumers, as with the older arsenicals. But the new petrochemicals were vastly more disruptive of ecosystems, from the immediate vicinity of the farm all the way to the oceans, where pelicans and sardines died.

A reaction was not long in coming, though at the time it seemed far too slow. The FDA adopted the strict Delaney clause on food testing in 1958; scientists like Rachel Carson began to raise a hue and cry; and a Toxic Substances Control Act was enacted by Congress in 1976. Meanwhile, many an ordinary home owner and farmer began to notice the "silent springs" all around.[71] By the 1970s, many entomologists and other scientists within the bowels of agribusiness were raising their voices in concern. At the UC Berkeley College of Agriculture, entomologist Robert van den Bosch (a student of Ralph Smith) led the charge, and he and his dissenting colleagues were harried and harangued for their trouble. Stepping out of the company line in the College of Agriculture has never been good for one's professional health.[72]

By the 1980s, despite the fierce opposition put up by the whole pesticide-agribusiness complex, the tide was turning, thanks to the combination of rising costs, increased regulation, bad publicity from poisonings, public protest, and scientific evidence; and alternatives to the petro-pesticide treadmill were being sought. Today, some kind of integrated pest management—the radical alternative espoused by van den Bosch and his ilk—is now the norm, and the organo-chlorine and organo-phosphate poisons are largely a thing of the past. Nonetheless, hundreds of poorly tested chemicals are still widely used, and tons of pesticides are still applied (over seventy tons in 1991, when reporting was made mandatory by the state). Fresno County leads the country in pesticide applications, just as it does in farm output. Levels of air and water pollution from agriculture in California are still very high. Regulators are only just beginning to clamp down on the farm sector, since the turn of the millennium.[73]

Environmental concerns forced Congress to begin putting strictures on agricultural research and the agricultural extension system at the land-grant colleges. Retrenchment at the agricultural experiment stations was made worse by the 1965 reapportionment in the legislature, reducing the power of rural counties in Northern California. Major cutbacks occurred as early as 1966, and continued through the belt-tightening state and county budgeting of the 1980s. A major reorganization came in 1974, combining the experiment stations and extension service under a single head. One of the striking developments over the last generation has been the greater commercialization of science in agriculture. The land-grant university research system has lost ground to private industry, thanks to neoliberalism in general and biotechnology in particular.[74]

Ironically, one of the most vital inputs to agriculture is a much beloved insect, the European honeybee. Bees are used by the millions for pollination of crops. They are indispensable for some crops, like almonds and cucumbers (which are not self-pollinating), and helpful for increasing fruiting on almost all orchard and truck crops. Honeybees are hired guns, moving from farm to farm through the season, and are not resident on large farms. Indeed, because of large-scale monoculture, they cannot survive past the flowering period of each specialized crop and starve to death if left in one place too long. Honeybees are provided by a little-known subsector of agribusiness, beekeepers, who keep hundreds of hives, contract their hives to growers, and breed new colonies. California developed the largest bee-breeding industry in the Americas, supplying not only its own beekeepers but those across North America. UC Davis apiculturalists perfected artificial insemination of queen bees, mimicking the success of poultry and livestock scientists, and the campus has the most comprehensive bee biology facility in the country.

Only lately have the scientists done much research into alternatives to European honeybees, themselves a monoculture introduced from abroad. Rarely have other pollinators been tried, as in the case of the fig wasp from Turkey (which is specific to certain species of fig). California's many native bee species have largely been ignored, even though they can be better pollinators than honeybees (especially the several bumblebee species). They have never been encouraged or commercialized—perhaps because they don't hive in such large, easily handled numbers—and many are dying out from loss of habitat and overuse of pesticides. Honeybees, by contrast, are threatened not by pesticides so much as by mites and diseases, which devastated hives across the state in the late 1990s. The reason? It is not so much the spread of a rapacious killer, the apian mite, but the failure to inspect hives for infestation because of cutbacks in county farm agents due to state fiscal woes after Proposition 13.[75]

In the 1990s, the focus of innovation and concern shifted to biotechnology, which is, so far, tightly linked to pesticides and pest control. Breeders have long been aware of pest control as a potential within plants and disease resistance in animals, and this became a matter of scientific urgency in the 20th century. As early as 1905, Spreckels Sugar gave money to the university to breed against

Almonds in blossom with beehives, San Joaquin Valley, 1985

curly-top disease in beets. Another example is the introduction of several rust- and bunt-resistant melon varieties by university researchers in the 1930s. In the 1950s, the ag scientists helped discover a variety of alfalfa resistant to the rampaging spotted aphid and a lettuce seed free of the mosaic virus. An ironic twist to this kind of breeding was a search for plants that would not suffer damage from chemical pesticides, as in Ferry-Morse's sulfur-resistant melons, introduced in the 1930s.[76]

In the case of the new biotech, the chief genetically modified crops are of two kinds. One type carries a gene from the *Bacillus thurengis* bacteria, BT, which produces a toxin deadly to caterpillars. It replaces what was previously a method of biocontrol (a natural pesticide) applied by spray. Both BT corn and New Leaf potatoes are in this group. The other main type are Roundup Ready crops, mostly corn, soybeans, and rapeseed. Roundup is the trade name for glyphosate, a popular plant killer sprayed for weed control. Roundup Ready crops have had a bacterial gene added that makes them immune to the weed killer's baneful effects. The result is that fields do not need to be plowed or harrowed for weed removal before planting or during the growing season.

These GM crops offer less insecticide use and soil disturbance, both of which lower production costs, so farmers have thus far been favorably disposed to them. But they have a downside. More weed killer is used than ever before, and its effects are felt at great distances, as in the die-off of plankton at the mouth of the Mississippi River. And the Roundup resistant gene can spread into surrounding areas, with the possibility of a superweed's emerging through natural crosses. The BT in GM crops lingers months longer in the environment than sprayed BT, and BT is now the only pesticide used over large territories, meaning that there is a strong likelihood that insects will develop resistance to it. It also spreads by pollination to nonfarm flora, where it kills caterpillars of desirable insects like Monarch butterflies.[77] Fears of such adverse effects are widespread; they lie behind the considerable political opposition to GM and behind government restrictions around the world. Apparently the lessons of *Silent Spring* and the chemical age have not been entirely forgotten.

The (il)logical end of petrofarming is to eliminate the land as the basis of production. This goes beyond the fungibility of the leasehold, mentioned earlier, to the reduction of the soil complex as a growth medium. Two aspects of soil are dispensed with: the nutrient qualities, which are replaced by fertilizers, and the microorganisms, which are killed off by pesticide applications. Michael Pollan tells a story of the Russet Burbank potatoes grown in Idaho for McDonald's french fries, in which he compares the soil of an industrial grower with that of an organic farmer; the former is thin and gray, the latter moist and alive. This could just as well be a California agribusiness operation. California strawberries are now grown on soil that is first fumigated with methyl bromide to kill every living thing, then covered with plastic and planted through small holes. By such means, all the strawberries for the whole country can effectively be grown on a few thousand acres in the Pajaro Valley.[78] With greenhouse hydroponic tomatoes and flowers, growers can dispense altogether with soil and climatic influences.

Agro-industrialization took a sharp turn toward chemicals as it continued a mad dash into the future. E.W. Hilgard helped jump-start this process, but a modern strawberry field would no doubt send the old man spinning in his beloved sod. The biological foundation of farming is seemingly under wholesale assault, as capital tries to mount the barricades of a reluctant nature. As Pollan goes on to say, "To put the matter baldly, an [organic farmer] is working very hard to adjust his fields and his crops to the nature of nature, while [industrial farmers] are working equally hard to adjust nature in their fields to the requirement of monoculture and, beyond that, to the needs of the industrial food chain."[79] This kind of agriculture is a crude bludgeon that cannot be long sustained. One sees the limits of petrofarming on all sides, from degraded soil to toxic runoff. The chemical companies know this, and it has propelled the

Field covered in plastic, 2003

likes of Monsanto and Dow into biotechnology—as a way out using more technology to solve the problems of past technological fixes.

The organic movement is in large part a direct response to the crisis of petrofarming. The well-grounded farmer takes offense at the harm done to the precious soil, degraded by industrial fertilizers and fumigants and never allowed to lie fallow and recover. For the consumer, the key element is the toxicity of pesticides and fertilizer contaminants. For the more philosophical, it is the unsustainability of the whole petrocomplex in a world of limited resources. Whatever the impulse, organic farming has thrived in California since the 1970s. But there are two difficulties facing the organic path. On the one hand, deeply organic farming requires few external inputs, hence few profits for chemical companies and other manufacturers. As Michael Pollan puts it, "[This] type of agriculture doesn't leave much room for the Monsantos of this world: organic farmers buy remarkably little—some seed, a few tons of compost, maybe a few gallons of ladybugs." On the other hand, as Julie Guthman demonstrates, big-time organic farming runs the risk of replicating the system it pretends to replace: intensive exploitation of land and labor and large-scale integration of the commodity chain.[80]

A CHICKEN-AND-EGG STORY

Animal feeding has been a fundamental mode of intensification in California agriculture, yet has received little attention compared to mechanization, pesticide use, or irrigation. In the long history of domestic livestock, animals have grazed on grasses, forbs, and roots in pasture, savanna, and woodland, but with the coming of modern capitalist agriculture in Britain and the Netherlands, animal husbandry began to be supplemented by systematic feeding, confinement, and rationalization of care. The first agrarian revolution brought enclosure of pasturelands for sheep and cattle and "the new husbandry" of the 18th century, which meant planting legumes and root crops instead of leaving bare fallows, and "yard feeding" to supplement grazing. The second agricultural revolution meant the introduction of industrial feeds like oilseed cakes in the early 19th century.[81]

California has played a decisive role in modern livestock feeding in the American West and in the United States generally. The earliest measures were supplemental feeding with oats and barley. Irrigated pasture on a large scale was first introduced to stabilize cattle feeding after the calamitous drought of the 1860s. This was undertaken by the modernizing cattlemen Miller and Lux and Kern County Land Company, and by other big ranchers as the era of the open range came to an end. They were the first American commercial operations to begin adapting the unique conditions of Mediterranean California and the erratic rainfall of the western range. This occurred well before the great droughts of the northern Great Plains in the 1880s and Southwest in the 1890s would reveal again the fallacy of overstocking in boom times.[82]

About the same time, new varieties of grasses and legumes were introduced to support intensive stocking better than the native grasses, which suffered dearly from overgrazing under drought conditions. The USDA and university scientists were eager participants in the rapid transformation of the California landscape to nonnative forage, bringing in everything from Bermuda grass to Ladino clover and Australian salt bush. In the early 20th century, led by Arthur Sampson, they vigorously supported heavier stocking under the guidance of Clementsian theories of climax vegetation, carrying capacity, and reversibility of damage. Sampson studied under Frederick Clements at the University of Nebraska before coming to Berkeley, and he translated his mentor's ideas from forests to grasslands. In the process, he founded a new field of range science. His peers also engaged in attempts to breed better grasses. But under the influence of Clementsian theory, they paid too little heed to California's unique climatic conditions.[83]

California agribusiness took mass feeding of livestock to a whole new level in the early 20th century, but in an unexpected arena, poultry. Up to that point, chickens had lagged in the degree of commercialization and industrialization in the United States. They were an everyday part of barnyard life, and little more. California changed all that, and in the process propelled Petaluma in Sonoma County to the world's biggest chicken hatchery for over fifty years. Never before had there been such a concentration of poultry in one place: half a million in 1890, 1 million by 1904, 2 million by 1930, 7 million in 1950. Their purpose was not to generate meat but to lay eggs for the commercial market. Petaluma proclaimed itself the "Egg Basket to the World," hitting over 40 million dozen shipped in 1919 and 100 million dozen in 1930. (Alameda County was another center, with 1.5 million chickens in 1930.) For the first time, a poultry *industry* had come into being, decades before Tyson and Perdue appeared in the eastern United States.[84]

This degree of mass production and spatial concentration required confinement, artificial incubation, and intensive feeding. Confinement took place in large chicken houses, which still litter the Sonoma landscape. Electricity was

Chickens and sheds, near Petaluma, circa 1910

vital to artificial incubation and better henhouse conditions. Incubation was improved dramatically by Lyman Bryce's homegrown invention of the incubator in 1879; he and Christopher Nisson were the first in the United States to sell chicks commercially. By the 1920s one Petaluma incubator could hold almost 2 million eggs at a time (California had over 500 hatcheries by 1940). For feed supplies, Petaluma was within easy range of the North Bay dairies for skim and dry milk and the Delta for corn, but especially important were the nearby coastal fishing centers. Petaluma developed specialized merchants, breeders, and feed distributors, along with its 3,000 chicken farmers.

The role of fish in animal feed has been little appreciated, but it was significant, as shown by Kate Davis. Around 1900, California developed the first industrial fishery in history, using modern boats, internal combustion engines, and manufactured nets. Fish by the thousands of tons were brought in to shore to be processed in canneries and rendering plants. The most notable fish was the sardine, which supported the largest fishery in the United States, but tuna were important in Southern California, and salmon and sharks in the north. Most of the plants were in Monterey, on Cannery Row, though there were five along the upper San Francisco Bay and many in San Pedro, as well. While one thinks of canned sardines, salmon, and tuna as being for human consumption, a large proportion of the sardines, almost all sharks, and a good deal of tuna and miscellaneous catch were turned into fish meal and fish oil for farm animals and fertilizer. This amounted to tens of thousands of tons a year in the interwar period, with a value of around $3 billion in 1935 alone.[85]

The health of the chickens was the biggest obstacle to mass production of eggs. Confined birds are susceptible to a number of maladies, such as "roup," caused by a lack of vitamin A, and vitamin D deficiency, due to lack of sunshine. It turns out that fish oil is a source of these vitamins, and it was being using locally before the scientific link to vitamin deficiencies was discovered in the 1910s. Chickens fed with soybeans in the Midwest did not fare as well as California poultry. Petaluma's chicken industry got outside advice from a UC Poultry Experiment Station established there in 1904 (though it was moved back to Berkeley a few years later). Not surprisingly, the College of Agriculture was the first in the country to have a full-time poultry pathologist and poultry extension specialist. While the University of Wisconsin led the way in vitamin research, researchers at Berkeley and Davis, such as V.A. Asmundson and T.H. Jukes, would make major contributions with the discovery of Vitamin K (B12) and the B-complex in the 1930s. Ironically, such advances in nutrition science were applied to poultry and livestock earlier than to humans.[86]

Petaluma's days as cock of the walk came to an end in the 1950s. Local chicken farmers fell prey to the next agro-poultry revolution, and the geography of production shifted dramatically. The national center of broiler production moved to the South under the leadership of Tyson Foods and Perdue Farms. The modern mass-production system relies on confinement in wire-bottomed cages, artificial lighting, hormones, vaccines and antibiotics, computerized feeding, increased scale, and feed-to-foodstore integration. California did not fall much behind, however. It has remained among the top ten states in broiler output from the 1930s to today and remains the number-one egg-producing state in the nation (with over 700 million dozen eggs sold, mostly within state, in 2000). The San Joaquin Valley took over as the state's main poultry region after the 1950s.[87]

Feeding and breeding in poultry were closely aligned. In 1921 university extension scientists began working with the Sonoma County Farm Bureau on a hatcheries project to improve chicken breeds, which bore fruit in the 1930s. The university people were not alone in their efforts: one of the most important chicken breeders of the middle of the 20th century, Charles Vantress, worked in a private breeding operation near Marysville (before moving to the South). Vantress was a contributor to the national Chicken of Tomorrow breeding contest launched in 1945 by A&P stores, which utterly transformed the chicken into the full-breasted beast of today. Turkey breeding also advanced, with the help of California breeders such as Nicholas Farms of Sonoma. By the 1950s, however, breeders had plumped up turkey breasts so much that the poor beasts could no longer copulate, so Frank Ogasawara and Fred Lorenz at UC Davis developed artificial insemination methods for turkeys. Ogasawara became known as one of the world's foremost poultry scientists as a result.[88]

Dairies—always a major player in the state's agribusiness—were the next to move to intensive feeding, after the chicken industry. They did so in the 1930s and '40s, realizing significant increases in milk output and keeping California well ahead of the rest of the country in milking productivity. Southern California dairies led the way by introducing confined or "dry-lot" feeding, using cut alfalfa and hay, fish meal, oilseed meal, and a variety of by-products such as citrus pulp. While most of the feed stock was local, some feeds were imported from as far away as the southern United States and the Philippines. San Joaquin dairies followed this pattern, but with less confinement and more local agricultural by-products, such as cottonseed meal, sugar beet pulp, and rice stalks. The explosive growth of sugar beets, cotton, and rice after 1910 provided cheap surpluses of feed, and in the depressed markets of the 1930s, Sun-Maid even sold off raisins as animal fodder. The introduction of field choppers after World War II made the harvest of alfalfa for dry-lot feeding easier.[89]

Naturally, the university and agricultural extension scientists were busy assisting in the science of feeding and breeding larger livestock as well as poultry. They jumped into the dairy business in 1901, mostly through careful monitoring and advising of private breeding programs. In 1921, extension agents announced a program to double milk productivity—which succeeded on 100,000 monitored animals, keeping California far ahead of the rest of the country. Beginning in the 1910s, researchers made extensive studies of types of feed and feeding regimens for all the main categories of livestock, including sheep; the first such experiments were held, not surprisingly, on a Miller and Lux ranch near Gilroy. Extension also circulated information on feeding practices and demonstrated the virtue of better breeds and feeds at the university farm in Davis. University researchers later added to the mix with scientific investigations into animal nutrition and vitamins, metabolism, hormones, and reproduction. But much of the improvement came from stockmen themselves, not the scientists.[90]

Even more startling were the San Joaquin and Imperial valley beef feedlots developed in the late 1940s and early 1950s, which revolutionized beef finishing. Although small feedlots were common in the Midwest for finishing beef before slaughter, and custom feed yards for finishing cattle were first estab-

Dairy farm near Linden, 2003

lished in California in the 1920s, confined feeding on a mammoth scale was radically new. California feedlots soared from 125,000 head to over 1 million head from 1945 to 1960, bringing in cattle for finishing from rangelands all over the West. The feedlot revolution swept out of California, transforming beef production, bringing down the Chicago packers, and shifting the location of the meatpacking industry. Feedlots and packers moved rapidly into the Great Plains, which caught up and passed California in beef production in the 1960s.[91]

In 1951 antibiotics were introduced into animal feeds, greatly bolstering immunity against disease and allowing more animals to be packed into hen-houses, dairies, and holding pens. Cattle were found to fatten more readily on antibiotics, as well. In feedlots, cattle are stuffed with corn and other feed in order to fatten them up and develop softer, marbled meat (embedded fat)—a heightening of trends begun in the 19th century. The feedlot revolution lowered beef prices and raised beef consumption markedly in the postwar era, until beef was surpassed by chicken meat mass-produced in the gigantic operations of the southern and San Joaquin operators. Beef intake has suffered a decline over the last generation with growing health concerns and fear of heart disease among American consumers. What has only recently become clear, however, is that corn-fed beef is more to blame for the detrimental health impacts than meat, per se.[92]

To amplify the feedlot effect, cattle, poultry, and dairy cows began to be fed hormones in the 1950s. The pharmaceutical industry had entered the agricultural fold as a major supplier and promoter of its products. Milk output went up as dramatically as any other animal product in this era, increasing by two-fifths in California with roughly the same number of cows between 1950 and 1964 (and doubling across the country). The extension service was pleased to have introduced stilbestrol. One of the first hormones, DES, was later banned by the Food and Drug Administration because of its malignant effects on human mothers and fetuses—but only after years of contention.[93] A debate has broken out today over biotech-engineered bovine growth hormone, BGH, which has still not been approved by the FDA after a decade of conflict.

Even more bizarre things are on the horizon, now that animals and pharmaceuticals have become so closely entwined. One of the latest tacks of the biotechnology revolution is to utilize bacteria, plants, and animals as biological generators of pharmaceuticals. Indicative of this, controversy recently broke out in Santa Cruz County over a zoning decision to allow transgenic goats to graze on land reserved for farming. Since these animals were producing a biomedical product in their reengineered bodies, the question that plagued the community was whether this should be considered a "farm" or a sort of bioindustrial "pharm."[94]

In sum, the supply of agricultural inputs has been deeply commodified over time, and the quantity and quality have risen, revolutionizing agrarian production. The supply of inputs has been a vital site of the emergence of California agribusiness. That is, it has been an arena of advance in the division of labor, commercialization, technological innovation, improved farm management, and state involvement over an ever-widening compass of activities. The complexity of the input system is such that it has come to look nothing like traditional farming. No doubt California's irrigation system is its most stupendous achievement, but the nurseries, fertilizer plants, machinery sector, dairies, feedlots, poultry districts, and research apparatus have all had their day in the sun and have touched many other areas of agro-production practices around the world. Today, it appears to be the biotechnology companies that are going to be the key movers and shakers of agribusiness in the foreseeable future, with yet more revolutionary impacts on California and the world's food systems.

6

The Harvest of Agribusiness

California agriculture appears as an immense accumulation of commodities, to be sure, but the crops cannot pile up in the fields. Successful farming—like all industry—depends crucially on moving produce forward to market. Agrarian capitalists—like all capitalists—have therefore to attend to the downstream stages of their commodity chains. Our narrative must, therefore, move far beyond the farm and crop production to capture the full extent of the commodity (food) chain and to grasp the full scope of agro-industrialization. Indeed, the merchants, processors, and retailers have came to dominate the food system as a whole over the last century, dwarfing farming and input provision in value added and employment, as well as in the number of large, well-known corporations. We can divide this downstream side of the commodity chain into four parts: merchandising, processing, retailing, and consumption. Each will be discussed in turn, and the narrative will unfold both analytically and historically, since the most dramatic developments in each area have come in a certain (if imperfect) chronological sequence.

First, the commercial pipeline that leads from farmers to consumers runs through a complex set of buyers, wholesalers, and carriers. At the center of these sales networks stand the commission merchants, packers, cooperatives, and grower-shippers. California gained significant competitive advantage over agriculture in other states by being well organized early in this domain, and its marketing cooperatives became the model for the rest of the country. A second stage of agro-commodity chains is processing. Without the industrial processors, such as canners, ginners, and meatpackers, many crops would be inedible, unusable, or untransportable. Here again, California was in the vanguard of the production of industrial foods, particularly canning, and it boasted some of the largest food corporations in the country, especially Cal-Pak (Del Monte).[1]

Third, at the retail end of food circuits stand the grocery stores and restaurants. Food retailing (particularly fast foods) has become a topic of considerable national attention recently, but has been almost wholly overlooked in California agrarian studies. It turns out that the state has been a leader in the evolution of supermarkets and restaurants from early in the 20th century. Fourth, at the end of the food chain stands consumption. This is not simply ingestion, but the process by which consumers come to want certain foods and not others. In this, they have been deeply influenced by promotion: advertising, food redesign, and packaging. California has seen some significant innovations in the way food is presented, prepared, and marketed. California has also provided a lively demand for its own agriculture produce and industrial foodstuffs. Finally, even though California agribusiness contributed mightily to mass consumption of bland American cuisine for many decades, it also, in the end, has been involved in a significant shift of the country's eating habits toward fresher, exotic, and more refined cuisine.[2]

One aspect of the capitalist order of agribusiness that must be kept in mind throughout the discussion is that a prime logic of capitalism is the sales imperative: finding outlets for the ever-increasing volume of production. This is no less true in agriculture and the food system than in automobiles or clothing. The fecundity of California farming has been a mixed blessing in that overproduction has repeatedly plagued every crop segment as more land has been brought into production and productivity has increased. This has forced growers to seek out ways of getting produce to market. But the imperative extends beyond growers to those who wholesale, process, and retail foodstuffs. Their rates of profit and capital accumulation depend on three things: acquiring good-quality crops cheaply and dependably, moving products along the pipeline efficiently (keeping down their own labor and other costs of merchandising), and adding value in further processing steps in factories and restaurants. Thus, the downstream capitalists have been every bit as innovative and modernizing as the growers and input suppliers.[3]

Another key concept in treating downstream stages of the food chain is *integration* of the division of labor by companies, markets, associations, transportation systems, and so forth. Vertical integration of this sort has been a central idea in business history and theory since Alfred Chandler's pioneering explorations of the modern industrial corporation. Although nearly everyone once thought that the large corporation was the answer to all questions of integrating commodity-production-capital circuits, over the last generation the puzzle of industrial organization has been discovered to have more than one solution, like Rubik's Cube. Large corporations such as Del Monte and DiGiorgio have sometimes risen to prominence in California agribusiness, but to this day it is impossible to point one's finger at *the* dominant players, as one might for General Mills or Cargill in the grain economy of the Midwest. It is decidedly not true that the citrus industry or any other part of California agribusiness has ever "typified managerial corporate capitalism," as some claim,[4] or is easily fitted to existing models of business organization and its history. Hence an investigation of the full complement of stages and linkages in the food chain is essential for fleshing out the beast called agribusiness.

FACTORS IN THE FIELDS

California's agrarian capitalists were always keenly aware of the need to expand the market for their produce. Agricultural marketing evolved across the board in the 19th century, and a workable fabric of modern marketing and distribution was knit together by many hands: railroads, packers, canners, bankers, and governments. It would be far advanced by the time of the state-sponsored marketing boards of the New Deal. These actors in the distribution network are what I am calling the factors in the fields—to use the old English term for merchant agents. The story of farm cooperatives in California is the best-known piece of this tale, and the names of the big co-ops still resonate: Sunkist, Sun-Maid, Sunsweet, Blue Diamond. Their history has usually been told as a heroic struggle of small growers to wrest control of marketing away from the merchants, packers, and shippers. While the co-ops' achievements were extraordinary, they were neither solitary, unexpected, nor particularly noble. They were just good business. We need to reframe the history to capture the full evolution of downstream organization of the commodity chain, starting with the wholesale merchants.

In order to reach potential markets, California's agro-capitalists had to address a highly fragmented division of labor. In part, they faced their own size limitations and the need for *horizontal integration* to reduce competition, keep prices up, and maintain a common front vis-à-vis the markets. But the more important goal, just as in steel or automobiles (as Chandler discovered), was to pull together the strands up and down the crop commodity chains.[5] That is, produce had to move from the farm to be packed, shipped, distributed, and sold. This was a matter of concern not just to the growers, but to everyone up and down the line. Who would take charge of the commodity chain depended on the product and the outcome of intercapitalist conflicts and maneuvers. But out of this struggle came the modern integrated growing-shipping-marketing system.

The story of large-scale agro-marketing begins, like everything else, with grain. As the wheat fields spread, the wheat moved by boat down to the rivers and the bay, passing through dozens of little ports, like Alviso, Petaluma, and Colusa. Grain merchants clustered, above all, in San Francisco's commercial district. That was home to the greatest wheat baron of all, Isaac Friedlander, who grew nary a stalk but cornered the flour market in 1852, then invested in flour mills and began exporting to China. As railroads reached into the Central Valley in the 1870s, Isaac founded Port Costa at the Carquinez Straits (where the Delta meets the Bay) as the principal trans-shipment point for grain. He chartered a fleet of ships and orchestrated the trade to England until his bankruptcy and death in 1877. Friedlander's passing did nothing to dampen the wheat market, as others like George McNear (organizer of Sperry Flour) stepped in to take his place as miller-merchants. The success of California wheat exports depended on Liverpool and Hong Kong grain merchants, as well. About one-third of California's crop was exported, generating tens of millions from the trade, and overnight the state became a global player.[6]

California's second agro-industry, ranching, was mostly focused on the regional market. The early cattle business was organized around auctions at cattle yards, usually next to the slaughterhouses in San Francisco, Oakland, and Sacramento. Miller and Lux, who controlled the holding grounds at Baden (south San Francisco), cobbled together the first integrated agribusiness company in American history. Their strategy was efficiency through vertical integration, not simply land acquisition. Nonetheless, for all Miller and Lux's 2 million acres, they never controlled the butchering and tanning end of the business, nor did they eliminate independent ranchers.[7]

In fruit, business organization began with packers, jobbers, and commission merchants. In local markets at San Francisco or Los Angeles, this job was handled by wholesale merchants, often with their own distribution system through the big produce markets or fruit stands. Italians such as Angelo Giurliani of Star Foods were prominent in the wholesale trade. Before founding Bank of Italy, A.P. Giannini ran his father-in-law's wholesale produce business in San Francisco. But as California began to export fruit (fresh, dried, and canned), the task of managing the commodity chain became more difficult; it included packing for shipment, orchestrating transport, timing at both ends, targeting distant markets, developing efficient wholesale distribution, and maintaining a forceful sales effort. Large merchant-packer firms, such as Rosenberg Brothers of San Francisco, emerged to handle the job of merchandising across the country. Some dried fruit was already going abroad, as well, in the late 19th century.[8]

A key component of the distribution chain was, of course, transportation.

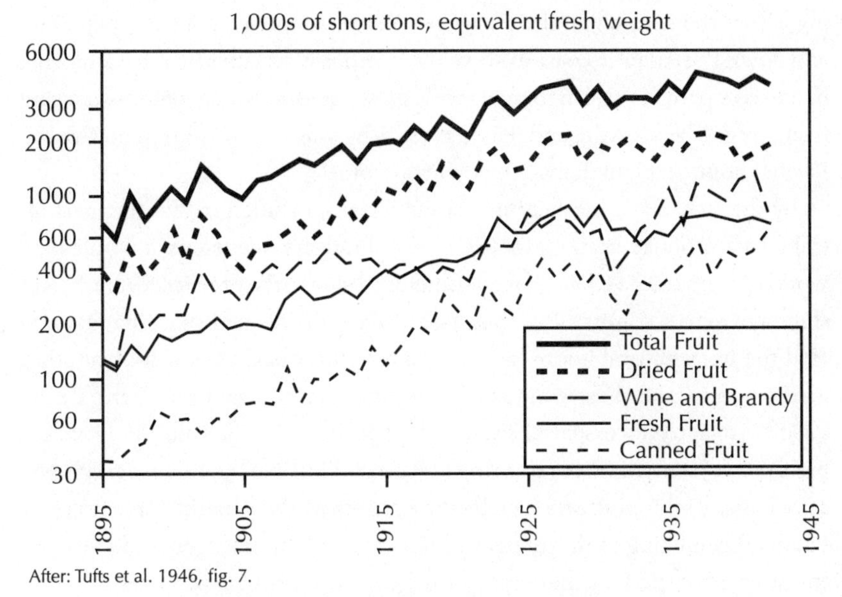

After: Tufts et al. 1946, fig. 7.

California fruit, fresh and processed, 1895–1945

But the mythology of transportation-led development in California agriculture needs to be rethought. Transportation is only one link in the distribution chain, and not inherently more important than the development of merchandising, finance, or retailing. For example, California's first food exports did not go by railroad, as is often supposed, but by ship. That would be the wheat trade, of course. Moreover, the California-Liverpool wheat trade was a primary cause behind the introduction of clipper ships and behind their continued use well after the era of steamships arrived (the dry California wheat did not rot on long voyages). Fruit also went by ship if it was canned or dried, and this continued after the arrival of the transcontinental railroad in 1869. Certainly, the railroad opened up new markets to the east, giving a vital boost to the horticulturalists, but California was not as isolated as we imagine. In any case, the same stimulus of an expanding rail network was felt on farm marketing across the country in the post–Civil War era, and a farm belt opened up and down the Atlantic coast to serve the northeastern cities by rail. California had to compete in eastern markets with these regions, which were also raising fruits, vegetables, cattle, and dairy cows. Why was it victorious? The Central Pacific cannot explain that competitive outcome.

By the late 1880s, another innovation in transportation made an impact on California produce marketing. Refrigerated railcars opened up a whole new world of sales after a Chicago fruit dealer introduced the practice for fresh fruit shipment across country. But this, too, was not the deus ex machina that created the horticultural boom (as writer after writer declares). It took another generation for the Western railroads—Santa Fe, Southern Pacific, and Union Pacific—to introduce special trains (like Southern Pacific's Pacific Fruit Express) and systematize ice reloading and precooling. Nor should we forget the advantages to shipping provided by the opening of the Panama Canal in 1915. Finally, the building of the interstate highways and the emergence of cheap air transport have sped the marketing of delicate fruit in our time.[9]

More important than the railroad to California's agricultural success were the marketing cooperatives that arose in the first two decades of the 20th century.[10] The first successful marketing associations were the California Fruit Union (1885–94), organized by fresh fruit growers in the north, and the Orange Growers Protective Union (1885–93), organized by citrus growers in the south. John Anderson, Santa Clara County pear orchardist, was first president of the California Fruit Union (CFU) and Harris Weinstock, a Sacramento vineyardist and dry-goods merchant, its guiding light. In its nine-year existence, the fruit union introduced auctions at Eastern markets, won lower railway rates by contracting in bulk, and was shipping roughly two-thirds of California's fresh fruit exports. The depression of 1893–95 ended both pioneering grower's unions, leaving California awash in surplus produce from the planting boom of the 1880s.[11]

The crisis sent growers into a frenzy of activity to control markets, find outlets for fruit, and restore prices. The citrus growers of Southern California, who felt overproduction most severely, were the first to hit on the right model

Freight boats on the Sacramento, circa 1915

by organizing around district exchanges which then banded together as the Southern California Fruit Growers' Exchange in 1893 (centered in Riverside, heart of citrus country). Some 40 percent of the state's citrus was under this umbrella by the time it morphed into the regionwide California Fruit Growers' Exchange (CFGE) in 1905, expanding to roughly three-quarters of citrus output between the wars. The CFGE ran its own packing houses, introduced standardization and grading of fruit, and adopted a lasting brand name, *Sunkist*. These would become common industry practices across all the fruit sectors.

Nor were the deciduous fruit growers idle after the CFU collapsed. The Placer County Fruit Growers' Protective Association (1896–99) was launched in Newcastle. When this fell apart, the Placer growers called for a mass meeting of orchardists in Sacramento in 1901. With the citrus growers there to advise and the raisin king, Theo Kearney, there to exhort, they founded a long-lasting body, first called the California Fresh Fruit Exchange and, after 1907, the California Fruit Exchange (CFE), when it morphed into a joint-stock company (it went back to being a membership organization in 1937). Using the blue anchor trademark, CFE was marketing about one-fifth of fresh deciduous fruit in 1930.

Running parallel to the growers' organizations were associations of the independent shippers, who did not want to be outflanked by their grower rivals. The first was the California Fruit Growers and Shippers Association, formed in 1895 to serve as a clearinghouse for fresh fruit shippers to improve information on markets in the East and open up auction houses there. This was supplanted by California Fruit Distributors (1902–27), which coordinated shipments, bargained for lower railroad rates, and hired its own agents to work the eastern terminals and open up new markets. Up to 1912, it handled 75 to 90 percent of all fresh fruit shipments by rail out of California; by 1917 this was down to 50 percent (with CFE picking up the slack). Even before the turn of the century, fruit packers, especially in Placer County, were leasing land to Chinese and Japanese farmers under exclusive consignment contracts—an integration of production and marketing as profound as anything the co-ops or sugar companies were doing at the same time. Of signal importance was the move to standardization (grading of fruit quality) by shippers—paralleling Sunkist's practices and encouraged by the State Board of Horticultural Commissioners.[12]

For dried fruit, the vanguard organization was the California Raisin Growers Association (1898–1904), incorporated under capitalist Theo Kearney, who had made a fortune as a promoter of Fresno-area colonies and brought the capital, business savvy, and ruthlessness to mount such an effort. His goal was to create a "trust" that controlled at least 75 percent of raisin output. The association died amidst good prices and bad blood, and Kearney followed it to an early grave. The price squeeze in 1908–09 revived growers' interest, however, and they created the California Associated Raisin Company in 1912, which would market successfully under the Sun-Maid label for the rest of the century. True to Kearney's vision, this was not an actual co-op, but a joint-stock corporation without membership control, "a distant species of cooperative organization, a hybrid of cooperative principles and corporate trust devices," as Virginia Woeste puts it.[13] In the Santa Clara Valley, where prunes and apricots predominated, cooperative associations started springing up around 1890 as collective drying yards. These came under the aegis of the Santa Clara County Fruit Exchange (1892–1916), which packed and shipped the dried fruit on a model similar to the citrus growers. Equally important was the Campbell Fruit Growers Union (1892–1913), a grower-owned enterprise. A statewide dried fruit co-op, the California Cured Fruit Association, was also put together in San Jose in 1900 but liquidated in 1903. Cooperation faded in the good years up to 1910, beset by grower reluctance to join and refusal to sell to their co-op if a better price could be had elsewhere. But pressures to do something about overproduction and stagnant prices built up again. In 1909, the California branch of the Farmers Cooperative and Educational Union established an office in San Francisco and drying plants around the Bay Area. Meanwhile, the California Cured Fruit Exchange was formed in the Sacramento Valley in 1912.

As in fresh deciduous fruit, a parallel organization of the packers came into being, the Dried Fruit Association of California (1908). They performed the service to the whole industry of gaining uniform sales contracts with eastern buyers and in negotiating with export merchants (including inspections on the dock). The growers, in a spat with packers, in 1917 finally came up with a unified front, the California Prune and Apricot Growers, which would prove to be a lasting institution. Based in San Jose, it chose the Sunsweet brand name for its products and commanded about 75 percent of the prune and apricot market in the 1920s (changing its name to Sunsweet Growers in 1923). Struggles with the packers would continue, including a coup in hiring packer O.A. Harlan to save the growers' co-op in 1928.[14]

Other sectors followed their own circuitous routes to cooperative organization of marketing. The two most important, not yet mentioned, are the California Almond Growers Exchange (1910) and the California Walnut Growers

(Blue Diamond brand, 1912), both of which controlled around 75 percent of their crops by 1917. After 1915, legal changes facilitated the rapid organizing of many more crops, such as in the California Bean Growers (1916), California Associated Olive Growers (1915), Central California Berry Growers (1917), California Pear Growers (1918), and the California Rice Growers (1920). Nor were livestock left out, for there were the Dairymen's Cooperative Creamery Association (1909) and the Poultry Producers of Central California (1907). By 1915 there were almost 200 cooperative marketing and purchasing groups in the state. By the 1920s co-ops and associations marketed well over 50 percent of state agricultural output.[15]

The co-ops were thoroughly practical organizations of businesspeople, not socialist or utopian projects. Economist Ira Cross observed at the time, "[Co-ops] seem to have been organized with but one object in mind, and that is to make money for their stockholders. They are mere profit-making associations." But the cooperatives were more than this; as Woeste documents exhaustively, they were a thoroughly original form of business organization that melded agrarian concerns with modern corporate form, and changed American business and antitrust law in the process. Co-op managers also discovered, as did manufacturers, that more important to long-term profits than monopoly prices was expanded throughput by gaining better control of distribution and integrating the flow of information and commodities. Co-ops helped fix in place the modern system of contract farming, winning the right to demand delivery from growers and performance by merchants down the line. California's cooperatives broke the ground for a new era of agribusiness with a far higher degree of integration of most commodity chains than ever before.[16]

The state was a helpful partner in the orchestration of cooperative marketing and production control. In 1915 California became the first state with its own legislated marketing commission, charged with fostering cooperatives and other associations, and a state marketing director, Harris Weinstock of the old California Fruit Union. This was followed by the Standardization Act of 1917, which gave the state powers of inspection and quality control in accord with the co-op model. The California Fruit Standards Act of 1927 further ramped up the level of collective control, empowering grower-run marketing boards for every major fruit crop. This policy was expanded to all crops in the federal Prorate Act of 1933 and the Agricultural Marketing Agreement Act of 1937, which gave legal authority to commodity groups to self-regulate quality, price, sales, packaging, competition, promotion, and research. California cooperatives had become the model for the whole country. From this point, agribusiness was neither "cooperative" nor "state managed," but a fine amalgam of the two, the remarkable public-private partnership that has always distinguished the American polity.[17]

The cooperatives, for all their success, were only one piece of the puzzle of marketing and agribusiness organization in California. The co-ops never fully displaced packers and merchants, but co-existed in a kind of creative tension, engaging in parallel practices of brand-naming, standardization, promotions,

Silos, near Colusa, 2002

advertising, and improved transactions in Eastern markets. The biggest shipper of all, the Southern Pacific, while an adversary on prices, was a vigorous ally in marketing California produce.

In certain sectors, the key actors have been a combination of grower and shipper—large growers who successfully unified production and marketing, as well as handling the produce of smaller farms. Examples can be found in the 19th century, in figures such as the Horst brothers, hop merchants in San Francisco and New York who operated a dozen hop ranches from California north to Canada. The largest and most well known of the modern merchant shipper-growers are the DiGorgio Brothers (incorporated in the 1930s), who began with a produce auction house in Baltimore and moved into various segments of the commodity chain, including farm production (especially fresh vegetables and table grapes) in the San Joaquin Valley. Integrated grower-shippers were always the major players in markets for seeds, lettuce, cotton, and other crops that were never brought into the cooperative fold. They prospered with the great increase in California fresh vegetable shipments after World War II. Strawberries provide an example of mixed approaches by a troika of intermediaries: large grower-shippers, co-ops (representing the middle-sized growers), and independent shippers (handling goods chiefly for the smallest growers).[18]

A most important competitive strategy for shippers of California produce, as well as growers, is to use the state's climate to maximum advantage. Especially as the southern desert of the Imperial Valley came into production, California dealers could offer eastern merchants a supply of fruits, nuts, and vegetables through a greatly elongated season. As Harvey Levenstein observes, "By the 1930s [Americans] were well on the way toward eradicating perhaps the oldest distinction in human diet: seasonality. . . . [T]he expansion of fruit and vegetable growing in California, the Southwest and Florida had made many fruits and vegetables available for much longer periods than before. . . . Practically the entire nation was blanketed with immature citrus fruits and indestructible iceberg lettuce from southern California."[19] After World War II, California would dominate the market by this means.

SUNSHINE IN A CAN

Food processors have been key players in the broad scheme of agricultural advance in California. This group has been led by the canners, but includes meatpackers, millers, vintners, and more. Unprocessed foods have always been a minority of the output of California agribusiness. Instead, wheat was ground for flour and baked into bread, barley and hops brewed into beer, grapes dried and fermented, peaches and apricots canned, almonds and walnuts roasted, sugar boiled and crystallized, prunes dried, tomatoes puréed, butter churned, beef butchered, rice polished, strawberries frozen, and orange juice concentrated. Only a handful of crops, such as kiwis, lettuce, and nectarines, have mostly escaped this fate to be delivered fresh and intact. In deciduous fruit, for example, the quantity dried, canned, and fermented was triple that sold fresh through the first half of the 20th century.[20]

The fruit and vegetable canning industry has been the most prominent of all processing sectors. In 1925 it was second only to oil refining in output value among California manufacturing industries, and it was the largest employer by far, other than timber and lumber. But other processing sectors have loomed large, as well. Meatpacking was third in output in 1925; bread and bakery products were ninth—ahead of motion pictures!—and butter and cheese just behind that, followed by flour and feed. In other words, not only did the industrialization of agriculture enlarge the agrarian sector, but agro-industry, especially processing, was a huge part of *manufacturing* in California. This contrasts with the rest of the country, where food processing fell out of the top ten industrial sectors by 1900.[21]

The first dramatic impact of agro-industrialization involved wheat. A large portion of the wheat for export was converted into flour. Well over one hundred mills were in place around the Bay and Delta regions by the early 1860s, producing almost a million barrels of flour. With the coming of steam-powered roller mills, perfected in Minneapolis in 1871, the industry restructured to fewer, larger mills, such as the Sperry, Golden Gate, Albers, and Starr, in Stockton, San Francisco, Oakland, and Vallejo, respectively.[22] In 1892, with prices and markets under pressure, several milling companies consolidated under the Sperry name (later bought out by General Mills around 1916).

Sugar mills were no less significant. The first sugar beet factory in the United States was built in 1870 by E.H. Dyer in Alvarado (Union City). Spreckels's San Francisco mill on Potrero Point (1878) was the largest factory in the West until he built an even bigger one near Salinas (around 1897). Meanwhile, California and Hawaiian sugar moved into an old Starr flour mill at Crockett (next to Port Costa and across from Starr's Vallejo mills) and converted it into one of the largest sugar refineries in the world. Other sugar mills could be found at San Francisco (American Sugar), Alameda (Alameda Sugar), and Lathrop (Holly Sugar). Big business was the mode of operation from the outset, dominated by integrated grower-processors. Diverging from the plantation system used in Hawaii, Spreckels became a pioneer in (sub)contracting with California sugar beet growers in the 1890s, including tenants on Spreckels's own 20,000 acres; by 1927 Spreckels ceased to produce any beets itself (most other processors preferred to farm their own land, however). Spreckels created the sugar trust by merging with American Sugar around 1897, and created the most profitable beet sugar complex in the country by combining horizontal *and* vertical integration. Even so, the trust broke up, leaving four companies operating in California after World War II; sugar is no longer a leading commodity, and Spreckels Sugar (later Amstar) did not outlast its competitors, closing down in 1992.[23]

Cattle were brought in from the far-flung grazing areas to be finished and slaughtered close to market in the 19th century. Concentrated "butcher towns" arose in San Francisco, Emeryville, East L.A., and other urban centers to supply the retail trade. Tanneries and rendering plants sprouted up in close association to the slaughterhouses. When the Chicago packers, led by Swift and Company, brought their factory-based meat cutting to the West Coast, they put the ax to both little butchers and Miller and Lux, but they eliminated neither all independent butchers nor independent ranchers (indeed, the degree of vertical integration on the supply side went down). Large and small joined together in the Pacific Coast Meat Association. Meatpacking in California finally disappeared under the onslaught of the boxed-beef revolution coming out of the Midwest in the 1970s.[24]

In dairying, processing was a critical complement to fresh milk production. Butter and cheese were the main long-lived products in the 19th century, with Merced and Stanislaus counties supplying most of the butter, while southern Santa Clara and Sonoma counties concentrated on cheeses. In the early 20th century, the manufacture of condensed and evaporated milk allowed the San Joaquin Valley to become the state's largest dairyland, while selling little fresh milk (nationally, condensed milk became the largest single canned-good item, as well). The refining of dairy products was aided by machinery such as the centrifugal cream separator and the power butter churn, followed by milk condensers, evaporators, and driers. Some of this was produced by local companies, such as Jensen Creamery Machinery Company of Oakland. Pipeline milkers, refrigerated tanks, and milk tanker trucks contributed to the modernization of fresh milk production around World War II.[25]

The principal method of preserving deciduous fruit was drying. Having quickly saturated local markets, deciduous fruit growers found their savior in drying in the 1870s. Preserved fruit could travel long distances by ship and transcontinental railroad. Drying was dominant for raisins around Fresno and prunes in Santa Clara and Sonoma, and was popular in apricots, peaches, and apples. California raisins and prunes took over North American markets, and made inroads into Europe, as well. Drying fruit in California's summer sun proved to be easy, done in large drying yards alongside packinghouses.[26]

Fruit and nuts were usually boxed and shipped to market by merchant packers and cooperatives, occasionally by commission merchants. Packing became a major intermediate step in the overall labor process, and packinghouses sprang up throughout all the fruit-growing regions and near railheads and ports, usually in smaller farm towns like Anaheim, Sebastapol, Lincoln, and Mountain View. In 1933 there were some 200 packinghouses around the state, and hundreds more small packing sheds on farms. Most of these came to be run by the grower cooperatives, but independent packers still played an essen-

Drying fruit, Turlock, Stanislaus County, circa 1900

tial role. Packers and co-ops employed most of their labor to handle fruit cleaning, cutting, drying, sorting, and packaging. Packing was labor intensive, but to speed things along they developed machinery such as automatic graders, dippers, and box nailers. Fully automated drying lines were achieved by 1890. A machine for packing oranges, for example, was invented in Riverside by Fred Stebler and George Parker.[27]

Because of its machine methods and high value added, canning was the king of agro-processing in California for a century, from 1875 to 1975. The state's first cannery, in 1857, was San Francisco's Cutting Fruit Packing, which was also the first to put up fruit in cans rather than jars. Other pioneering canners were Josiah Lusk in Oakland (1868) and James Dawson's San Jose Fruit Packing Company (1871). Canneries began to proliferate in the 1870s, after the invention of steam-pressure kettle canning made it possible to control temperature, reduce cooking time, and prevent cans' bursting. Major breakthroughs followed in the closing of cans, with the Cox capper in the 1880s and crimping in the 1890s. The fruit pack grew from 80,000 cases in 1869 to 2.8 million in 1900. Fish canning was a major subsector, starting with salmon along the Sacramento River, then shifting to sardines circa 1900, followed by tuna.[28]

Canning methods continued to improve with machines to prepare the produce, such as pea podders, corn huskers, fruit pitters and peelers, and various slicers and dicers. A stream of innovative machinery came out of local firms such as Anderson-Barngrover and Food Machinery Corporation. More efficient pressure cookers evolved in the early 20th century, along with more rapid cooling of cans. Cans improved, as well, with double-seamed cans, open-topped cans for crimping, and enamel-lined cans all emerging. Continuous cooking and canning systems were commonplace by the 1920s, along with conveyors. Productivity would go up repeatedly over the next half century. Fordist mass production was by no means restricted to the automobile industry in this age of robust American industrialization. Nor was automation driven by an "inner compulsion to achieve higher standards," as two food historians hold, but by the usual laws of capitalist accumulation: competition, profit seeking, and cost reduction. In fact, it often went further in California, thanks to the scale of canning and its supporting machine makers. For example, California tomato canners became much more mechanized than competitors in the eastern United States though mechanical peeling was not achieved until the 1950s.[29]

Canned vegetables—chiefly tomatoes and asparagus—entered the arena in a big way after 1895, leaping to 8 million of the 17 million cases produced in the state by 1918–19. Tomato products would surge ahead to become the leading sector among California canned goods by the 1930s. Canned-good output began to rise rapidly in the 1920s, hitting 30 million cases in 1929, then, after slumping in the early years of the Great Depression, rocketing upward from 1935 through World War II. By 1948 California's fruit and vegetable pack was 66 million cases, and canneries employed around 100,000 people at peak sea-

Cutting room, George Hyde cannery, San Jose, 1920

son.[30] Along with canning came a bevy of companies supplying the cans, jars, crates, and cartons to store and ship produce; they included the National Can, American Can, Eagle Box, and California Barrel companies. Some were locally owned and some national, but all had large factories in the Bay Area and Los Angeles.

The canning companies consolidated over the same time period as the growers and shippers, and for much the same reasons—initially to control competition but ultimately to better integrate the commodity chain. The largest canner of the 1880s and '90s, J. Lusk of Oakland, was a creature of the well-connected financial and marketing firm of William Coleman. In 1899 a dozen companies, including Cutting and Lusk and controlling roughly one-half the total fruit pack, merged into the California Fruit Canners Association, headquartered in Oakland. That association morphed into California Packing Corporation (CalPak) in 1916, with the addition of Griffin and Skelley, Central California Canners (product of an earlier Central Valley merger), and Alaska Packers, all engineered by the merchant George Armsby. The new company moved its headquarters to San Francisco.

CalPak can rightfully be considered the second national food conglomerate, a decade ahead of General Mills and General Foods (only Nabisco, formed in 1898, was earlier). It vied with Campbell's, Heinz, and Libby-McNeil-Libby to be the largest canning company in the country. CalPak, with over fifty canneries across the state, and Libby together controlled half of California's canning output.[31] While scores of small companies disappeared, several independents flourished, including Hunt Brothers, Stewart Fruit, and Virden.

The new generation of large canners contracted with farmers to secure produce, as had the sugar mills, fruit co-ops, and packer-shippers before them. This practice began here (and across the country) at the turn of the century. (Small canners in the 19th century had often grown their own produce.) Three-fourths of canning tomatoes were under contract in 1922. Del Monte was contracting with 6,600 different growers in 1937 (though Del Monte also owned farms of its own, including the 4,000-acre Fancher Ranch near Merced, which produced peaches). Grower-processor contracts were not arm's-length deals. In the processing-tomato industry, for example, the processors could control the variety planted, the planting and delivery times, hauling, and other factors. They went so far, as the industry developed, to provide seeds and plants, to offer technical assistance, and to supervise farm practices. The relation between processors and growers was by no means egalitarian, and the economic cut between canners and growers was determined through a vigorous jockeying for position between them and their collective organizations. Canner contracts helped spur peach, pear, and other farmers to organize into trade associations. Such relations were not entirely contrary to the interest of growers, however, because contracts could improve the stability of markets, the performance of farming operations, and payment flows and access to credit from canners or banks. A notable example occurred when the canners stepped in to rectify the massive overproduction of peaches in the 1920s.[32]

The kind of business integration that the contracting system represents was long ignored by students of industrial organization, who were in thrall to the modern corporation. Yet a study of the tomato industry undertaken in the 1950s sounds remarkably prescient in light of today's theories of vertical integration and disintegration, recognizing that the grower-processor relations were open, mutually beneficial, and dependent on particulars of crop type and markets. Remarkably, the industry "deverticalized" at a time when integrated corporations were expanding all through American industry. To this day, the contract growing system perfected by canners predominates in every crop that is heavily processed, such as canning tomatoes, frozen potatoes and strawberries, sugar beets, cotton, and juice oranges.[33]

California canners hit their peak around World War II. California led the other states in fruit and vegetable canning by 1900, and it packed over half the national total in 1946. California employed one-quarter of the national food-processing labor force in both 1945 and 1958. In 1958, the state accounted for over one-third of value added in canned fruits and vegetables; led in the production of seven of the top twelve fruit and vegetable products; and provided over 90 percent of three of them—as well as 100 percent of fruit cocktail, naturally. This dominance is staggering when one considers that U.S. canneries accounted for three-quarters of *world* output in that year. Most of those canned goods were consumed in the United States, but exports did creep up, from 3 to 6 percent between 1950 and 1963.[34]

Not surprisingly, California was headquarters to four of the top eight canning companies (CalPak, California Canners, Hunt, and Dole). CalPak's sales tripled from 1939 to 1950, and by the 1960s the company had grown into the world's largest canning company and one of the biggest agribusiness multinationals, with subsidiaries in twenty countries.[35] The labor process kept changing in canning after the war, with the introduction of several new mechanical devices that improved peach and cherry pitting, automated the filling of cans with peaches and apricots, and improved tomato canning. This and union-

California Canning and Food Processing 1899–1992

CANNING (fruit and vegetables only)

	Firms	Employees	Output Value
1899	136	7,486	$13,082,000
1909	196	7,757	32,915,000
1919	303	19,575	189,956,000
1929	389	29,958	220,465,000
1947	218	26,379	393,445,000
1963	192	25,060	856,002,000
1972	160	24,500	1,252,900,000
1987	124	19,100	3,609,709,000
1992	120	18,700	4,238,600,000

FOOD PROCESSING*

	Firms	Employees	Output Value
1899	1,476	16,149	$89,110,000
1909	2,034	21,921	189,927,000
1919	3,035	51,233	695,535,000
1929	3,572	72,924	1,812,626,000
1947	2,803	120,510	1,104,038,000
1963	3,067	155,731	7,140,000,000**
1982	2,536	173,500	31,263,000,000
1987	2,521	162,200	35,450,700,000
1992	2,751	170,400	44,296,600,000

* Food processing as defined by U.S. Census Bureau after 1940. For earlier years, I have added all relevant categories. Sugar refining had to be estimated.

** Includes some double counting.

Source: U.S. Census of Manufactures, various years.

NB: California ranked #1 among the states in canning after circa 1905 and #1 in food processing after circa 1950.

ization virtually ended piecework, which was replaced by hourly pay. But it also reduced the labor force, which fell rapidly to only 26,000 by 1958—one-quarter of what it had been a decade earlier. After a grace period provided by the robust postwar economy, canned goods began to decline as a share of the food market. While output was still rising into the 1960s, per capita consumption of canned goods remained flat (consumption of frozen foods was only one-sixth that of canned goods as late as 1963, but it had risen 150 percent in the preceding fifteen years).[36]

Frozen foods moved to the forefront, surpassing canned goods and dry cereals as the vanguard of American food processing. Californians were shipping barrels of frozen strawberries east as early as 1922. Clarence Birdseye made a key contribution to frozen foods in the late 1920s with his waxed-paper and cellophane packaging. But most early frozen food was sold only to processors and institutional kitchens, even as output doubled during the war. It took the postwar introduction of large, open-top freezer chests in supermarkets and substantial freezer compartments in home refrigerators to make frozen foods practicable on the mass market. The breakthrough product after World War II was, somewhat unexpectedly, frozen orange juice concentrate, which became a breakfast staple and constituted one-fifth of all frozen foods sold by 1950. Frozen concentrate was developed in Florida in 1942 and commercially perfected in Los Angeles in 1945 (frozen orange juice was first introduced by Minute Maid in Florida, but the key to mass acceptance was concentrate). Safeway was a pioneer in fast-freezing of peas under its Whitney subsidiary and Bel-Air brand name in the 1940s. William and David Berelson of San Francisco packaged frozen strawberries in 1941, which soon became the third most common frozen product, and then founded Ore-Ida Potatoes, which introduced Tater Tots in 1952. The UC College of Agriculture was a pioneer in frozen pack research.[37] In the long run, however, the shift toward freezing favored the Northwest, with its greater crops of berries and peas.

The California canners were gradually eclipsed, and their numbers began to shrink through bankruptcy and consolidation. By the late 20th century, only a few canners, such as Del Monte, Tri Valley Growers, and Sunkist, remained. As canners ran into trouble, some were bought up by grower co-ops to keep their markets open and by supermarket chains to provide house brands. (By contrast, Campbell's does almost all its tomato processing today near Stockton.) Del Monte ceased to be a major player in corporate America. Like so many other food-processing companies, it was no longer at the cutting edge of innovation in products, processing, or organization. It hit the wall in the early 1980s, along with the rest of the U.S. agrarian sector. It was gobbled up by the R.J. Reynolds conglomerate, along with Nabisco and Standard Brands, just as Green Giant fell to Pillsbury. The engineer of the takeover was Kohlberg Kravis Roberts, an investment banking firm that spans the San Francisco–London–New York financial circuits. Del Monte emerged from this smothering financial embrace as a lean, independent company based in San Francisco in the 1990s, and its brand name quietly remains a standard on supermarket shelves. Tri Valley went belly-up in 1999.[38]

Agro-processing's most enduring icon has not been the can but the bottle of spirits. Europeans brought the art of wine to the mining frontier in the 1850s, and the names of the founding fathers still resonate in local legend: Haraszthy, Krug, Schram, Beringer, Niebaum. The fledgling wine industry was flying high by the 1880s, granted breathing room in the market by the *phylloxera* epidemic across France in the preceding decade. Californians won thirty-five medals at the 1889 Paris Exhibition. But most California wine was still made from lowly Mission grapes, until growers replanted with better varieties after *phylloxera* came back to bite them in the 1890s. Napa was hardest hit, and its vintners fell one by one before the invading armies (not to mention economic depression and overplanting). Sonoma climbed over the wreckage to regain supremacy, mostly on the strength of a rising group of Italian grower-vintners, such as Samuele Sebastiani. A new generation of vintners put down roots in Napa as well, led by Georges de Latour (Beaulieu) and Louis Martini.[39]

Producer organization got its first leg up in the grape and wine sector. The California Winegrower's Association was paralleled by the California Wine Makers Association (both in 1862), the two bodies contracting with each other. A State Viticulture Society was established in 1875 to muster legislative and university support for the industry. Early winemakers were as much a producer network as today's are, and spinoff wineries were common: Krug trained with Haraszthy, Beringer with Krug (as did Carl Wente of Livermore). But the industry's center of gravity at this time was San Francisco wine merchants, not the vintners. Most wine came downriver in bulk, was stored in city cellars, and was bottled and sold under merchant labels. In response to a glutted market from 1886 to 1894, the merchants put together the California Wine Association (1892) to stabilize prices and peddle California wines and brandies in London. After the 1906 earthquake destroyed millions of gallons of wine in San Francisco warehouses, a merchant-led consortium created Winehaven at Point Molate (Richmond); it was purported to be the largest winery in the world.[40]

Four-fifths of the state's wineries went down before the drought of sanctity from 1919 to 1934. Ironically, total output continued to be robust. Catholicism saved the wineries from Protestant sledgehammers by the exemption granted to sacramental wines; rabbis were also to be found searching for wines to bless. In Napa only Louis De Latour's Beaulieu survived, thanks to his monopoly grant from the archbishop of San Francisco. Italians sold wines to the churches and the pharmacists, and grapes to everyone else. The second Viticulture Commission, created in 1913, helped the growers shift to table grapes when Prohibition hit.

After Prohibition, California wine was mostly manufactured in bulk in the San Joaquin Valley; the Gallo brothers were the supreme purveyors of plonk, gaining over half of California's total wine production in the postwar era. The rediscovery of quality wines brought a transformation of the North Bay in the 1970s (along with some parts of the Central Coast). The number of wineries grew from a handful to nearly 1,000, as new vintners entered the business rapidly, out of passion and/or riches made elsewhere. This time around, Napa emerged as the elite province, with Sonoma and the rest gradually catching up. Total output value of the industry hit an astounding $12 billion by the end of the millennium—roughly equal to half of all on-farm crop output for the state.[41]

Brewing and distilling are major forms of processing, but they are generally

Giant wine cask, Lachman Cellars, 1898

held to be beyond the pale of histories of food and the food industry. These processes have absorbed a large quantity of barley, corn, and hops grown in California since the 1850s. Beer and ale, in particular, were the foodstuff of choice for most manual workers in the 19th century, and nearly every block of downtown San Francisco had its own brewery or saloon, while every small town bristled with drinking establishments and sported a local brewery. The business and labor history of brewing and distilling are almost always treated as something entirely urban and outside the compass of agribusiness. Yet the most famous labor incident in the state's history, the Wheatland Riot, took place on a hop farm.[42]

HEADING DOWN THE AISLE

Agribusiness is a moving target. No sooner had the big canners and co-ops hit stride than the organizational contours of agricultural processing and marketing reconfigured again. California growers, shippers, and processors were not indifferent to the retail stage of marketing, but they could not transform it by themselves. That was done by a different set of capitalists, who moved front and center in the agrarian circuit of capital by the 1930s. These were the grocery chains and supermarkets.

Grocery stores were an innovation of the post–Civil War era, representing a more specialized form of retailing than previous general stores. The 20th century ushered in a new mode of selling groceries, the chain and discount stores, which would sweep away most of the independents. They would, moreover, revolutionize retailing generally, leading the way to everything from Montgomery Ward to Wal-Mart and McDonald's. This is not so surprising if one remembers that food and drink were long the biggest items in popular consumption, absorbing 40 percent of disposal income in 1900 and still roughly 25 percent in 1950. Groceries were still the biggest retail business in the United States at mid-century, with sales over $30 billion. Chain grocers became the key merchant-capitalists in the food system by World War I, selling almost one-third of all food products nationally by 1930 (with only 16 percent of all stores) and two-fifths by 1960. A&P was the largest retailer in the world in the interwar period, and it would not be overtaken by Sears Roebuck until after World War II. A&P was the fifth-largest nonfinancial corporation in the country in the 1920s, and the second-largest corporation after General Motors in 1950.[43]

The chains bested independent grocers by means of several innovative tactics. The essential ones were simplicity in selection (open shelf display and self-service), no credit (cash and carry), economies of scale (buying in bulk), and economies of throughput (quicker inventory turnover). The 1920s brought the larger "combination store," which featured groceries, produce, meat, dairy, and baked goods. In the 1930s, the giant stand-alone discount store—the supermarket—swept across the land, offering even greater selection, mass checkout, shopping carts, and parking lots. Supermarkets soon numbered in the thousands. It only remained for the large store to be absorbed into the chain in the 1940s. After a brief outburst of opposition to chains and supermarkets in the 1930s—which almost led to congressional action—the juggernaut was allowed to roll onward. The independents fought back with larger stores, self-service, and affiliated wholesalers, and they were able to stem the tide of control by the national giants. Nonetheless, by 1960, 69 percent of grocery sales were through supermarkets. While the share of chains was still 40 percent in that year, it would take off again and reach 60 percent by 1975.[44]

California was in the forefront of developments in grocery marketing. Los Angeles, with Ralph's, Von's, Seelig, and Alpha Beta, was in the advance wave of self-service and the "supers." Los Angeles had self-service stores by 1912, four years ahead of the pioneering Piggly-Wiggly in the South (though Piggly-Wiggly joined in the action in Southern California in a big way in the 1920s), and it had huge combination stores in the late '20s, before King Kullen and Big Bear hit New York (bringing the supermarket idea home to the rest of the country). Big stores in Los Angeles were already over 10,000 square feet in the early 1930s, over 20,000 in the late 1930s. As Richard Longstreth documents, "No other part of the United States came close to matching the proliferation of these emporia that occurred [in Los Angeles] before World War II. The supermarket was not wholly invented in Los Angeles as were the super service station and the drive-in market. . . . Yet a greater contribution was made to the development process in the Los Angeles metropolitan area than in any other part of the country."[45] Around 1950, supermarkets were handling almost two-thirds of grocery sales in California, compared to just under half for the country as a whole (only Florida and New Jersey boasted higher figures).

In Northern California, Safeway became the largest grocery chain in the West by 1930 and the second largest in the United States thereafter. Safeway began as a chain of local groceries put together by M.B. Skaggs of Idaho between 1915 and 1925 (there were 428 in ten states). Skaggs moved to Oakland in 1926, after he bought Safeway, the largest regional chain in Los Angeles (which had been assembled by Sam Seelig); he added MacMarr stores of Portland in 1931. At that point, Safeway passed Kroger in size, peaking at 3,500 stores, and had the highest sales per store of any of the chains. Safeway stuck with the neighborhood grocery model until World War II, targeting areas by income and ethnicity, but afterward opened hundreds of supermarkets. In the 1940s it closed most of its smaller outlets, reducing its number of stores to around 2,000 while building a host of new, large stores. Safeway became a leading symbol of mass marketing in the postwar era, from its moderne S logo (1951) to the Corbusier-inspired store design of William Wurster (for the Marina District store, 1959). Sales passed $1 billion after the war, $2 billion by the mid-1950s, and $3 billion by the mid-1960s, and came to nearly $5 billion in 1970. The company was at the top of the competitive heap in terms of profits. It went international in the 1960s, to Britain, Germany, Canada, and Australia, but remained overwhelmingly a western U.S. company. Safeway finally reached the pinnacle of grocery chains in the United States in the 1970s. Lucky Stores, another Oakland-based chain, rose quickly through the ranks with the help of buyouts of smaller chains, and made number 4 on the national grocery rankings from the mid-seventies to mid-eighties.[46]

The chain grocery stores altered the agro-food chain in fundamental ways. At the sales end, they created a new aesthetic and practice of shopping. Chain stores were more open, standardized, and hygienic than mon-and-pop stores, with displays featuring great pyramids of goods (with the variety of items numbering in the hundreds) to tempt consumers with visions of abundance.

Canned goods were at the center of such displays. Chains also ushered in the self-service revolution, which gave consumers the right to shop on wit and whim—the quintessential freedom of American consumerism. This appealed especially to women and immigrants during the interwar era, as Tracey Deutsch has pointed out, because it offered a degree of relief from male shopkeepers, sales pitches, ethnic pressures, and credit relations that must have seemed an immense liberation to many people.[47]

With the supermarket, the sense of freedom to shop was amped up. Along with it came the symptomatic layout of retail modernity: high, vaulted ceilings; long, wide aisles; multilayered shelves; and gleaming refrigeration cases. Outside came massive parking lots, sleek modern buildings, and huge signs to

Safeway store interior, circa 1954

shout at motorists. These retail landscapes were perfected, above all, in Los Angeles between the wars. The cornucopia took a quantum leap forward, with the average number of items for sale hitting 3,000 by 1946 and 6,000 by 1960. Stores added freezer compartments, air-conditioning, stadium lighting, and chrome on everything from produce bins to shopping carts. Parking lots grew to playing-field size. When Khrushchev came to America in 1962, where better to take him to prove the superiority of capitalism than the Safeway designed by Wurster in San Francisco? Without question, supermarkets changed forever the look of American retailing—and American cities.[48]

On the supply side, grocery retailers were "redrawing the geography of product movement," in Richard Tedlow's phrase. An observer of the 1930s called it "the rapid obliteration of recognized channels of distribution."[49] What this meant was that the chains bypassed wholesale merchants, on whom the small grocers relied to do their own purchasing, and built their own warehouses and trucking systems. In 1930 Safeway already had eighty warehouses and claimed the world's largest grocery wholesaling operation; after the war it shifted to forty gigantic, single-floor regional warehouses equipped with advanced computer tracking systems. By the mid-1930s, many chains' direct purchases from suppliers exceeded those from wholesalers (for Safeway that point was reached by 1940); by 1960 it was up to 80 percent for all supermarkets.[50] Just as the co-ops were cutting out the middleman from behind, the grocery capitalists were squeezing them out from the retail end of the distribution chain.

In many commodity sectors, the supermarkets revolutionized agrarian production by contracting directly, demanding higher and more consistent quality and steady product flow over the year, and in some cases taking over suppliers or establishing new processing subsidiaries. Safeway, in particular, favored direct procurement from farmers, and especially from grower cooperatives like Pacific Egg Producers and Land O' Lakes. It had specialized buyers going from farm to farm, and it generally preferred fair price and high quality over bargain buying. Safeway's provisioning networks stretched across the nation, but California growers provided half of all produce handled by car lots in 1940. By 1960, farm output under contract for all the United States was 15 percent, but for California sectors it reached very high levels, such as 50 percent in fresh vegetables, 75 percent for citrus, 90 percent for processing vegetables, and 95 percent for milk, broilers, and sugar beets.[51] Much of this must have been with supermarkets.

Chains also manufactured their own house-brand products. By 1930, 12 percent of grocery sales were provided by the chains' subsidiaries (this did not increase significantly over the next thirty years). Safeway was the most active acquirer of suppliers and had over fifty processing plants by 1950, for baking, milk bottling, coffee roasting, canning, cheese making, fish processing, preserving, and candy making. These were located all over the West and Midwest.[52]

Self-service dairy products were pioneered in the 1920s by modernizing grocers such as Oakland's Mutual Stores (bought out by Safeway in 1931). Supermarkets were critical to the introduction of milk, butter, ice cream (in cartons), and prepackaged cheese, presented in modern electric cooling systems. By 1950, almost all dairy departments were self-service. Route dairies, with their thousands of home customers, could not match the low costs of the grocers, and they rapidly disappeared from the urban scene in the 1950s.[53]

Beef became a pivotal good in the jostling for position between supermarkets, chains, and independent grocers. By the mid-1930s, Safeway and other chains were already chafing over indifferent quality control by meatpacking firms, and moved to age and cut carcasses in their own warehouses. California beef was of notoriously poor quality up to this time, and both indifferent consumer taste and the lenient grading system maintained this low standard. In 1940, the supermarkets adopted the new USDA grading system, a wedge against the meatpackers, and moved to improve the supply of grain-fed beef. In a move that would revolutionize meat production, Safeway vice president Dwight Cochran moved over to the Kern County Land Company after the war and established the first large-scale cattle feedlot, capable of handling 50,000 head. The supermarkets then promoted precut, prepackaged meat for self-service, which swept away the in-store butchers by the early 1950s. A generation later, Iowa Beef Packers would revolutionize meatpacking again with boxed beef, precut at the factory and shipped directly to supermarkets.[54]

Poultry production was dramatically altered under the auspices of the supermarkets, as well. The watershed came with the breeding of a broad-breasted chicken, which was spurred by Howard (Doc) Pierce, national poultry research director for A&P stores. Pierce organized the poultry trades, the USDA, the state agricultural colleges, and the poultry press into a National Chicken-of-Tomorrow Committee, which ran a contest for the best double-breasted breed in 1946–47. California breeders were in the forefront of the contestants.[55]

Refrigeration was central to the evolution of the supermarket and the introduction of frozen foods, packaged meat, and fresh dairy products. Frigidaire company's coil system replaced ice after 1925, and the air compressor followed in 1930. After that, cold cases became the streamlined centerpieces of stores, and products such as meat, beer, and milk could be displayed and sold without rapid deterioration. Frozen vegetables and frozen meat became firmly established during the war (paper packaging was not rationed like metal for canned goods). Prepackaging also depended on postwar advances by Dow, W.R. Grace, and others in creating strong, clear plastic wrap and acceptance by once-reluctant consumers of prepackaged foods.[56]

Chains were the first to bring fresh produce into the grocery store, though it was typically durable items like potatoes and onions in the early years. Produce lagged in the shift to supermarket sales, and produce departments only became a central attraction of the supermarket in the 1950s (self-service was slower to take over in produce than other departments, as well, appearing in less than half the stores in 1953). As fresh produce sections grew, chains and supermarkets contracted directly with co-ops and growers in large quantities.[57] This favored big grower-shippers who could both meet the quantity and quality demands of the retailers and supply produce year-round (from their own holdings and via subcontracts with smaller operators). The web of contracting spread its net over an increasingly far-flung geography of production from diversified areas across California and the Southwest (although the core growing areas could be remarkably small and intensive, as in Salinas Valley lettuce, artichokes, and strawberries).[58]

It would be a mistake, however, to think that supermarkets and large

Bunching broccoli, Los Osos, 1974

grower-shippers finally converted California agribusiness into a few vertically integrated, corporate entities. For example, the top dozen grower-shippers handled two-thirds of all Salinas Valley iceberg lettuce in 1972, but they grew less than this on their own and shipped a smaller proportion of all California/Arizona production; they also did not cross over to dominate other fresh produce lines in the same way. As Friedland et al. observe, grower-shipper integration is high in California lettuce and table grapes, but not in tomatoes or raisin grapes. These shippers sold not only to supermarkets but also to terminal (wholesale) markets in distant cities and to large institutional buyers, such as the Department of Defense.[59] The tendency toward vertical integration in agrarian capitalism does not run a straight course, but follows many braided currents of commodity production and distribution. As Tedlow concludes, "there is no single answer to this problem for all historical eras, for all industries, or even for all companies within an industry."[60]

As the century wore on, managers discovered a new angle on the organization of distribution and retailing that would prove every bit as important as scale or direct control of supply: inventory control and flow management. Supermarkets introduced automated checkout and inventory tracking using barcodes and scanners circa 1980, revolutionizing systems of supply and selling throughout the retail sector. This represents a shift in flow management as significant as Japanese just-in-time production methods in car assembly, but much less ballyhooed.

Following just behind the burgeoning grocery business in the 20th century has been the restaurant trade. Fast-food chains, in particular, have become the most important in volume markets for agribusiness. California has always been a place for eating out (unlike most of the country outside the big cities) and for innovation in restaurant formats. The everyday restaurant spread markedly in the 1920s and '30s, when diners and cafeterias thrived and the first drive-ins and chains put in an appearance. Los Angeles was the birthplace of the drive-in market and restaurant, and Sacramento the home of the first drive-in chain, A&W Root Beer (followed by Bob's Big Boy in Los Angeles).

The chain restaurants were hit hard in the Depression, falling to only one-seventh of the trade, but came back with a roar in the 1950s. Southern California gave the world McDonald's (the mother of them all), along with Denney's, Sambo's, and Taco Bell. Northern California could claim the A&W, Shakey's Pizza, Round Table Pizza, and Doggy Diner. With the fast-food revolution of the 1960s, national food sales outside the home rose from 20 to 30 percent between 1960 and 1990. Almost all of this growth was captured by the chains. Fast-food purveyors gained market share by the same mass-throughput principles as supermarkets: standardization, volume, and rationalization of work flow. They were both famously self-service (Ray Kroc got rid of server girls early on to keep away the teenage boys). Not surprisingly, California's per capita restaurant food and drink sales were high (fifth among the states in 1987).[61]

Fast foods made new, high-volume demands on growers in California and elsewhere. Frozen potatoes (precut for french fries) exploded after McDonald's made this the standard. Iceberg lettuce benefited immensely from the hamburger stands, as did fresh tomatoes, onions, and mushrooms. McDonald's is the largest single buyer of fresh tomatoes in the country. Fast-food chains have also been the most important buyers from the meatpacking and industrial chicken industries.[62]

Hidden by the fast-food explosion was the fact that total food expenditures per capita were flat after 1950 and would eventually pinch retailers. It was not until the profit crisis of American capitalism after 1970 that grocery and restaurant chains felt the heat. As their fortunes began to turn down, the supermarket chains lost their position on the cutting edge of American business, as had the canners and co-ops before them. A&P went into steep decline and was sold to German investors in 1978. Safeway hit the skids, but recovered smartly—only to be targeted by a financial raider (a common occurrence in the 1980s). It was saved by a whisker by means of a leveraged buyout engineered by investment bankers Kohlberg Kravis Roberts in 1986 (not coincidentally, George Roberts is based on the San Francisco peninsula). In the 1990s, Lucky's fell to Idaho-based Albertson's, Ralph's was bought by rising L.A. star Food4Less, Alpha Beta was grabbed by American Stores of Utah, and Von's was purchased by Safeway on the rebound.[63]

In tandem, grocery "superstores" came on line, averaging 25,000 different items and 45,000 square feet by 1980 (up from 6,000 items and 19,000 square feet in 1970); sales per store increased by half over the decade. Supermarkets stayed open longer hours to respond to the competition from convenience stores and the extended working time of women and families. Fast-food chains hit a major bump in the recession of the early 1970s, which was particularly strong in Southern California, but then rebounded—often after being bought out by old-line food processors such as Pepsi, Pillsbury, and General Foods.[64]

Food retailing split wide open in the 1980s, following the growing schism in income and wealth distribution of the last two decades. Fast-food joints and big-box discounters tracked the fall of working-class wages, as well as the movement of women into the labor force. Overall, the torch of innovation in retailing passed from the supermarkets to big-box discounters, such as Wal-Mart, Kmart, Price Club (Costco), and Target, which grabbed some of the traditional supermarket product lines. Kmart led the pack in the 1960s but stumbled badly in the 1980s. Wal-Marts averaged 50,000 square feet and 50,000 items by 1980, and the really big boxes were twice that large. Their formula was just an extension of the supermarkets': size and volume, regional warehouses, subcontracting supply, and tracking merchandise by barcodes and computers. But the volume was vastly greater and supplemented by global supply chains for the lowest-cost production. Wal-Mart, which had sales less than the big supermarket chains until 1980 and less than Kmart up to 1990, became the world's biggest retailer in the 1990s and, like A&P before it, one of the biggest corporations in the world.[65]

On the upper end of the market, the expanding niche of exotic, luxury, and organic products were sold through a new generation of outlets that cut into the supermarkets from the top. In the 1980s and 1990s, these included luxury supermarkets (Andronico's, Whole Earth, Trader Joe's); specialty shops (wine shops, brew pubs, organic grocers); and nouvelle cuisine restaurants. Farmer's markets also enjoyed a renaissance in California as purveyors of freshness, wholesomeness, and all-important authenticity.

FRUIT COCKTAIL

The last step in the food chain is the uptake of foodstuffs by the consumer. The history of consumption has become a prime topic in cultural and economic studies in recent years, enriching the understanding of how industrial capitalism has unfolded. Food, in particular, has come in for serious study as an item of mass consumption. But the cultural approach can fall prey to the same error as conventional economic theory: giving undue priority to consumer taste as the driver of economic development. It is essential to see that the connections along the commodity chain run in both directions. Production and consumption patterns have marched in close formation, with processors and merchants actively stimulating the juices of consumer taste at the same time as they try to anticipate consumer demands.

On the one side are the consumers: the buying public whose demands must be met if the food commodities are to sell. California food products have reached deeply into the broad American market, and even into foreign lands. They have had to confront American foodways, where mainstream tastes have emphasized volume over variety, or what Michael Pollan calls "a monoculture of taste" (meat, potatoes, white bread, sugar, gravy, apples, onions, tomatoes). Americans have never ventured far from their culinary origins in three limited cuisines of Northern Europe—English, Irish, and German—and had settled into a broad national consumption pattern by at least the 1820s.[66] Yet the core consumers for California agribusiness have almost always been the people of this state—a fact often overlooked in the rush to tell the tale of agribusiness exports. California tastes have sometimes moved in regional counterflow to the rest of the country. California food producers and marketers have thus had to evolve with national patterns of food habits, promoting the blandest of the bland, while at other times they have played a more discriminating part in bucking national trends—and even creating them.[67]

On the other side are the food purveyors, who have pursued several strategies to promote the consumption of their products. One strategy is to put new and desirable foodstuffs before consumers to whet their appetites. This has meant the redesign of food and food products, such as the introduction of exotic, improved, or hybrid plants; the development of new breeds and animal husbandry practices to create more agreeable meat; the milling and mixing of grains and sugar; the manufacturing and freezing of entire meals; and the enhancement of food color and flavor.[68] Along with this industrialization of food has come an enormous investment in promotion, including better packaging, media advertising, retail display, and architecture. To some extent this has been a cultivation of taste, although the tendencies in American consumption and production of food have been quite mutually reinforcing in the elimination of strong, strange, or original flavors.

The logic for capital to engage in market innovation, promotion, and elaboration of manufactured food is clear, just as it is in the making of new, better, or more elaborate automobiles and toys: sell more goods, capture greater market share, employ more labor and capital, and generate new surplus value (value added). While it is vital to acknowledge the force of consumer culture, as theorists such as Daniel Miller and Jackson Lears do, it will not do to recycle a version of consumer sovereignty guiding the market. It is essential to see that production and consumption evolve in tandem in modern economies—personal computers, for example, were not something anyone knew they wanted until they were invented and then mass-produced at a reasonable price. While social developments affect the course of food consumption in profound ways—think of changes in women's homemaking role—the course of domestic history cannot easily be severed from the flow of commodities and wage labor in and out of the household. Furthermore, producers weigh heavily in the balance of determination because they have an entrenched interest in creating new products, lowering costs of production, and expanding markets, while consumers are far more diffuse in their interests, desires, and power. The result is that producers have had immense impact on American foodways, from breakfast cereals to frozen pizzas. Moreover, they have had a radical effect on business in general, since food processors and retailers were among the first modern corporations, pioneers of mass advertising, and early proponents of brand naming.[69]

The story of food consumption in California begins with its cities. From the outset, San Francisco and the gold towns supported a varied local agriculture, including wineries, hop farms, and dairies (not to mention commercial fishing, oystering, and hunting). The success of beef, dairy, fruit, and truck crops since the early days cannot be separated from the rise of an urban market, and it would always give California farming more in common with Atlantic coast agriculture than Midwestern or Southern staples. Moreover, the notorious lust of the forty-niners for good food and strong beverage—backed by ample gold dust—established San Francisco as a city where eating out has always been a normal, not exceptional, part of life. If ever a place did not have an authentic, traditional cuisine, it's California; yet San Francisco developed into one of the three reliable centers of gourmet eating in America, along with New York and New Orleans. This did not apply to everyone, of course, and a century before nouvelle cuisine, E.W. Hilgard was arguing that cooking schools should be established to raise the standard of California dining.[70]

In the early days, California wheat and flour were highly valued abroad for biological and environmental reasons—always a key factor in food distribution and consumption. The state's wheat was known for its dryness, whiteness, and a gluten content favored by bakers from China to Germany. The wheat's quality was the combined result of climate, variety, and modern milling. In a sense, flour was the first industrially processed foodstuff, especially once it went to steel rollers, which produce a finer, whiter flour than millstones. Industrial sugar soon followed, and California exporters benefited as consumer tastes across America and Europe shifted from the more natural (molasses, honey, and brown sugar) to the processed (pure, white, granular sugar), the latter becoming cheap and abundant thanks to advances in processing in the late 1800s. On the other hand, California olive oil suffered a body blow from the introduction of cheap, tasteless, industrial cottonseed oil (Wesson oil) that emerged from the Mississippi delta at the turn of the last century.[71]

Turning to livestock, the cattle industry grew on the bellies of the state's city folk, as Californians joined the national turn to beef and away from pork and mutton after the Civil War. This came with a shift toward stock raising, corn finishing, and industrial beef dressing, when the character of the meat changed significantly toward fattier ("marbled"), more tender cuts. Thus, Mexican cattle in California and the Southwest lost out to the Yankee stock not just because of drought in the 1860s but in part because Anglo-Americans turned away from the former's tougher, gamier meat. Hereford, Angus, and other British breeds were carefully imported and blended with older stock to meet—and create—this more "refined" taste. Milk also became a national drink in this time, with the invention of canned and dried milk (by Borden and Nestlé) and the cultivation of better milch cows. California was, if anything, ahead of the curve in milk production and consumption—both fresh milk delivered locally and industrial milk products. Pasteurization, introduced around 1900, reassured urbanites that their milk was free of bacteria. California led the way in mass-producing eggs, which became favorites in eastern markets because they were gleaming white and hence particularly popular with consumers, who associated them with good hygiene—an increasingly critical aspect of American foodways in the 20th century.[72]

Another contribution of California consumers was to bolster the California wine industry (which means the American fine wine industry, since New York State wines continued to be made with native grapes, a much inferior product). They also downed goodly quantities of local beer and whiskey, mostly produced in small batches in the cities themselves. California wine was also exported by the boatload, and Winehaven's "Big Tree" wine label became a fixture in English pubs in the early 20th century. While the quality of wine from Mission grapes was never very good, the vintners supported the State Viticulture Board's efforts to promote their product at home and abroad and backed the Pure Wine Bill of 1887 to stop adulteration that undercut California wine's reputation.[73]

The earliest market for fruit was strictly local, and apples were king because they kept well, Yankee miners knew how to grow them, and Americans already had a robust appetite for hard cider. As stone fruit took off after the Civil War, especially as an export crop, growers and merchants were quick to pay attention to quality and appearance in the eyes of the consumer. Sun drying was the principal means of preserving fruit for shipment, a method as old as civilization; but Californians quickly seized on sulfuring dried fruit—a process developed by San Jose's Henry Coe in the 1870s—because it gave the product greater consumer appeal by preventing peaches, apricots, and pears from turning black (in the 20th century, oven drying returned because it yielded moister fruit). Early canners experimented with new fruit varieties, such as the Elberta peach; figured out how to can unprecedented crops (asparagus canning was worked out by Tom Foon of Alviso); and discovered additives to preserve good form and color in the final product.[74]

Nonetheless, Americans still did not eat many fruits or vegetables in the 19th century, regarding fresh produce with suspicion. California fruit cooperatives must be given considerable credit for turning this around, as they ramped up efforts at pleasing and seducing consumers. The Southern California Fruit Growers Exchange was a pioneer, thanks to its brilliant advertising manager, Don Francisco. He launched a sales campaign for fresh oranges in the Midwest in 1907, inventing the catchy Sunkist trademark, elaborate grocery-store displays, and glamorous labels on shipping crates. Sunkist's most effective sales tactic was probably the simplest: grading. Because Sunkist culled small and flawed fruit, buyers could be convinced that its oranges, lemons, and grapefruit were the largest, ripest, and sweetest on earth. This tactic was soon imitated by every other fruit co-op. Thanks to such efforts, enormous, overwatered, tasteless fruit would become something of an American fetish—for instance the modern strawberry—as would the avoidance of every blemish or sign of bug life, in the same way that marbled beef and bland olives became popular favorites across the country.[75]

An array of sales strategies were pursued by the fruit merchants. Packaging to please the consumer was one, and Sunsweet and other co-ops began to offer a range of carton sizes for dried fruit more suitable to retail sales. Another was recipes geared to the product. Sunsweet was promulgating prune recipes invented by its staff in the 1910s, a decade before General Mills's Betty Crocker came on the scene to usher in the age of Bisquick. California citrus was further promoted as the bearer of good health—whether through bogus claims about stopping "acidosis" in the 1920s or through exaggerations of the benefits of Vitamin C, discovered around 1930. Fresno's Raisin Day, Petaluma's Egg Festival, and other civic promotions were aimed at the regional market. The marketing efforts of fresh fruit and vegetable growers and shippers paid off as consumption rose and California became the principal purveyor of mass-produced freshness. But it did not always work. California's exotic crops went against the deep conservatism of American tastes, as the case of the avocado shows. Introduced around 1900, it languished for decades despite high hopes and feverish promotional efforts (including the hired advice of Don Francisco)—and what market it had was almost entirely in California and the Southwest.[76]

The attempt to capture consumer markets moved into manufactured foods by the late 19th century. These went beyond simple milling or canning to making the product something quite distinct from the crops that went into it. The first manufactured foods in this sense were Van Camp's pork and beans, Campbell's condensed soups, and H.J. Heinz's line of sauces and condiments in the late 1880s, followed by breakfast cereals in the 1890s. Though usually overlooked in histories of American food, California processors deserve credit for several contributions to manufactured foods, including canned black olives (ca. 1900), vacuum-packed ground coffee (ca. 1900), and fruit cocktail (ca. 1917). Canned olives were engineered with the help of university researchers, canned coffee by Hills Brothers of San Francisco, and fruit cocktail at Libby's plant in Sunnyvale. The canned olive industry was later bailed out when UC researchers in the Food Products Laboratory headed by Professor William Cruess solved the botulism problem for canned goods in the early 1920s, after a notorious poisoning in Ohio by Mrs. Freda Ehmann's famous California black olives.[77] Such manufactured foodstuffs changed the way Americans cooked and ate, from dropping meat at breakfast in favor of cereal to opening a can of fruit cocktail for dessert. The state's vast output of canned tomato juice, tomato paste, catsup, and other new tomato products would, during the interwar period, become the most ubiquitous processed foodstuffs in the country, and they no doubt played a role in Americans' acceptance of spaghetti with tomato sauce—the only immigrant food to enter the national diet up to that time—not to mention the postwar pizza craze (with the help of prechopped, canned black olives developed by Bell Packing company in the 1930s).[78]

The big canners were eager promoters of the mass market for California goods at the turn of the last century. Canning was increasingly standardized, rendering products indistinguishable without brand names and advertising. The Del Monte brand, introduced by the Lusk Company of Oakland in 1891, was only one of many early brands (Big Bear and Coleman Flag had been introduced in the 1880s). This was the dawn of food brand naming across the country, famously associated with Kellogg's Corn Flakes, Post Grape-Nuts, and Nabisco's Uneeda Biscuits. The CalPak canning empire inherited the Del

Sunnyside Fruit Salad label, circa 1920

Monte trademark through the corporate merger of 1916, in a strategy paralleling that of W.F. Sloan's assemblage of General Motors from several known car companies in the 1910s and prefiguring E.F. Hutton's creation of General Foods out of dozens of trademarked product companies, like Jell-O, in 1929. CalPak promoted the Del Monte brand through aggressive advertising, including the first industrial movie ever made (1917), ads targeting jobbers and grocers in the *Saturday Evening Post,* and appeals in the six major women's magazines that featured colored pictures of food for the first time. In this CalPak was keeping pace with the other food processors, such as Kellogg, which were the leaders in national advertising up to 1930. CalPak would change its corporate moniker to match its brand name in the 1950s.[79]

Canners and other food processors leapt on the nutritional bandwagon in the 1920s, making exaggerated claims for the healthfulness of their products. Vitamin research was taken up with a vengeance by the new field of nutrition science. University of Wisconsin agricultural chemists figured out how to add vitamin D to pasteurized milk in 1928, and this became standard practice even though milk did not naturally contain that vitamin. In the late 1930s, Agnes Fay Morgan, the head of the Department of Food Science and Technology at UC Berkeley, made pioneering studies into the role of vitamin G (B_2), leading to claims for its miraculous effects in preventing aging. Southern California food faddists—who gave the world the Hollywood eighteen-day diet and

Gayelord Hauser, food advisor to the stars—led the charge to vitamin pills in the 1930s. By the 1940s, food processors were introducing "fortified" foods (with vitamin additives) to bolster their image against the criticism that processing eliminated delicate nutrients. Eating was now something one did for good health, not pleasure—rather like jogging and weight lifting in our own time.[80]

Supermarkets were invaluable in the introduction of name-brand processed foods, especially canned goods, which proliferated in the 1920s. Canned-goods displays at state fairs and national expositions were other early methods of insinuating California products and brands into popular consciousness. All this effort in product change, publicity, and promotion, combined with cheap mass production and low prices, must have worked, because the national consumption of canned and processed foods shot up. Canned and processed foods were also readily accepted because they meant greater convenience in household food preparation as middle-class families lost their servants and women began moving into the labor force in number. The two world wars gave a further boost to sales and to consumer acceptance of canned foodstuffs, as the military piled on the processed food for the troops. Food scientists like Emil Mrak helped perfect army rations (Mrak's fame in this regard propelled him to the post of first chancellor at the new UC Davis campus). Canned goods amounted to only about one-tenth of household consumption in 1910, but thirty years later Americans were eating almost as much processed as fresh food.[81]

After World War II, food processing entered a new stage, with freezing and more elaborate prepared foods hitting the grocery shelves. Canned goods would begin to look old-fashioned as storage, cooking, and eating habits shifted again. Along with frozen orange juice, vegetables, and berries, a raft of increasingly elaborate "convenience" foods appeared, beginning with cake mixes and the Swanson TV dinner. The latter relied on the arrival of large freezing units in suburban homes (split-level refrigerators were introduced in the 1960s, by which time one-quarter of American households had separate freezer units). There was also an explosion in the number of soft drinks sold.

Over half the products on grocery shelves in 1960 had not existed in 1945, with high value-added prepared foods in the vanguard.[82]

There was an extraordinary boost in per capita food consumption during and after World War II.[83] Americans spent more on food, both absolutely and as a percentage of household income. No doubt they were eating better, but one wonders how much went to pay for processing, since the consumption of prepared foods went up by half between 1939 and 1954. Rising incomes explain part of the increase, but housewives were more keen than ever on reducing food preparation time at home, and the young were moving out to their own apartments and looking to Betty Crocker and frozen pizza to replace Mom's cooking. Restaurants also started purchasing more frozen and canned supplies to speed food preparation in their kitchens.

Another dimension in the manufacture of foodstuffs began to unfold in the 1950s: better foods through chemistry—the twin of petrofarming. In the 1950s, hundreds of new additives were introduced for color, taste, texture, and shelf life. This blossomed into fully "derived foods" by the late 1960s. Derived foods are constituted from component starches, proteins, sugars, and flavors of natural crops, which are remixed into unnatural but desirable forms, from hamburger helper to instant soup. Chemical additives became staples of the American diet at the time, before the health hazards of certain dyes, preservatives, and stabilizers were recognized in the 1970s. These manufactured foods only nod toward the farm and the crop, not to mention taste. The lead in this development came from the chemical, petrochemical, and pharmaceutical industries. UC Davis researchers in the Department of Food Science and Technology helped along the deconstruction of food with advanced work on milk components, enzymes, and flavors, even before federal support was ratcheted up by the 1946 Research and Marketing Act. Nonetheless, there was still room for entrepreneurism in the burgeoning snack food arena. Mervyn Wangenheim founded Granny Goose in 1948 in Oakland, after graduating from Stanford in economics in 1938 and helping to develop a slow-cooling method to prevent spoilage in canned fruits and vegetables sent to the troops while working for California Processing Company.[84]

Fast-food chains had by the 1960s become the biggest sellers of processed foods after supermarkets. These restaurants won over kids and busy moms with their brand of popular foodstuffs like hamburgers and fries, produced cheaply, reliably, and repetitively. Fast foods were not completely revolutionary. Americans have always eaten on the run. As one foreign observer noted in the early nineteenth century, "Abundance seems to breed a vague indifference to food, manifested in a tendency to eat and run, rather than to dine and savor." A century later John Steinbeck would observe that Americans' sense of taste had disappeared through lack of stimulation. Perhaps most importantly, fast-food restaurants served up meals in squeaky-clean environments, satisfying the American fetish for hygiene in food establishments.

Food historian Donna Gabaccia sums up the modern history of American eating thus: "For better or worse, American foods have more often been products of American industry than of American kitchens."[85] Gabaccia goes on to show that mainstream industrial foodways were created almost exclusively by Anglo-American capitalists. Immigrants were prominent only in meatpacking, brewing, and wine making. This dovetailed with the general rejection of non-whitebread foods during the high tide of industrial processing. California bucked this trend to some degree, with its Italian vintners like Sebastiani and canners like Contadini, German sugar and cattle barons Claus Spreckels and Henry Miller, and Sicilian produce merchant Giuseppi DiGorgio. But its divergence from the Midwestern norm has become most marked in recent years.

CALIFORNIA CUISINE

At the end of the 20th century, changes in food habits among Americans had a significant effect on California agribusiness. These changes were mostly instigated by Californians searching for better restaurants, new ways of life, and better health. Though California agriculture had played a leading role in the mass production and marketing of food, there was a distinctive middle-brow,

Midwestern stamp on American foodways during most of the 20th century. Since 1975, California (and especially the Bay Area) has come into its own as the beacon of changing foodways, no doubt reflecting its ascendancy as the largest state and the engine of American growth. Specifically, it led the way in fresh, organic, exotic, and refined foods for the country, thanks to its massive immigration and enrichment of the upper classes. In the ascent, the market for four food groups increased: fresh produce, organics, luxuries, and exotics. These have already reoriented the state's crop mix, providing growers and shippers with high-value niche markets to replace those glutted by the sheer productivity of California and global agriculture.[86]

Fresh fruit and vegetable consumption began to rise in the United States in the first quarter of the 20th century, thanks to the fruit merchants and changing ideas about nutrition. Californians were ahead of the curve on this, eating more fruits and vegetables than in the East. After all, they were the principal consumers of the state's varied crops. They were also exhorted by the scientists. Hilgard had long been encouraging Californians to eat more fruit, and he carried out some of the earliest laboratory studies to show that fruit was nutritious and healthful. Later, Berkeley nutritionists added their professional voices to the chorus. After World War II, however, fresh fruit and vegetable consumption plummeted nationally in the face of the processed food binge (down 64 percent and 17 percent, respectively, between 1947 and 1963); as a result, fresh produce declined from 88 percent to 65 percent of per capita consumption, by weight. At that time, a typical store carried about 65 fresh produce items.[87] Worse, the definition of "fresh" had changed. In keeping with the rest of the array of chemicals added to foodstuffs, ethylene and hormone sprays had come into general use to allow fruits and tomatoes to be picked early and force-ripened in storage on their way to the supermarkets. Another move encouraged by the supermarkets was to have growers field-chill and wrap vegetables to reduce spoilage in transit and display.

Yet within a decade, the fresh segment was rising again. By 2000, fresh vegetable consumption per person had nearly tripled over the preceding twenty-five years, fresh fruit intake had nearly doubled, and stores were typically carrying more than 300 produce items.[88] The top-selling fruits were bananas, apples, grapes, melons, and berries, and the top vegetables were lettuce, potatoes, tomatoes, onions, and carrots. In California, the amount and variety were even higher. Demand for fresh produce was, as before, more robust here, thanks to close proximity to the nation's largest supply of fruits and vegetables. Equally important were the eccentricities of California consumers, with their health consciousness, Hollywood-induced search for fresh faces, countercultures, and large numbers of vegetarians and vegans. But most important, growers and retailers realized that fresh produce had become a higher earner than processed foods, reversing a long trend in value-added food manufacturing.

Adding to the force of fresh produce in California was a turn to organically grown crops, free of pesticides and chemical fertilizers. Beginning as a fringe alternative, organic produce had a firm foothold in the state by the end of the 1960s. It was promoted by the counterculture and the burgeoning environmentalism of the region, and even got star billing through the San Francisco Zen Center's restaurant, Green's. But a taste for organics entered the mainstream more through the avenues of health concerns than consciousness raising about the earth. In any case, organics found a regular place at the table of prosperous Californians by the 1980s, with their status duly certified at the farm gate by an extensive state and private regulatory apparatus. By the end of the century, organics, which had spread even to field crops such as corn and potatoes, inched to over 2 percent of California crop production.[89]

Elite foodstuffs have proliferated wildly in California in recent years. Many of these have been headed for restaurant tables first. The new California cuisine, cultivated at Chez Panisse, Stars, and a growing circle of Berkeley, San Francisco, and Los Angeles restaurants, radically changed tastes across the state and the country. Haute cuisine has always risen and fallen with the strength of the high bourgeoisie, virtually dying out in the 1930s with the stock crash and higher income taxes. With a recrudescence of conspicuous and elegant consumption among the Reagan-era rich, the 1980s and '90s became fertile ground for a revival of fine dining. High average incomes of California professionals and technical workers added to the brew. Not that serving the rich was necessarily the intent of Alice Waters (who started Chez Panisse as a local operation in 1972) and the food revolutionaries, who wanted mostly to improve American eating and cooking along French lines (certainly a meritorious goal).[90] But that's how it has worked out.

Nouvelle cuisine altered farm supply chains close to the big cities. Spurred by Waters, who made a campaign out of encouraging alternative suppliers, a whole new segment of foods hit the racks in farmer's markets, in specialty shops, and on restaurant plates. The trend bolstered small producers of vegetables, oysters, cheese, and fowl, among other things. It also led to a revival of fine olive oil making after a century-long hiatus.[91]

The boom in exotic produce began around 1980, starting with the kiwi craze and the Americanization of guacamole.[92] There was soon a cascade of things such as raddichio, star fruit, bok choy, and lychees. These filled more than niche markets for elite consumption. A new era of immigration brought with it a phalanx of exotic foodways and foodstuffs. Immigrants, especially those coming from Asia, introduced a host of previously little-known crops, such as bitter melon and daikon, to California markets and menus; and immigrants usually were the pioneer growers of these exotics, as exemplified by the Hmong farmers of Fresno County. They even awakened Americans to things under their noses, such as wild mushrooms gathered in California and Northwest forests by Asian and Mexican workers. Long-ignored varieties of chilies, potatoes, rice, and citrus put in an appearance as the burgeoning Latin American and Asian populace demanded things familiar to their palates. Along with new crops came a host of locally made tortillas, salsas, and sauces, along with some incursions by Mexican companies such as Pan Bimbo and Manseca.[93]

Of course, this movement echoed off the flowering of high restaurant cuisine, with the search for fusion dishes to represent the new cosmopolitanism, as Americans grew more accepting of the unusual. Conversely, wider experiences of travel in exotic (often tropical) climes made Californians, and Americans in general, more open to new tastes and experiences. Thus, the landscape of food and drink changed dramatically in the last years of the 20th century.

But the landscape of California agribusiness changed less, as Julie Guthman has shown, because high-value cropping fit comfortably into the time-tested patterns of California's specialty-crop agriculture, industrial processing, and sophisticated marketing networks. Indeed, it offered salvation to growers faced with low profits in the agricultural doldrums of the 1980s. They were more than willing to try anything with niche profit margins—even if the niches were saturated within a few years.[94]

The most valuable of the elite products of late 20th century were fine wine and beer. Always California strengths, wine and beer underwent a mammoth expansion and reconfiguration after 1975. Collective memory for quality wine disappeared for the most part during Prohibition and had to be painfully reconstructed after 1934, among both producers and consumers (most post-Prohibition wines were "dessert wines" made from grapes such as the Thompson Seedless). The arduous task of restoring fine wine was carried out by a small contingent of Napa Valley winemakers; members of the Napa Valley Grape-Growers Association; the staff of the Wine Institute; and university faculty, such as A.J. Winkler and Maynard Amerine, at Berkeley's Fruit Products Laboratory and Davis's Department of Viticulture. Of particular importance were planting of better grape varieties, cold fermentation, use of stainless steel, control of malolactic fermentation, introduction of pure yeast tablets, and mechanized sterile bottling.

The North Coast wine industry began to promote its products vigorously through tastings, winery visits, labeling, varietal designations, and, finally, local appellations. Recognition at a key Paris wine tasting in 1973 gave California wines worldwide visibility. But the domestic market had to be expanded through persistent advertising, comparative tastings, and efforts to link wine with the good life. This was carried out not just by individual wineries, but with the invaluable assistance of the Wine Advisory Board and Premium Wine Producers of California. Table wines finally passed dessert wines in the late 1960s, and now premium wines from the North Bay region dominate the industry. Despite significant increases in exports, the demand for high-end California wine is still concentrated in the state itself, and the Bay Area drinks more wine by far than any other part of America—as it always has. Los Angeles is not far behind. The economic boom of the 1990s drove the wine industry into a frenzy of expansion, with wine grape acreage ballooning from 300,000 to almost 500,000 acres; the subsequent bust has hit the subpremium market hard, as evidenced by the marketing ploy of selling a house brand affectionately known as "Two Buck Chuck" (Charles Shaw) by discount grocery chain Trader Joe's.[95]

Saloons and small breweries were equally decimated by Prohibition, and afterward the industry became concentrated in a few major brewers selling mostly bottled beer via supermarkets and liquor stores for home consumption. By the 1960s even the few remaining regional brewers, such as Lucky Lager and Falstaff, were knocked out by the national brands. But a revulsion against banal brewing kicked in by the 1970s, and microbreweries and brew pubs began to sprout up around California. Northern California was the birthplace of the new craft brewing, for reasons of both taste and a rebellious culture. A Berkeley home brewer broke the legal ban on microbrewing in the late 1970s. Today California supports more microbrewers than anyplace else in the country. Brands such as Anchor Steam, Sierra Nevada, and Gordon Biersch have become household names among California beer drinkers of post-Bud sensibilities.[96]

While the market for food was changing at home, California growers and processors turned more than ever to global markets to absorb their mass-produced output. This is driven principally by market saturation at home, which hit hard in the general economic downtown of 1979–84 and drove many American farmers to the brink. That episode ended the halcyon postwar era of rising prosperity. As grocery sales stagnated and margins were cut by the discounters, agribusiness sought new outlets. In this, they benefited from the lower dollar of 1985–95 and the rapid growth of Japan and East Asia. Agricultural and food exports from California expanded from $4 billion in 1980 to $12 billion in 1994. By the early 2000s, California was exporting 20 percent of its crops. It is the leading exporter among the states, and its share of exports (20 percent) exceeds its share of U.S. farm production (12 percent). The chief

California's Leading Export Crops, 2002

(in millions of dollars)

Rank	Commodity	Value	% U.S.
1	Almonds	$829	100
2	Cotton	513	25
3	Wine	486	91
4	Table Grapes	367	93
5	Oranges	303	57
6	Dairy Products	301	30
7	Tomatoes, processed	215	93
8	Walnuts	184	100
9	Rice	183	24
10	Beef & products	168	30
11	Strawberries	157	93
12	Raisins	152	100
13	Lettuce	136	61
14	Pistachios	131	100
15	Prunes	128	100
16	Peaches/Nect.	107	81
17	Hay	104	23
18	Broccoli	95	89
19	Lemons	84	92
20	Carrots	71	83
Total exports		$5,374	26

Source: University of California, Davis, Agricultural Issues Center, Issues Brief no. 23, 2003.

exports in 1980 were cotton, rice, and wheat; about a quarter of fruits and nuts were exported and perhaps one-tenth of vegetables. Subsequently, there have been marked increases in the export of wine, table grapes, dairy products, and tomatoes. Agricultural exporting in this day runs smack dab into the global furies, as third world nations have become more militant about rich countries' protective barriers for their farm sectors. Paradoxically, California agribusiness sees free trade as a good thing, even as it utilizes the Export Trading Company Act of 1982 to market crops such as rice and cherries abroad through government-sanctioned monopolies that keep prices up. On the other hand, the state's prime exports today are unaffected by federal farm-support programs, which were phased out in the 1970s.[97]

7

Capital in the Countryside

Agribusiness in California has always been defined by more than large farms, integrated commodity chains, and industrialized production; more than factories in the fields, rivers of empire, or Cadillacs in the desert. It has meant capital in the countryside: rivers of finance and investment across the land. Money is the purest form of capital, its starting point and finish line. This is not money as mere means of exchange and measure of value, but money as storehouse of value, means of investment, and measure of return. Money as profit, the measure of all things; money as accumulated capital, the lifeline of economic prosperity. If California agriculture was predominantly capitalist from 1850 onward, as is our claim, then it ought to be well saturated with that most abstract form of capital, money. And it has been.

But capital is more than money, more than doing business. Behind the skin of commodities, arteries of exchange, and muscle of production beats a human heart. Capitalism is ever and always a social order of rank and power. On the one side stand the capitalists—owners of money, property, land—and, on the other side, workers—those who labor in field and factory. In the world of California agribusiness, we should find landowners, operators, canners, ginners, and their ilk facing off against workers in field, shed, supermarket, and dairy. And we do.

This is the hardest part of the agrarian story to grasp, because money and class are two of the most slippery categories in the economic lexicon. As George Henderson says, "As simple as the question may seem—'how ought capitalism in agriculture to be recognized?'—the answer is rife with complexity."[1] Indeed, not all money is capital, and financial relations are often well hidden by the discretion of banks and accountants. Classes are not parading about in name tags at conventions of The Bourgeoisie or The Proletariat, nor are they doing battle in the street every Sunday. Nonetheless, capital and class are not impossible to spot; they are every bit as real as grades of oranges or soil horizons, if just as subtle—which is to say, they are hard to mark off with sharp lines, but no less tangible for all that.

One of the ways to spot capital and class at work is to look for the webs they leave, like the assiduous spider who keeps a discreet listening post beside his handiwork until the time is right to pounce. One type of interweaving of agribusiness is through flows of investment and credit, and that will occupy the first section of this chapter. Another kind of networking is organizational and political: business associations, business alliances, and partnerships with the state. These will be the subject of the second section. Last, we shall look for capital's antithesis: a class of agro-industrial workers trying to organize themselves into a coherent opposition force to agribusiness. What the third section of the chapter will reveal is less the unity of the working class than further proof of the impressive power of the capitalist class to keep the social order of the countryside in place. In these ways we can begin to see how wide the net of capital and class has been cast across the California landscape, and why it is essential to speak of agribusiness as a class system that goes beyond farmers and rural society.

CAPITAL FLOWS OVER THE LAND

From the beginning of the American era, California enjoyed a surfeit of capital from mining and used it to good effect to invest in agrarian production. This internal circulation of funds continued long after the mining era, and

California financiers developed some of the most innovative forms of credit and banking in order to facilitate such investment. As Henderson observes, "California was the primary financier of its own agricultural development," which "excelled at the circulation of money capital through the farm."[2] Indeed, money coursed through the veins of the whole agro-industrial system, with its many parts, commodity circuits, and production chains. This can be seen in any number of cases throughout California's agrarian development.

The entry of capital into agriculture occurred immediately with the buying up of wheat and cattle lands using fortunes made in the mines and in commerce. The first great surge of investment came in the 1860s and early '70s, using capital accrued during the Gold Rush (1848–55) and the great Comstock silver rush (1859–75). Contemporaries roundly denounced the "land monopoly" of the time, but land was no more concentrated than the capital that purchased it. San Francisco was the main place this money piled up, and hence the source of most investment in outlying land holdings. Wheat barons such as William Chapman were really just San Francisco financiers. Some miners took their money and bought land without going through San Francisco, as in the cases of Martin Murphy in Santa Clara County or Charles Reed in Yolo County. The cattle empire of Miller and Lux was a partnership of a butcher, Henry Miller, who never went near San Francisco, and a city banker, Charles Lux, who was allied with the Parrott banking family. Kern County Land Company was another such successful partnership, of Lloyd Tevis and Ben Ali Haggin.[3]

San Francisco capitalists were also present at the birth of the wine industry. Agoston Haraszthy, the father of California wine, was a notorious promoter whose Buena Vista Viticultural Society was founded in 1863 with the backing of Billy Ralston, San Francisco's premier financial operator. The free-spending Haraszthy was thrown out by the stockholders in 1866, and the society went under along with Ralston in the deluge of 1875. Southern California's Anaheim Colony, another wine enterprise of this period, was put together in San Francisco, and so was Italian-Swiss Colony in Sonoma, founded by banker Andrea Sbarbaro. In the wine boom of the 1880s, many city capitalists jumped on board: George Hearst (mining) set up winery operations near the city of Sonoma, James Fair (mining) near Petaluma, and Alexander Duval (railroads) and Francis Smith (borax) in the Livermore Valley; Leland Stanford (railroads) planted the largest vineyard in the world in the eastern Sacramento Valley with the largest winery to accompany it.[4]

The story of agrarian investment is not confined to the biggest of bourgeoisie, however. City banks began lending in the countryside at this time, as well. Such channels of credit were no less important a part of the advancing agribusiness infrastructure than railroads or irrigation ditches. San Francisco banks, which controlled the vast majority of funds in the state in the 19th century, were very active lenders. For example, they held about one-third of the real estate loans to Stanislaus and Merced counties in 1879 and almost two-thirds in 1899. San Francisco banks were active in Southern California, as well, lending almost 40 percent of the real estate funds to San Bernardino County in 1899 (while Los Angeles banks lent another third). Most important were the savings banks, which focused on real estate lending and held 45 percent of California farm mortgages by 1914.[5]

Money did not just flow from the center. As California agriculture pushed farther into the countryside, hundreds of banks sprang up in rural counties. Most of these country banks were started by local growers and their allies and were vigorously boosterist; they recycled local wealth into trade, water supplies, and processing industries, as well as farming. They even transgressed racial lines, with Anglo bankers busily financing Chinese tenant farmers in the Delta in the late 19th century. The assets of banks outside San Francisco had caught up with the city by the time of the earthquake. The country banks were not isolated from the pool of capital in San Francisco, however. Correspondence relations were developed with city banks almost immediately, and clearinghouse centers were established after 1906 in the larger outlying cities like Stockton, San Jose, and Fresno.[6]

With an improved banking system, interest rates fell. Paul Rhode argues that the shifts from mining to agriculture and wheat to horticulture were facilitated by lower interest rates that benefited capital-intensive tree planting in the 1880s. The next upsurge of investment took place after 1895 and continued through the mid-1920s, despite a downturn just before World War I. This was a period of great prosperity generally and a robust time for American agriculture, which enjoyed its highest price levels of all time; so money could and did flow toward agriculture. Farm mortgage debt ballooned by five times between 1890 and 1925, and "No other state attracted as much [farm mortgage] investment as quickly as California." The result was that California had the second-highest state total of farm mortgage loans in 1925 and by far the highest value of mortgage debt per acre, $202 (farm debt did not peak until 1932, nine years after the rest of the country). The capitalization of California farms, which had been impressive from early on, was strengthened as more agrarian credit became available. And the buildup of debt would become an essential pressure on growers to force up productivity and returns in subsequent years.[7]

Banking reforms greatly increased the circulation of capital, helping lending activity rise dramatically after 1900. Regulatory changes included revisions in national bank rules, followed by changes in state bank charters and the establishment of the Federal Reserve Bank in 1913 (with San Francisco as headquarters of the Pacific States district). Federal land banks were established by the Farm Loan Act of 1916. The most striking development came in 1908 when California legalized branch banking for the first time in U.S. history, permitting the extraordinary growth of A.P. Giannini's Bank of Italy. Giannini expressly targeted agriculture and the interior, and his first branch was in his native San Jose. Other banks followed suit, like the Mercantile Trust Bank of San Francisco and the Bank of Oakland (later merged into American Trust), and the Security Bank of Los Angeles and Los Angeles Trust and Savings Bank.[8]

Bank of Italy had two dozen branches by 1919 and almost 300 by 1927, the year it metamorphosed into Bank of America; by 1930 it controlled almost half of all California bank units, mostly country banks it had purchased. Giannini held almost one in every ten farm mortgages in the state by then. One could argue that branch banking allowed San Francisco and Los Angeles to rein in the growth of country banks and maintain their hegemony over California capital circulation. Nonetheless, branch banking was a brilliant solution to the problem of time and space discontinuities in agricultural production up and down California.[9]

Short-term credit was critical to farmers' success, as well. Banks provided some of this, especially for purchase of equipment, but a large segment of "trade credit" was forthcoming from suppliers or vendors, such as irrigation districts, seed companies, veterinarians, and machinery dealers, and from buyers such as commission merchants, shippers, packers, canners, and grower-shippers, who might provide seeds, fertilizer, or cash advances. There are, unfortunately, no estimates of this sort of credit, which consists in delayed or advanced payment schemes rather than formal loans at interest. But it bridged the temporal gaps in production and commodity sales. Most of this credit consisted of compensatory fluxes of money within the agro-industrial complex of California, so the expansion of the supplier and shipper nodes of the whole production system meant more money available to smooth out the circulation of capital.

In all the histories of crop sectors, one sees the benefit to growers of suppliers who could wait for payment and buyers who could pay more up front. A decided advantage of the big millers, ginners, packer-shippers, canners, and grower-shippers that emerged in the late 19th and early 20th centuries is that they had the financial wherewithal to pay up on time. A big reason that the contracting system was expanded was to assure price and payment schedules to growers. The co-ops succeeded, as well, because they acted as finance capitalists, assembling capital, making payments, and extending bridging credit more effectively than most independent merchant shippers. On the one hand, they secured better terms of sale, year-round markets, and more rapid payment back to the growers, while on the other, they could pool growers funds, attract bank loans, and withhold payment to members in order to have investment funds and working capital for packing sheds, agents, harvest labor, etc.[10]

Capital also flowed into off-farm operations. On the input side, banks and city investors were crucial to the provision of water, power, feeds, machinery, and genetic research. California was richly supplied with electric power long before most farm states, thanks to heavy investments by San Francisco capitalists in hydroelectric dams and transmission systems in the early 20th century. In water supply, the first big irrigation canal was built in the 1870s by the biggest city capitalists of the day: Ralston, Friedlander, Lux, and Chapman. Ditch companies and irrigation districts were financed by selling bonds to city investors—though only when district legislation was amended in 1911 did capital flow in quantity (San Francisco financiers actively promoted the new legislation in Sacramento). Giannini became an enthusiastic buyer of irrigation bonds in the 1920s.[11]

On the output end of the agro-industrial system, finance capital played an equally vital role in the expansion and consolidation of California agribusiness. For example, Spreckels and the sugar companies pumped huge amounts of capital into milling operations, as did McNear and the flour companies. Later, CalPak was assembled by food merchant George Armsby with the help of New York financiers, his brother, banker James Armsby, and the Bank of California. Giannini was an enthusiastic lender to canners and packers, as well as to wineries. Safeway was put together under the auspices of Charles Merrill, New York investment banker, who had cut his teeth on retail chains like Kresge before that. Merrill served as a director; Merrill and Lynch held a fifth of Safeway stock; and after Skaggs stepped down in 1934, the presidency passed to Merrill's protégé, Lingan Warren, then to Merrill's son-in-law, Robert Magowan, and on to the latter's son, Peter.[12]

Following World War II, agricultural credit again expanded rapidly in California, helping to finance the increased inputs and mechanization of capital-intensive petrofarming. Total farm debt shot up from $650 million in 1950 to $3.5 billion in 1970, and then skyrocketed to over $10 billion by 1980. Credit was ample and interest rates low in part because of the favorable economic conditions of the times and in part because of New Deal agricultural reforms that augmented the role of the Federal Land Banks for mortgages, the Federal Intermediate Credit Bank system and its local Production Credit Associations for short-term funds, and the Farmers Home Administration for high-risk borrowers. Commercial credit also played a role. Land sales on contract, where the buyer picks up prior mortgages and payments, became a significant way of avoiding return to the banks and new long-term debt burdens. A different strategy for reducing capital needs which became widespread in this era was to acquire inputs through leasing and contract services, which turn fixed capital into circulating capital. Nonetheless, growers in California became even more financially overextended than farmers nationwide during the 1970s, and were squeezed severely by the high interest rates and diminished markets of the great agricultural recession of the early 1980s.[13] This helped trigger the latest round of agrarian restructuring.

Similarly, finance capital played a big role in nonfarm portions of the agro-industrial division of labor after the war. For example, credit from banks and supermarkets was essential to the rise of the feedlot system in California. City investors bought up billions of dollars of state water bonds for the State Water Project. The recovery of the North Coast wine industry was helped along by Bank of America and Wells Fargo's active branches in the Napa Valley, as well as by personal fortunes invested in boutique vineyards and wineries and by some outside corporate capital, such as Seagram's, Coca-Cola, and Moet et Chandon. Campbell's Soup Company provided the funding for Calgene's first biotech tomato and the Dean Witter investment bank directed stock buyers to the buoyant prosperity of the Del Monte Corporation.[14]

George Henderson's great insight in *California and the Fictions of Capital* is that finance capital looks upon the blockages and interruptions caused by natural conditions of agrarian production as opportunities as much as obstacles. For example, the extended time and distance of transport to market creates an opening for capital to invest in railroads, warehouses, and depots. Similarly, a crop failure due to drought presents the opportunity to invest in irrigation works; a pest problem a chance to create a spraying company; a soil problem a possibility of selling fertilizer or gypsum. Of course, capital cannot just profit from failure and being locked up in fixed investments; agricultural production must succeed and capital turn over, the more prodigiously the better. It is always a matter of balance.[15]

Furthermore, gaps in the flow of capital are to be found everywhere in the economy, cutting across time, space, and the division of labor. The genius of capitalist finance has been the development of banking and credit systems that leap over such gaps, that make them into new fields of play for profitable lending and investment. One capital's temporary paralysis is another's opening. Here the operative term is not fixed capital but "fictitious capital"—loan capital created on the hope of future returns. Such capital is no mere fiction but a bet against the future; it becomes real as soon as profits accrue and the loan is paid back with interest. And it makes the imagined unities of the agrarian circuit of capital real, as the gaps are plugged and the agro-production system functions smoothly. Thus, in the circulation of money we find the advance legions of capital in the agrarian system of California, the great unifier of disparate parts of the division of labor beyond even the organization matrix of agribusiness in commodity production and circulation.[16]

Finally, finance is the oxygen pumped into the engine of accumulation to make it run faster. Credit provides extra funds for investment and profit seeking, for the whole speculative process of anticipating future prospects and making them come true. But finance is also implicated in the excesses of accumulation. The drive to accumulate and internecine competition stimulate businesses to keep investing even as profit conditions start to deteriorate at the peak of a boom period, and the collective result is to push accumulation over the cliff of profitability and into the next vale of recession. This cycle of overinvestment is quite general across sectors, countries, and the global market, though never in one synchronous movement. Finance is a major propellant in the overheating of business booms, because it so readily fans the flames of enthusiasm and excess for the latest, greatest opportunity to make money. But then it steps in to bridge the gulfs of recessionary despair and to lead the way into new fields of fortune, as Henderson also argues.[17]

Financial force feeding can be seen very clearly in the history and geography of California agriculture. Back at the beginning of our story, in chapter 1, a picture was drawn of the rolling waves of crops over time and space as California agriculture grew. An essential push behind those wavelike surges of planting and stocking was finance. Finance not only helped agribusiness's explosive spread across the landscape, financial excess also led to overplanting and price drops of a catastrophic kind. One of the first financial bubbles in California drove the wheat and sheep booms of 1862 to 1875, and created the vastly overbuilt Mendota Canal. The crisis that followed encouraged the general turn toward fruit to diversify the agrarian economy, as well as the passage of the Wright Act to stabilize irrigation. Horticulture became the next object of financial fancies, with the planting boom of the 1880s ending in the deep panic of the early 1890s (along with the final wheat crash). This crisis stimulated the organization of the fruit cooperatives and the first consolidation of canneries, both with the help of outside financiers.

In the 20th century, boom-and-bust cycles continue to rearrange the landscape of agribusiness, both literally and figuratively. The prewar boom brought investments in new lines of produce, such as fresh vegetables and berries, canning tomatoes, and cotton in Imperial Valley. That bubble popped in 1914, taking down the land speculations of William Tevis and his partners in the Delta, forcing the Santa Clara Valley prune growers to finally get themselves organized as Sunsweet, and triggering the formation of CalPak, among other events. The next wave, from 1918 to 1929, built on new foundations and places. The two most dramatic examples are the Imperial and the Southern San Joaquin valleys. Imperial Valley received a tremendous impulse of finance into its farms and irrigation district from the Los Angeles area—chiefly because Los Angeles was getting rich on oil, but also because of limits on citrus expansion beginning to be felt there. A similar flood of money pushed the accelerator on the cotton boom in the San Joaquin Valley in the 1920s. Everyone jumped on the cotton bandwagon—outside land developers, electric utilities, and local chambers of commerce—but most important was Bank of America, which was financing half the cotton crop by 1929, at a cost of $10 million a year; less evident but equally essential was bond financing of several irrigation and water districts in the cotton region.[18]

For all the ins and outs of finance and the work done by capital flows to bind together places and projects, the human element cannot be overlooked in knitting up the raveled sleeve of capitalism in the countryside. For this side of things, we turn to the question of class and capitalist class formation.

MAKING CLASS

Agrarian class analysis in the United States is forever stuck in the Jeffersonian-Populist line of thought, the touchstone of which is farm size and family ownership. Even in California, where mass agrarian settlement was never the order of the day, the question of farm size has dominated the debate over agribusiness and obscured the nature of the capitalist social order in the countryside. We have already seen that while there were both large and small farms in California, they almost all shared three fundamental conditions: they were businesslike, they were well capitalized, and they employed wage labor. These growers, as capitalist producers, are the linchpins of the agribusiness circuit of capital. But agrarian class analysis has to find a wider field of play than the farm.

I propose to expand the understanding of the agribusiness class in three directions. One is to see how growers organized themselves into associations in pursuit of their common interests, and found leaders capable of advancing their cause. Another is to look over the farm gate to growers' relations with financiers, suppliers, processors, grocers, and scientists in the towns and big cities—that is, to see the geographical integration of class interest and the formation of business alliances. A third is to revisit the question of growers and the mobilization of state powers.

These are not only matters of an agrarian class pursuing its economic interests by various means—organizing, networking, politics—but ways in which an agribusiness class constituted itself—or what is known in sociology as "class formation." In the spirit of Edward Thompson's *Making of the English Working Class,* we are in search of the making of California's agrarian capitalist class.[19] The key to California agribusiness is ultimately not commodities, industry, or science, but the people behind them all. They have been the shapers of their own destiny, and responsible for the success of their historical project of agrarian development. A class analysis of California's countryside therefore has to take into account the collective action of the growers; their unity across the entire financial, organizational, and social fabric of agribusiness and across the landscape of city and country; and, finally, how the key powers of the state have been mobilized by and for agribusiness.

The agrarian capitalists are not easy to pin down, of course. They are a mobile and variable target, like Heisenberg's electrons. Not only are they both large and small, family and corporate, Anglo and immigrant, as previously discussed, they don't always have a fixed position relative to capital or the division of labor. As Henderson demonstrates, for example, the capitalist farmers are caught between superiority to the wage worker and subordinance to finance capital, and their sense of class position flips back and forth depending on business conditions and personal success or failure. Henderson does not put a name on this class position, but it is the ineluctable dilemma of the small proprietor, the petit bourgeois of classical sociology.[20]

Goodman, Sorj, and Wilkinson make a powerful argument against the idea of a natural division of labor between country and city, which inflects so much rural sociology. Yet in making their case for taking in the whole agro-industrial system beyond the farm gate, they overstate the centrifugal forces breaking up the farm and give the impression that industry has taken over entirely from the grower. Nevertheless, growers here have not been reduced to semi-proletarians whose money, inputs, management directives, and market outlets are all commanded by outside capital, in the manner of poultry farmers under the thumb of Tyson and Perdue in the South, as some observers think is the inevitable fate of all agriculture under capitalism.[21]

On the contrary, California farming, processing, and supply industries have evolved in tandem, and have been knit together by a variety of means. Indeed, California agriculture has always been a fluid business system and a combined class of growers, industrialists, and financiers in constant formation and re-formation. In keeping with their nonpeasant, business origins, agribusiness capitalists in California have rarely been just growers or merchants for life.

They were promiscuous actors who thought nothing of transgressing boundaries between sectors. George McNear, the flour king, founded First National Bank of Oakland, which later branched into rural communities throughout the East Bay. The Horst brothers were not only hop growers and merchants, they were fruit canners and driers. Equipment maker Ben Holt of Stockton tested his first crawler tractor on his family's Delta farm. C.W. Reed, the state's largest nursery operator in the 1860s, later became a leading grower and shipper of fruit. A.P. Giannini's father was a farmer and Amadeo started out as a fruit merchant. Theo Kearney was both land developer and raisin grower. Harris Weinstock was a grape grower, a city merchant, and a state marketing director. Eastern fruit merchant Giuseppi DiGorgio moved to California and became a grower, with huge holdings in grapes. Southern cotton merchant J.G. Boswell followed a similar path, and grew to be the largest cotton grower and cotton ginner in the world. G. Harold Powell moved from the USDA to head of the citrus co-op and became one of the largest lemon growers in Southern California. Of course, there were small farmers who were not this mobile and never moved along these networks of power, who struggled to keep up, and who often failed. This has been the case in every line of capitalist business since the first industrial revolution. Why should we expect any different in agrarian capitalism?

A key characteristic of the farmer-growers of California is that they have been great self-organizers in pursuit of collective interests, a dynamic more important than their differences and conflicts. San Jose's Pioneer Horticultural Society was founded in 1853 and the first state nurserymen's convention was held in 1858. The State Agricultural Society was formed in 1854 for self-education and promotion. The California Wine Growers' Association and the California Wine Makers' Association, established in 1962, were two of the earliest trade groups; the merchants organized as the California Wine Association thirty years later. The first ranchers' organization was the California Wool Growers' Association in 1860, followed later by the California Cattlemen's Association and the California Dairy Association. The fruit growers began to organize as soon as they became a force in California agriculture in the 1870s. The Farmers Union Corporation, established in 1874 in San Jose, may have been the earliest such effort, though it was short-lived. Hop growers tried to organize as early as 1877. Fruit growers held annual horticultural conventions at Sacramento from 1881 to 1931.

At the dawn of the 20th century, new associations came into being that reflected the shifting mix of crops. The U.S. Sugar Beet Association, the Lima Bean Growers' Association, the Citrus Association, the Olive Growers Association, the California Avocado Association, and others sprouted up. The processors set up the California Creamery Operators Association, the California Canners' League, the California Fruit Growers and Shippers Association, the Dried Fruit Association of California, the League of California Milk Producers, and the California Meat Association, among others. The modern California Nurserymen's Association was established, as well. A flurry of crops got themselves organized after World War I, like the California Canning Peach Growers, the California Pear Growers, the California-Arizona Ginners and Crushers, and the San Joaquin Cotton Growers.[22]

California growers also organized across sectors. This got started with the Grangers in the 1860s and the Farmers' Alliance in the 1880s, national farmers' movements that started in the Midwest and South but had a following in California, as well. As Populism died down by 1900, the growers turned toward marketing cooperatives. The cooperative spirit of California growers has been much celebrated in local histories, but Californians were, frankly, not much different than their farm brethren around the rest of the country in this regard. Moreover, by 1900 industrial trade associations were the standard of the day in American big business, and California agriculture was in the thick of this movement. Where it diverged from the rest of American agriculture was in the success of the marketing cooperatives after 1900, in which California would establish a new national model for businesslike farm organization. The line between co-ops and trade associations was blurred in the process.[23]

With the help of the Progressives and the Hiram Johnson administration, the 1910s were the time of a great leap forward in grower and agribusiness organization, a development that would be strengthened in the New Deal and thereafter, but never really improved upon. From then on, most growers could turn to their cooperatives or marketing boards as instruments of collective action, education, and order. Just after World War I, growers came together in the various county Farm Bureaus, and statewide as the California Farm Bureau Federation—the dominant voice of the growers as a class through the rest of the 20th century. Growers would also organize to secure labor and to control workers as Farmers' Protective Leagues in the 1910s, Labor Bureaus in the 1920s, and the Associated Farmers in the 1930s. They did the same in 1980 with the formation of the United Agribusiness League to oppose the United Farm Workers Union, and it is still going strong today.[24]

Nor should we forget the ethnic associations of the various immigrant farmers. For example, Japanese success relied greatly on associations, which helped new farmers find land, negotiate leases and prices, arbitrate disputes with owners, market the crops, and spread scientific knowledge through Japanese-language publications. After their internment in World War II, Japanese farmers joined white organizations in an effort to conform. The current secretary of Food and Agriculture is A.G. Kawamura.[25]

We must be equally cognizant of the connections between town and country and linkages that brought together different kinds of capitalists across the division of labor. The presently popular term "networking" can be used to describe the interweaving of farmers and businessmen into a capitalist class across geographic space and boundaries. The term conveys at once the sense of social interaction in the process and the necessary linkages to be forged between different segments of the capital chain.[26]

As the local banking system spread through Northern California after the Civil War, farmers and businessmen worked together to create a financial infrastructure for local development, mostly at the county level. Prominent growers, shippers, millers, and other notables composed the boards of every country bank. In the case of the Bank of Woodland, a group of landowners (wheat growers and cattlemen) gathered behind the prominent Stephens family, who covered other bases, such as the local gas and water companies. The Bank of Suisun was founded by a group led by Reuel Robbins, a lumberman turned shipper, farmer, and developer; a prominent sheep rancher; a commission merchant-flour miller; and a landowner who was also the son of a Supreme Court justice. Such connections via banks did not end at the county line. Reuel Robbins also headed up the nearby Bank of Vacaville, this time in league with fruit growers, including one who was director of the Vaca Valley Railroad and another who was prominent in the California Fruit Union. An extreme case is the Bank of Benecia, which brought together a far-flung network that included a banker from Colusa, a land agent for the Southern Pacific, and a Marysville attorney, and that shared three directors with the Bank of Haywards in Alameda County. The solidarity of these capitalist clusters was reinforced by ethnic and regional affiliations; for example, the Bank of Woodland was run by Missourians, the Bank of Colusa by Scots and Yankees.[27]

Another key local institution in which a range of capitalists came together was the marketing cooperative. These were fought for, organized, and run by prominent townsmen as well as growers. In every case, farmers worked hand in glove with other agribusinessmen, and the boards of grower co-ops were frequently "dominated by merchants, lawyers and bankers." In San Jose, for example, Thomas Montgomery, a realtor, banker, and part-time grower, played a leading role in a succession of organizing efforts that ultimately created Sunsweet. The trustees of the Sun-Maid raisin cooperative were a cross section of Fresno's business leaders, most of whom were also growers. Yet another local site of collective action was the irrigation district. Here, too, prominent businessmen joined with growers to lobby for and administer these critical institutions, which, while legally creatures of the state, were firmly in the hands of rural landowners and town developers.[28] Finally, every county had its chamber of commerce, the quintessential booster of local interests and modernization, and agribusinessmen cycled in and out of county government, a critical level of political organization for capital improvements and collective voice. There were, of course, fierce rivalries and notorious fallings out among local elites, as among the founders of the Imperial Valley, but what is striking is the degree of integration and stability over the years.

The building up of towns and cities in the agrarian districts of California was part and parcel of the formation of the agrarian capitalist class. Towns, like farmlands, were investment opportunities and speculations, then were settled during the general process of agricultural development in the valleys, as in the case of Brawley and Calexico in the Imperial Valley, Lockeford and Linden east of the Delta, and Pixley and Earlimart in the San Joaquin Valley. Some were model towns, like Patterson, other acts of spite among competitors, like El Centro. These towns grew through the residential congregation of farmers and the commercial clustering of suppliers, co-ops, banks, merchants, and more. Or frequently less. While some, like Fresno, San Jose, and Salinas, swelled into important second-rank cities, others, like Los Banos, King City, and Colusa, remain small to this day. Some, like Linden, are easily missed by the hurried driver on the way to somewhere else. Surveying the overall landscape of urbanism in the valleys, one is struck by its modesty—if not impoverishment—given the wealth of California agriculture. Where is the dense palimpsest of towns, cities, and villages one finds all across the Midwest and the northeastern United States?

This is a puzzle that bothered Walter Goldschmidt, a Berkeley anthropology student in the 1930s and author of the much-bruited study *As Ye Sow.* In that book, Goldschmidt compares Dinuba and Arvin in the San Joaquin Valley, the former lying just south of Fresno and the latter east of Bakersfield. Goldschmidt concludes that the more robust small-town atmosphere of Dinuba comes from its surrounding intensity of small farms, while Arvin is bereft because it is in the midst of huge holdings, particularly those of the Kern County Land Company. Goldschmidt was quite right (and greatly resented for his conclusions, leading him to end his career studying Africa instead). Indeed, one can generalize his findings to say that the overall lack of a large class of family farmers deeply embedded in local circuits of commerce and social reproduc-

"God Bless America," Marysville, 2002

tion is an essential reason for the relative impoverishment of urban settlement. In this, California is nothing like Iowa, or even Texas.[29]

Conversely, Goldschmidt recognized that a substantial portion of the capital and the class engaged in agribusiness is located in the big cities of California. The countryside was developed by investors and promoters from the cities outward, rather than from the bottom up through the mass settlement of a family farm class, and dense supply lines and social networks still run back into the metropolitan areas. We have already seen how the leading farm counties in the state were, first of all, along the Bay-river axis, where the big cities such as San Francisco, Sacramento, and Stockton were prior implantations of the Gold Rush. By 1900 the agrarian hot spot of California was Los Angeles County, cheek by jowl with the country's second-fastest-growing urban center, and L.A. investors opened up the Riverside Colonies and the Imperial Valley. Fresno, in the San Joaquin Valley, was originally a creation of the San

Francisco colony investors. Fresno's own business leaders then became deeply instantiated in the grape and raisin sector's development.

More is involved than simply the flux of finance capital out of the cities to make the California countryside. The urban capitalists were also developers and builders who helped construct the agrarian landscape. One example from San Francisco is the deep involvement of Southern Pacific Railroad in developing the land near its rail lines by selling off its land grants, by financing farms, and by founding town sites; it also ran the Pacific Fruit Express, saved Imperial Valley from the Colorado River in 1905–07, and helped organize the raisin growers, among other things. William Chapman of San Francisco got Fresno off the ground with his Central California Colony after 1875, sending his agent Bernhard Marks (a former farmer) to clear and level the land, build an irrigation canal and roads, plant trees and vines, and provide credit to colonists. Farther north, the Stephens family of Yolo County cooked up a plan around 1900 to dam Clear Lake in order to irrigate lands along Cache Creek. The Stephenses tapped into San Francisco money via their brother-in-law, Comstock millionaire Charles Bonynge, called on San Francisco engineers (including Michael O'Shaunessey) and attorneys for expertise, and mustered the support of Roy Pike (of a paint manufacturing and oilfield supply and banking family) and George Armsby (fruit merchant) to support their scheme. Up in Sacramento, department-store merchant Harris Weinstock became the kingpin of the entire co-op movement; he used a team of San Francisco lawyers to advise fledgling co-ops. Weinstock was the uncle and business partner of Simon Lubin, chief of the CCIH.[30]

There are similar examples from Southern California in the early 20th century. The California Development Company, which established viable farming in the Imperial Valley by reengineering the Colorado River, was a partnership of land promoters, engineer George Chaffey, and bankers (including Chaffey's son) from Los Angeles. The Los Angeles Merchants and Manufacturers Association was an ardent backer of the Sunkist co-op, which had its headquarters in downtown L.A. The founders of Union Oil partnered with Nathan Blanchard, owner of the giant Limoneira ranch, which controlled half of Sunkist.

The cotton boom in the San Joaquin Valley was largely engineered by merchants who set up shop in Los Angeles. On a more personal level, J.G. Boswell married Ruth Chandler of the *L.A. Times* family and J.S. DiGorgio's brother-in-law was Hugh Jewett, a banker and grower from Bakersfield.[31]

The branch banking systems of the Bank of Italy and its competitors would establish another direct alliance between big-city and small-town capitalists. Giannini usually kept the management of acquired banks intact, to maintain contacts with local elites. In addition, as local unit banks were bought up and turned into branches, local businessmen on their boards were invited into the big-city fold. Some were elevated to prominent positions in major banks. For example, when Mercantile Trust Company acquired Garden City Bank in 1922, it brought onto its board of directors William Alexander and Thomas Montgomery of San Jose. Conversely, when the farmers faced militant labor in the early 1930s, Giannini helped finance the Associated Farmers to put down the rebellion out in the rural counties.[32]

A single major bank board could distill the whole of California agribusiness, as illustrated by the Mercantile Trust Company, one of San Francisco's top banks in the early 20th century (it merged into American Trust in the 1920s and thence into Wells Fargo in the 1950s). The founders of Mercantile Trust in 1909 were William Irwin, who had bought out Claus Spreckels's Hawaiian sugar lands, and John McKee, son of a San Francisco banker, also on the board of sugar plantations and PG&E. The bank's board would include, over the next thirty years, such agribusiness luminaries as James Armsby, brother of the founder of CalPak; John Mailliard, son of a food broker and rancher; Walter Rothschild of Rosenberg Brothers, dried fruit merchants; Peter Cook, landowner and banker from Rio Vista who sat on the state Board of Reclamation; George Montgomery, member of the Spreckels family who sat on the board of Kern County Land Company; and Paul Davies, later president of Food Machinery Corporation and a director of Caterpillar Tractor. A look at other boards, like that of Safeway, reveals similar ties to the broad spectrum of agribusiness and the agrarian capitalist class.[33]

All this organizational activity and networking is not just for social solidarity and the governance of markets. It is profoundly political, undertaken to influence public policy and the state. Agribusiness has a deep interest in government policy and in using the powers of the state to its advantage on everything from tariffs to technology. The hidden hand has maneuvered to get a good grip on the neck of the government. Growers not only influenced legislatures, governors, and congressmen, they organized to establish their own service agencies, such as the College of Agriculture, marketing boards, and the California Department of Agriculture. At the local level of county government, water districts, and town officials, they effectively have their own private state, in which the sheriffs, judges, and supervisors are themselves growers more often than not. Yet that easy formula of grower power over the state doesn't quite fit the facts, for the growers often did not want state interference or did not know what they wanted or what was in their best interests. So one finds the state being used by dynamic men (and a few women) to organize the growers and to push agribusiness forward for its own good. In short, agrarian capital is deeply involuted with government, in the classic manner of the American state—where "the business of America is business," as Calvin Coolidge so aptly put it. The preferred relation of state and capital is commonly known as the public-private partnership.[34]

California growers wasted no time in calling on the state for assistance in agrarian development. The State Agricultural Society was funded by the legislature in 1854, at the behest of wheat growers and nurserymen. The society vigorously advertised to stimulate immigration of farmers—hence the Cornucopia poster encountered earlier. The state forgave farmers their taxes for many years and subsidized several experimental programs in the 1860s, including silk farms and Haraszthy's doomed voyage to gather grape cuttings. The California Wool Growers' Association lobbied successfully for the federal wool tariff of 1867. In 1877 the legislature empowered irrigation districts—the first special district governments in America—allowing local growers to undertake water projects by securing the powers to tax, issue bonds, and condemn land, powers normally reserved to the state (reclamation and flood control districts would be empowered in the same way to assist in drainage and dam projects). A State Board of Viticultural Commissioners was created in 1880, at the behest of the wine industry. The Board of Horticultural Commissioners that followed in 1883 was put in place "by and for orchardists and vineyardists" to provide a forum for discussion, information exchange, and legislative relief. This was followed twenty years later by a state horticultural commissioner. Miller and Lux were the most accomplished integrators of capital and the state in this era, and they did so with little subtlety—keeping Land Office officials, tax assessors, and surveyors on the payroll even while those officials served in public office.[35]

We have already encountered the close relation between the University of California's College of Agriculture and agribusiness, as evidenced by research programs, extension services, and on-campus teaching. Relations between growers and professors were not always idyllic, however. Mutual distrust grew out of class and urban-rural differences, but more importantly out of understandable differences over the role of abstract knowledge versus practical arts. The first dean of the College of Agriculture, Ezra Carr, was sacked by university president Daniel Gilman for his sympathies with the Grangers, who wanted more practical education (Gilman left in a huff soon afterward). E.W. Hilgard was a better academic fit than Carr, as he had little regard for hands-on training compared to more scientific agriculture. Nor was he afraid to cross swords with the wine grape growers for their use of poor varieties, their bad placement of vineyards, their lack of knowledge of soils, their sloppy culling at harvest, and their poor enological practices. During his service on the Board of State Viticultural Commissioners, he clashed with growers so vehemently that the California Wine Association had the board decommissioned in 1895 just to be rid of him. Farmers railed against the university again after the turn of the century; they were led by Peter Shields of Yolo County and the State Agricultural Society, who disdained Hilgard's research emphasis, preferring more direct education of farm boys. Shields (a onetime law partner of Hiram Johnson) got the legislature to fund the new university farm at Davis, just as Hilgard was retiring.[36]

After 1915, however, California agriculture and the university marched in lockstep as never before. Emmett Fiske argues that this was a critical juncture in the history of the College of Agriculture, at which Thomas Forsyth Hunt and his colleagues sold their souls to agribusiness, making the pursuit of productivity take precedence over all else. This seems less a qualitative than a quantitative change, however. The purposes of the College of Agriculture had not really been altered, since, as Edward Wickson declared in 1918, "In California it was from the beginning . . . a question of doing as much as possible of whatever the public demanded." What changed was the college's financial resources and scientific capabilities, which made it possible to do much more for growers and agro-industry than it had been in the Hilgard-Wickson era.[37]

Nothing was more helpful to agribusiness than the organizational capacities of the Cooperative Extension Service and its director, B.J. Crocheron. Crocheron put farm advisors (Agricultural Commissioners) in every county, completed the system of county farm bureaus, and then, in 1919, created the California Farm Bureau Federation. The two organizations were so cozy that the federation's office was on the Berkeley campus until it moved into a downtown Berkeley headquarters in 1938 (it did not move to Sacramento until the 1970s). The Farm Bureau Federation worked hand in glove with the State Chamber of Commerce, formed in 1920 by California's industrial capitalists.[38]

The state of California and agribusiness worked hand in hand to come up with the system of co-ops and marketing orders that would reign over every crop sector for the next century. The growers came up with the first functioning co-ops, but then the state gave the cooperative movement its imprimatur (and the force of law) with the creation of a State Director of Markets in 1915, the Standardization Act of 1917, and the Fruit Standards Act of 1927. Agents of the state, such as Weinstock and Crocheron, were intimately involved in the establishment of several co-ops, including Sunsweet, the Pear Growers, and Calavo (Avocados). The co-ops then turned around and created a political arm, the California Agricultural Legislative Committee, in 1919. Joined by the Farm Bureau Federation, they lobbied the legislature for more powers for themselves, for legal protection of the marketing agreements against antitrust laws, and for the creation of a State Department of Agriculture. The federal government was also pushed by the California delegation in Congress to protect co-ops from antitrust prosecution through the Copper-Volstead Act of 1922 and to create the system of federal marketing orders in 1937, which were widely employed in the postwar era.

The state has engaged in any number of other aid and assistance programs for agribusiness, mostly cooked up by agrarian interests. One was promotion of exports. For example, Crocheron was sent on a tour of East Asia in 1929 to publicize California farm products, which were suffering a glut even before the Great Depression. Another was restriction of imports. For example, prune growers benefited from a tariff on foreign prunes approved by Congress in 1894. Avocado growers—a rather minor group in the overall mix—were able to keep Mexican avocados at bay through high tariffs and to have Florida avocados kept out of California (a rule that prevailed until it was thrown out as an interference in interstate commerce by the United States Supreme Court in 1961). A third, highly specific case of the use of state regulatory powers to support a single crop sector is the One Variety Act, passed by the legislature in 1925, to ban the growing of any variety other than Acala cotton in the San Joaquin Valley, so as to maintain uniformity of fibers coming off the gins.[39]

The state's most notorious beneficence has been water for irrigation. We have seen how irrigation districts, the city of Los Angeles, and the Federal Reclamation Service developed the power to make water flow to agribusiness. What about the State Water Project? The state of California took up the cause of the Sacramento Valley growers in their long battle with floods with the 1911 State Water Plan, which was funded in 1928—only to be taken over by the federal government in 1930 as the Central Valley Project, after the state budget collapsed. When southern and western San Joaquin Valley growers complained of falling water tables (thanks to the cotton boom), the state stepped in again with the State Water Project, concocted in 1940s, funded in 1960, and completed in 1972. And when that was not enough, the Department of Water Resources dreamed up the Peripheral Canal—which CalTrans started to dig as it built Interstate 5, even before the project was approved by the legislature in 1976.[40] When that was defeated, DWR still went ahead and added new pumps at the Delta to export more water. And in 1992, when the federal government threatened to reduce subsidies on the Central Valley Project, Governor Pete Wilson wanted the state to take the project back (he needn't have worried, as the reforms were defeated in Congress).

For the generations that came of age after the Korean War, it is hard to think of any group in California society better orchestrated to serve its own interests and more powerful in Sacramento than agribusiness. Quite simply, agriculture ruled the roost for almost a century and suffered no opposition in its own domains. Things came apart gradually between the 1960s and the 1980s, through an accumulation of contradictions, hubris, and enemies. The pesticide campaigners, opponents of the Peripheral Canal, and tax cutters underscored the environmental and fiscal costs of untrammeled agrarian development. National Land for the People, organic fanatics, and river rafters cut further into the mantle of legitimacy of corporate farming and its brutish production system. Meanwhile, other sectors of industry, such as aerospace, electronics, and urban construction, were overtaking agribusiness as a share of the state's economy and demanding more attention from the government. But the thing that best dramatized the excesses of the House of Agribusiness in California was the tremors coming from ground level—the stirrings of opposition from the people most directly suborned by the agrarian class, the workers. We turn to those upheavals now.

WOBBLY DREAMS

Standing across from the businesspeople in all lines of agribusiness are the workers offering up their wage labor in the fields, orchards, canneries, and supermarkets. They have formed a great stream of humanity numbering in the millions over the years. But have they constituted a working class? The story of the agrarian working class doesn't read like that of the capitalists, with its long record of unity across differences in pursuit of common goals. Instead, the workers have all too often been divided, disorganized, and left exposed to the vagaries of the market. They have had less access to the powers of the state than the grower-business class and more experience with the business end of police clubs and other heavy-handed means of social control.

Farmworkers have periodically joined together in militant opposition to their exploitation by growers, and this makes for a heroic history of class struggles in the countryside. But their efforts have been repeatedly stymied, making that history an American tragedy. Cannery, packinghouse, transport, and grocery workers have had better luck unionizing and winning significant improvements in their lives. But their gains, too, have often proved impermanent.

Unfortunately, a labor history that cuts across all the branches of California agribusiness has yet to be written. This is a testament to the power of the division of labor to separate people and obscure connections, as well as an understandable focus of radical sympathies on the extreme plight of the fieldworkers. Nonetheless, one looks in vain for histories of organizing in slaughterhouses, tanneries, breweries, wineries, sugar refineries, irrigation districts, and the like across the state.

The two streams of struggle, on and off the farm, have rarely converged. A sad fact of agricultural labor has been its isolation from the rest of the working class, here as well as across the United States. This is due in part to the rural nature of agriculture and urban condescension, yet rural workers were constantly circulating in and out of the cities. Another cause has been the pressure of the growers and their allies to keep farm and city rebellions safely disconnected. Not by accident was farm labor left out of the National Labor Relations Act in 1935—with disastrous effects on the rural workers. In California there are some inspiring tales of reaching out across the distances, especially by organizers of sweeping vision, but there have also been ample neglect and perfidiousness toward the rural working class.[41]

There is little record of 19th-century labor struggles in the California countryside. The Chinese were subject to tight control by labor contractors and white society in general, but they were not entirely passive in the face of their exploitation, and they formed fraternal institutions for self-protection and mutual aid. Unfortunately, they were openly refuted by the mass of the working class in the state when the Workingmen's Party fell under the spell of demagoguery and racism, abandoning its radical, inclusive charter. White urban workers led the charge to drive the Chinese out of California.[42]

Japanese and Mexican workers kicked off the 20th century on a promising note by launching the first strike in California agricultural history: the Oxnard sugar strike of 1903. The Japanese had the benefit of labor contractors from their midst who supported them in strikes. This solidarity was a source of irritation for the growers. But the Japanese strikers were rebuffed in the end by a white working class led by the American Federation of Labor (AFL). The Los Angeles and San Francisco labor councils made tentative steps toward promoting field organizing and racial inclusion, but fell under the sway of the AFL's rabid anti-Asian sentiment. It is not true, however, that animosity toward the Japanese was worse than to the Chinese—quite the contrary. The Japanese remained a force in farm labor and farming in selected areas, like Placer County and Santa Clara County, right up to the internment of World War II.[43]

Tight labor market conditions just before World War I facilitated a surge of organizing in the fields by the Industrial Workers of the World (IWW). This was a critical moment in California agrarian labor history because the harvest labor system had come of age and was in a position to be challenged by a vibrant American (and California) labor movement. The pivotal event was the Wheatland hop pickers strike of 1913 at the Durst Brothers ranch. At Wheatland, white tramp labor played a leading role in organizing an astonishing mixture of nationalities, who were fed up with grower hiring, housing, and sanitary abuses. The IWW organizers were the first to give public voice to the maladies of the California harvest system on behalf of all workers (since the Chinese and Japanese had been too much out of the white mainstream to be heard). Furthermore, the cross-ethnic solidarity at Wheatland ran against the grain of race prejudice, and it was consistent with Wobbly views of One Big Union and not entirely out of keeping with the California working-class outlook by this time. Racism is not an unalloyed quality of whiteness, but a nerve that can be agitated or calmed depending on circumstances and leadership (as we can see in our own time).

The growers won the battle at Wheatland (which they described, falsely, as a riot), thanks to local forces at their disposal in Marysville: a sheriff willing to attack strikers, courts happy to convict labor organizer Blackie Ford for murder because another worker shot two men, and hired gunmen working for the Durst-sponsored Farmers' Protective League—the first of the grower vigilante groups. They went on to win the war against the Wobblies, with the help of government spies from the CCIH, statewide police raids coordinated by the U.S. Department of Justice, and conservative courts that supported bogus convictions for sedition and other trumped-up charges. The follow-up to Wheatland established the terms for the future wars in the fields, as Don Mitchell has shown. Growers were not to be denied their cheap harvest labor system, but they had to keep a tight grip on a highly mobile labor force and stop all organizing in its tracks. Wheatland set a critical precedent for grower repression that would become a defining feature of California agrarianism thereafter. Growers showed themselves willing to resort to violence to keep the fields union free. This collective response of agrarian capital was an act of will and a political choice not determined by markets, land, nature, or racism.[44]

Wheatland made the growers suspicious of white tramp labor for a generation, thanks to fear of Wobbly agitation. Employers tried mobilizing more women and children during the war, and then relied heavily on an influx of Mexicans and Filipinos through the 1920s. But a decade of bad wages and working conditions, combined with the tightening noose on agricultural profits at the end of the boom, led predictably to a new round of labor conflict. When it came, beginning with cantaloupe workers in the Imperial Valley in 1928, the new wave of organizing and strikes was far beyond anything agribusiness had yet felt.

The Mexican workers who now comprised the bulk of the labor force were steeled by their work in mines, fields, and industries in northern Mexico and across the Southwest. These were not unschooled peons, but seasoned proletarians. They were strengthened by traditions of national pride, union struggles, and the revolution of 1910. Filipinos were equally militant, fueled by memories of island resistance to American imperialism in the 1900s. These workers' resources of righteousness were different from the free-labor, free-speech, and one-big-union cries of the IWW—although many belonged to the IWW in its prime. Many more were followers of the anarchist Partido Liberal Mexicano of Ricardo Flores Magón, who had fled prerevolutionary Mexico for Los Angeles. Others came out of the Communist Party of Mexico. Organizing efforts in the late 1920s were led by unions growing out of mutual aid societies, such as La Unión de Trabajadores del Valle Imperial (La Unión); the Confederación de Uniónes Obreros Mexicanos (CUOM); and the largest, formed in 1933, La Confederación de Uniónes de Campesinos y Obreros Mexicanos (CUCOM). As expected, the first efforts of La Unión and CUOM to improve wages and conditions in the cantaloupe fields of the Imperial Valley in 1928 met fierce resistance from growers. Growers had the police, media, courts, and INS—and even the Mexican consulate in Calexico—on their side.[45]

The farmworkers' cause was taken up by the American Communist Party and its newly minted Trade Union Unity League in 1930. The party attracted many of the most committed radicals of the day, most of whom were believers in a united working class across national, race, and gender lines. In California, unlike some parts of the country, these ecumenical commitments were more than words. Many of the key organizers were women, like Carolyn Decker and Dorothy Ray (Healey), while men like Pat Chambers had come out of the IWW with a broad vision of the working class. These white organizers eagerly pushed the cause of workers of color, and recruited them as co-organizers. This was not such a difficult proposition since, as Devra Weber has shown, the Mexican workers were already organized and militant on their own. A new initiative among lettuce pickers hit the Imperial Valley in 1930. Communist organizers came to their aid under the banner of the Agricultural Workers Industrial League (AWIL), bolstering the strike; but the growers called out all their troops: the Western Growers Protective Association, the conservative governor, the Los Angeles Police Department's Red Squad, and the judges. They threw the head organizers in jail and broke the spring cantaloupe strike.[46]

But the union organizers were not done by a long shot. Depression conditions were affecting agribusiness profitability and driving growers to cut wages sharply (by roughly half on average across California). Moreover, labor markets were tight because of a new U.S. policy to deport Mexicans. Union organizers Decker, Ray, and Chambers returned in late 1932 with a new mantle, the Cannery and Agricultural Workers Industrial Union (CAWIU). An unprecedented surge of rebellion swept through the fields in 1933, mostly under the CAWIU banner. There were thirty-seven strikes involving 40,000 to 50,000 workers; this represented more than half the farmworker strikes that year across the whole country. Orchard pruners went out in Yolo County, pea pickers in Alameda, cherry pickers in Santa Clara, berry pickers in El Monte (Los Angeles County), sugar beet workers in Oxnard (Ventura County), peach pickers in the Sacramento Valley, grape cutters in Lodi (San Joaquin County), lettuce pickers in Salinas (Monterey County), and cotton pickers down the length of the San Joaquin Valley. Lettuce workers in the Salinas Valley formed a Filipino Labor Union and repeatedly struck from 1932 to 1934. Amazingly, the farmworkers' explosion came a year before the more famous 1934 General Strike in San Francisco and the other labor rebellions around the country that year. The workers' demands were as revolutionary as those on the docks: closed shop and union-controlled hiring.[47]

The growers' reaction was swift. They organized a vicious counterattack using all the resources at their command. The Filipino Labor Union was broken, for example, when growers organized a company union and then sent vigilantes into the main labor camp and arrested the union head, Rufo Canete (a former labor contractor), in 1934. Vineyard workers near Lodi were attacked by deputized American Legionnaires, their leaders arrested in a raid on the union hall; and people on poor relief mustered into the fields. Finally strikers were rounded up in the camps at gunpoint and transported beyond the county line. In the cotton strike, the longest of all, growers and ginners formed a united front behind the Agricultural Labor Board, evicted cotton pickers from labor camps, and organized vigilantes under local Farmers' Protective Leagues. The vigilantes repeatedly threatened workers, until they succeeded in killing three near Arvin and Pixley as the California Highway Patrol stood by. Naturally, the local sheriff pinned the blame on other workers and on union leaders, who were jailed on trumped-up charges. The state of California and its officials were kept away from the fray by agribusiness hostility to any negotiated settlement, then the federal government stepped in and pushed through a slight wage increase without recognition of the union. The strike broke up after that.[48]

That winter the growers banded together as the Associated Farmers, the largest California vigilante movement since 1850s San Francisco. The Associated Farmers enlisted 40,000 members, with local chapters in farm towns ready to heed the call from local growers and sheriffs to get out the ax handles, guns, and baseball bats to apply to workers' hides. The Associated Farmers was supported by the American Legion, by Chambers of Commerce, and by local women's clubs, and funded in large part by CalPak, the Canners' League, PG&E, Southern Pacific, Bank of America, and other agribusiness capitalists. Whatever differences growers and processors, producers and financiers might have had, they were put aside to face the common class enemy. And "enemy" is the right word is this period of sharp class conflict; as John Steinbeck wrote in *Harvest Gypsies,* "In California we find a curious attitude toward a group that makes our agriculture successful. The migrants are needed, and they are hated." The Associated Farmers would continue to be active through the rest of the 1930s.[49]

Although strikes continued around the state into 1934, the CAWIU offensive had been stopped, and leaders Decker and Chambers were jailed for two years under the Criminal Syndicalism acts used a generation earlier against the IWW. The CAWIU was dissolved by the Communist Party in 1935. For the next couple of years, the scene of labor struggles shifted south to the Los Angeles basin, where the Mexican federation CUCOM was still active. The organization was part of the American Federation of Labor's Federation of

Agricultural Unions, started in 1936 to fend off radical organizing in the fields, but it went over to the rival Congress of Industrial Organizations (CIO) fold under the leadership of anarchist William Velarde. CUCOM led a series of strikes across Los Angeles, relying on the same kind of Mexican solidarity and militance as before. It was the first farmworkers' union to win recognition from growers. Yet overall wages were falling with the influx of Okies, adding to a labor surplus.[50]

When the CAWIU disappeared, organizers from the American Federation of Labor tried to form a joint agricultural and cannery union, only to be rebuffed by the AFL hierarchy. Instead, they broke away to form an independent union in 1937, called the United Cannery, Agricultural, Packing, and Allied Workers of America (UCAPAWA), affiliated with the emergent CIO. CUCOM merged into UCAPAWA. While the new union was the creature of labor radicals, it had little to do with the Communist Party, despite accusations to that effect in later public disputes. Dorothy Healey, a veteran of CAWIU, was one of the few direct links to the party. As with the IWW and CAWIU, union organizers depended on an ecumenical mix of Mexican, African, Asian, and European ethnicities in local Campaigns. They made a concerted effort to link processing workers to field-workers, and they had active support from unionized black ginners and compress workers. They also received a good deal of public support from urban liberals and unions, and more favorable treatment by the state of California and the federal government because of it. Carey McWilliams was made head of the Commission on Industrial Housing under left-leaning governor Culbert Olson, and the federal Farm Security Administration built model labor camps at Corcoran, Shafter, Arvin, and other sites around the cotton districts.[51]

UCAPAWA organized a dozen locals in the fields from 1937 to 1940, most of them in the cotton region of the San Joaquin Valley. But unionism was overshadowed during the first two years by concerns about distribution of relief funds, provision of housing, and the election of Olson as governor in 1938. Late in that year, wage cuts in cotton had again set off a round of strikes in Kern County, which moved up to the fast-growing cotton fields of Madera

County in 1939. UCAPAWA had some success in raising the cotton piece rate and signed the first labor contract in the history of California agriculture. But the rebellion in the fields was much more muted than in 1933 or even 1936, until jailing of pickets in Madera brought a unified response that stopped almost all picking in October. That brought a counterblow from a revived Associated Farmers, who brought out the vigilantes with their guns and pickaxes. After a vicious attack on a worker rally at the Madera city park, in which the California Highway Patrol sided with the assailants, the strike collapsed. Rain forced the growers to offer a higher wage, and work resumed.

A major success of the wave of militancy was in bringing the exploitation of California harvest workers to the attention of the nation through the masterful work of Paul Taylor, Dorothea Lange, Carey McWilliams, John Steinbeck, and the La Follette Committee in Congress. But despite national outrage against the California agrarian system, the plight of the workers was generally misunderstood as one of the triumph of land monopoly over the family farm rather than the need to gain legal protection for agricultural unions comparable to those in any other modern industry. Furthermore, the Okie inductees to the army of harvest labor proved to be reluctant warriors rather than the freedom-loving, true Americans many took them to be. The white workers had stronger leanings toward small property, individualism, and white supremacy than to unions. The backbone of the UCAPAWA was, again, Mexican. As the world war diverted attention and siphoned off the white workforce into the cities, the great agricultural valleys would return to an uneasy peace and be forgotten. As Don Mitchell trenchantly notes, the sublime picture of abundance that the rural landscape represents was restored to its pristine state, hiding the unrest and the pain of agricultural labor. That is "the lie of the land," which hides the social order from view just as surely as today the Nike swoosh hides the sweat of workers in the factories of China.[52]

Labor organizing has not been confined to the fields. The downstream sites of food processing—packinghouses, canneries, slaughterhouses, and sheds—

were all major employers. The overall numbers of workers in these realms was enormous—roughly half as many as the number of field-workers. California canneries and packinghouses employed at least 20,000 people by 1900, 70,000 by 1910, 100,000 in 1939 and again in 1948. Three-quarters of the food workers were women, and three-quarters of those were immigrants. Cannery and packinghouse work was hard labor: long hours and seven-day workweeks during peak season (followed by closure during the off-season); no time to sit down; heat, humidity, and slick floors; payment by the piece; and low wages (even lower because women were systematically underpaid compared to men).[53]

The factory workers were just as militant as field-workers, if not more so, and they were more successful in the end in gaining union recognition and improved wages. This might not have been predicted by those who think of women and immigrant workers as docile. It also might not have been predicted in terms of relative wages and conditions, since California cannery workers had it better than cannery workers elsewhere in the country. Yet the potential for success was greater, in that the big canning and food-making companies had market power through oligopoly and greater surplus value through factory production. The factory workers also had greater leverage than field-workers in that they were more rooted in local communities. Most were townsfolk with stable living arrangements, families, and social networks. As Glenna Matthews observes, "women could return to their work year after year and could experience it as a regular part of their lives and identities, despite its seasonal nature."[54]

The first recorded cannery action was led by Toilers of the World, an IWW-inspired local that organized in seven large canneries in 1917. Not surprisingly, this took place around San Jose, the largest canning center. Typically, the National Guard was called out by local authorities, but the union won pay raises and improvements for workers before disappearing a couple of years later. Savage wage cuts in 1931 triggered the next outburst in the Santa Clara Valley and stimulated the founding of the Cannery and Agricultural Workers' Industrial Union (CAWIU), whose first offices were in downtown San Jose. There was a riot in a downtown park when the vigilantes turned up to do battle with strikers, and San Jose quickly acquired a reputation for violent repression. The CAWIU saw a couple of its few successful actions in the strike wave of 1933 in the Santa Clara Valley, with the cherry and pear pickers.[55]

The unions finally gained recognition after a third wave of organizing in the late 1930s. In Southern California, the union advance unfolded under the CIO-affiliated UCAPAWA and the leadership of Dorothy Healey and Luisa Morena. The union had decided to get out of field organizing and try its hand with food processors. The first contract and closed-shop agreement in the history of California canning were achieved at a Los Angeles company in 1939. During the war, cannery workers won dozens of representation elections under the UCAPAWA banner, as did citrus packinghouse workers (though gaining contracts was another matter). The UCAPAWA was a democratic union with high levels of worker participation and a sharp left bent. The union changed its name to the Food, Tobacco, and Agriculture Allied Workers (FTA) in 1945, but was essentially the same organization.[56]

The union drive in the northern canneries followed a rather Byzantine course after 1937. On the one hand, the state AFL leadership was panicked by the growth of a radical labor movement and by the Communists around the Bay Area, and it wanted to keep UCAPAWA out of the canneries. On the other hand, the canners, organized under a new employers' group, California Processors and Growers (CPG), were ready to cut a deal; their interests had diverged from those of the growers because of the National Labor Relations Act, which recognized factory workers' rights but not those of farmworkers. So the AFL and the CPG worked out a sweetheart closed-shop agreement behind closed doors: the AFL would set up cannery locals, and the canners would do business with them. These were not part of a union, but rather temporary locals chartered by the state AFL, like Santa Clara's Local 20852. While the worker activists hated the deal, they succeeded in turning 20852 into an effective weapon to gain seniority hiring and wage increases by 1941.[57]

As the war ended, a fierce labor battle broke out between two unions, the Food, Tobacco, and Agriculture Allied Workers, moving in from the South, and the Teamsters, awarded the Northern California cannery organizing franchise by the compliant state AFL. The Teamsters defeated the FTA in 1946, using the power of the truckers to stop deliveries, the financial clout of the union, and some nefarious tactics and intimidation. The FTA, which had been one of the country's largest unions, declined rapidly and fell victim to red-baiting within a rightward-moving CIO (it was expelled from the CIO in 1950 to be replaced by the much smaller United Packing House Workers). Yet, despite the perfidiousness of the AFL and the Teamsters, workers ended up with solid union locals in place and improved wages and working conditions during and after World War II.[58]

The grocery chains became major sites of unionization in the 1930s and '40s, as well. The origins of the retail clerks' union go back to the 1890s in the eastern part of the country, but its modern story takes off in the 1930s after the union was flattened by the capitalist offensive of the 1920s and the onset of the Depression. In California, efforts by shoe-store and department-store locals to organize small groceries came to naught early in the century. But the rise of big chain stores and supermarkets gave unions a new opening, and the Depression an opportunity to win over workers suffering wage cuts and very long hours; moreover, the National Labor Relations Act of 1935 (upheld by the Supreme Court in 1937) gave them the legitimacy and legal muscle to do the job. The Retail Food Clerks Union was reconstituted around the state in the early 1930s. Not surprisingly, given the presence of Safeway and Lucky Stores in the area, Local 870 in Oakland and the East Bay led the pack. As one study observed, it "earned the title of 'the pace setter in the food industry' and a reputation as the most successful union in the retail field" well into the postwar era. Its able leader, James Suffridge, went on to the presidency of the national union from 1944 to 1968, during which time the Retail Clerks International grew from 60,000 to 500,000 members.[59]

Local 870 won its first recognition and closed-shop agreement with Lucky Stores in 1935, then with Safeway in 1937, after a membership strike vote. Locals in Los Angeles, San Francisco, San Jose, and around the state quickly followed up on this success. The chains recognized the Butchers' and Teamsters' unions, as well. Safeway, which had previously been staunchly antiunion, signed on with the AFL unions, in part to keep the more radical Longshoremen and the CIO out of its domains, and in part to put pressure on its smaller competitors, who were less able to afford union wages. Another reason was to win friends during the late 1930s, when special taxes and regulations on chain stores were widely bruited in California and around the country (A&P and other big chains also settled with the unions at the time).[60]

By 1939, the Retail Food Clerks Union had more than doubled wages, and Local 870 had the highest wages of any food local in the United States; these were used as the standard by the War Labor Board to bring up wages around the country. At war's end, Local 870 was the first food union to gain wages of $1 per hour, the forty-hour/five-day week, and fringe benefits; and by 1950 it had won full medical benefits and its clerks were earning more, on average, than manufacturing workers in the East Bay. As in the canneries, a bitter contest broke out between the Retail Food Clerks and the Teamsters after the war. The Dave Beck–led Teamsters tried to bust a strike by Los Angeles retail clerks in 1947 (as they had already done to the Oakland department-store saleswomen), but Local 770 held on to win anyway. The same scenario played out in 1949 up north, with Safeway trying to break a strike and sign a sweetheart deal with the Teamsters. The strike lasted almost a year, and Safeway tried to enjoin the union under the Taft-Hartley Act, but the Retail Food Clerks held out to win—backed by an effective boycott of Safeway stores that cost the company millions.[61]

The big chains were 88 percent unionized by 1955. The main grocery segment left nonunion was ethnic stores like those of the Chinese. The meat cutters were especially strong, with strict rules against part-time workers and Sunday work. But not all workers benefited. After all, the obsolete milk wagon drivers and bakery delivery unions were decimated over the 1950s. Then Safeway broke the meat cutters union in the 1960s by introducing precut meats. The Retail Food Clerks would merge with the Amalgamated Meat Cutters in 1979, as a result, forming the United Food and Commercial Workers (UFCW). Some packing-shed union locals were destroyed by the introduction of vacuum-cooled lettuce, an innovation in the 1950s that was introduced by grower-shippers who wanted to replace relatively expensive shed labor with unorganizable, cheap bracero labor, and that was welcomed by the supermarkets for ease of handling and durability.[62]

By the late 20th century the most dynamic segments of food retailing, the fast-food chains and the big-box stores, were resolutely nonunion, with few signs of a breakthrough. The Hotel Employees and Restaurant Employees Union (HERE) has had some success in organizing large restaurants, but never the fast-food chains. Safeway, following its restructuring in 1986, moved aggressively to lower its costs, selling almost half its stores and laying off almost half its workforce—and threatening to lay off the rest if they did not offer concessions. It also moved to break the union grip by subcontracting work to nonunion companies at its regional distribution centers, as it did by moving the Northern California center out of the Bay Area to Tracy. All of this was very much in the spirit of the union-busting Reagan era. Renewed pressure on grocery workers, due to competition for union-free Wal-Mart Superstores and the recession of 2000–2003, led to a regional strike by UFCW Local 770 in Southern California that is still being played out as this is written.[63]

Field organizing languished under the bracero regime. Ernesto Galarza and other dedicated souls tried to keep the flame alive after the war under the National Farm Labor Union and its successor, the National Agricultural Workers Union; they targeted the bracero program for elimination, knowing they could not organize in the face of such an unfree labor system. New hope for field-workers came when Cesar Chavez and the National Farm Workers Association (NFWA) joined together in the early 1960s. Chavez won the backing of the American Federation of Labor national leadership and merged with the AFL's Agricultural Workers' Organizing Committee to create the United Farm Workers' Organizing Committee (UFWOC) in 1966. Chavez, with the support of the AFL, won over liberal consumers in cities around the country in a boycott of nonunion lettuce and grapes. His biggest targets were the grape grower and merchant DiGorgio and the grocery chains Safeway and Lucky. The United Farm Workers of America, which was officially chartered in 1969, succeeded in bringing more than 50,000 workers under contract by 1970. Cesar Chavez was by this time a hero of radicals all over California, and the UFW had inspired the Brown Berets and Chicanismo more generally.[64]

Yet at the moment of triumph, the UFW's position started to disintegrate. The union was undermined by a powerful growers' counteroffensive, as growers stalled on signing contracts, harassed union organizers, and changed their production practices to reduce labor demand. A second blow was competition from the Teamsters union, which was outside the AFL at the time and not averse to raiding a fellow union's turf, as it had done before in killing off

"Vineyard March," woodcut by Richard V. (Dick) Correll, 1970

UCAPAWA. Chavez turned again to outside support, winning over the governor of California, Jerry Brown, and gaining a state law recognizing agricultural workers' right to organize under the purview of an Agricultural Labor Relations Board in 1975. Another flank attack on agribusiness came with the suit against the university by California Rural Legal Assistance for failing to consider the impacts of its research on mechanization for the welfare of farm labor.

But none of this was enough, and the union shrank until it had only a few thousand workers under contract by 1980. What went wrong? It has been hard to make an objective assessment of the UFW because of Chavez's larger-than-life persona; but with his recent death, some critical voices have been heard. According to one, Frank Bardacke, a major stumbling block was internal, owing to the limits of Chavez's organizing strategy. The UFW had always divided its energies between serving the workers in the fields and appealing to the urban consumer (mostly white and upper class) for support through boycotts or political intervention. Equally, there was a failure to build internal democracy among the farmworker membership, the heart of the union. The result was a neglect of basic organizing and an inability to hold workers' allegiance; the UFW thus started losing decertification fights right and left, quite apart from the Teamsters or grower antagonism.[65] California agriculture also began to experience a major restructuring of its harvest labor force, which once again shifted toward undocumented immigrants. This weakened the social basis of the union in more settled farmworker communities, and today labor organizing is greatly complicated by the transnational lives of a large part of the workforce.

Union organizing is at the center of every history of California farm labor. One expects that "working class formation" means unionization and solidarity, but that's by no means the long and short of it. Classes coalesce around many aspects of shared experience, such as community, housing, and collective consumption. They also break apart because of sheer forces induced by such things as nationalist fervor, racial supremacy, and gender. This has too often been the tragic course of the American working class. In California agribusiness, race and gender have dogged every effort at labor organizing and working-class solidarity.[66]

The "racial fault lines" in California, as Tomás Almaguer calls them, have been many and deep. They have certainly afflicted labor organizing in fields and packing sheds. Yet permanent racial hostilities have not beset California agribusiness workers as deeply as might be expected, if only because the turnover of racial and ethnic groups has been too rapid to freeze into clear lines of fracture. If anything, the racial boundary between growers and workers has built some solidarity among field-workers, and the ethnic solidarity of new immigrants has often been a profound reservoir of strength in organizing battles (as well as in daily life). The degree of racial inclusion and alliance in the agribusiness unions has been nothing short of remarkable, all things considered. Gender fault lines have also run deep, but mostly between women workers and the union leadership above them. Women have dominated in certain parts of the agribusiness division of labor, particularly canning and packing, and have been a large part of the grocery-store workforce. They have also been considered by their bosses, their families, and themselves to be part-time labor, adding to collective household income. At the same time, women have been some of the leading militants over the last century of labor organizing.[67]

The processing sites introduced important schisms into the workforce, as well. The detailed division of labor in the canneries was closely gendered and racialized. Before 1900, Chinese men were an important part of the cannery and packinghouse workforce. After 1900, it was mostly European immigrants in the north—Italian, Greek, and Portuguese—and mostly Mexicans in Southern California. But there were many others, including Armenians, Russians, Jews, and Slavs. Inside the plants, there was further slicing and dicing according to ethnicity and race (varying from plant to plant). Anglos and men normally held all the supervisory jobs, women the bulk of the ordinary production jobs.[68]

Another divisive force ought always to be mentioned in the American, and especially the California, case. Many workers see themselves not as working class, but as potential members of the capitalist, or at least small-owner, class. Farmworkers, cannery workers, and others have often pursued the agrarian dream of becoming farmers, and have left the workforce for greener pastures rather than fight for better wages and working conditions for someone else. Farm management expert R. L. Adams complained that white workers (mostly European immigrants) had a regrettable tendency to go into farm operation themselves. Equally did Asian immigrants seek to become farmers. The Chinese had to be content with tenant farming because of racial restrictions. The Japanese almost always saw farmwork as a stepping-stone to farm operation and ownership, and they became a substantial portion of the farming class between 1900 and 1940.[69]

Don Mitchell argues rightly that geographical mobility was a strength in labor-rights efforts and organizing, because workers could withdraw their labor strategically from employers and because they created networks of information and militancy (he is thinking especially of the IWW period). But why doesn't he mention the other kind of mobility that many workers sought, class mobility? They could be quite strategic about that, too, using immigrant networks for funds and mutual aid for support in the new enterprise. This petit bourgeois maneuver is still to be found today. It can be seen among the immigrant Hmong and Vietnamese who enter farming after a period of working for others in the Central Valley. It has also cropped up among Mexican workers in the strawberry fields, where class relations were thrown into turmoil by a system of sharecropping and subcontracting set up by grower-shippers. While this was clearly meant as a strategy to thwart UFW organizing in the 1970s, it appeals to many workers who desperately want to own a little land and be their own boss. This has made union organizing difficult to this day in the strawberry sector. Another variant on this tale of upward mobility is that of the many Mexican workers who toil in the vineyards or fields in California in order to save enough to buy a farm and plant an orchard back in Mexico.[70]

Quite the opposite of mobility, but equally important, is the settlement of workers into towns, neighborhoods, and cities. Settlement, and the sense of permanence and community it provides, has often been a resource for workers and union organizers. One sees this in the success of cannery unions in San Jose and grocery clerks in the East Bay and Los Angeles in the mid-20th century. Chavez understood this in his campaign against the bracero system. In the days of the Dirty Plate route and the Wobblies, workers were more often than not tramps or migrants who spent the off-season in Oakland, San Jose, or San Francisco. Early in the 20th century, Stockton became a major depot for farm labor in the north and Los Angeles in the south, particularly for Mexicans. Those cities also served as the main canning centers in large part because of ready access to the labor of women and children.

Over time the regularization of the California rural labor system brought more working-class settlements into being up and down the state. Brawley, Corcoran, Arvin, Pixley, and El Cajon were not just agribusiness towns; they became predominantly worker towns, and this explains their scruffy, impoverished appearance—reflecting the degraded state of farm labor. Such places were not necessarily sources of working-class strength as reminders—easily internalized by workers—of class inferiority and powerlessness. This geography of class separation is deeply etched on the landscape of California, and it is especially provocative where the classes faced off at close quarters, as in Yuba City (Marysville), West Sacramento, and West Fresno. Adding to the weight of oppression is, as always, the racial face of segregation, as with the Chinatowns from Walnut Grove to Los Angeles, the Manilatowns from Salinas to San Pedro, and the barrios of Santa Barbara, East San Jose, Alvarado, and a hundred other neglected places over 150 years of agribusiness triumphant.

A century of agro-industrial labor militance has yielded a poor harvest of rewards for its ever-new / ever-same working class. A bitter fact of agrarian development in California is that farm labor conditions are not significantly better today, and in some ways are worse than ever. As Devra Weber concludes, "The glaring disparity between the bounty of the agricultural industry and the conditions and wages of its workers remains as painfully evident as it has been for the past hundred years."[71] The mass of farmworkers have not been allowed to pull up to the table and join the movable feast of California agriculture. What that has meant, besides low wages and high rates of exploitation, is the untrammeled power of the agribusiness class over production, circulation, and politics. Capitalist class formation and solidarity could hardly have been more ideal. No wonder agrarian capitalism here has performed miracles with crops, soil, and sunshine, if not for the general welfare.

Conclusion

The Conquest of Bread

At the end of the 19th century, Peter Kropotkin, one of the leading geographers, social theorists, and anarchists of his time, wrote a liberationist manifesto, *The Conquest of Bread*. The title is tantalizing to the student of agrarian development, so I dug the book up from the library, dusted it off, and read it. What I found was not unexpected. Here is a classic revolutionary political tract of a kind going back to Tom Paine, Charles Fourier, and Robert Owen, with shades of Marx and Engels's *Communist Manifesto*. Like Marx, Kopotkin was an exile from his own country and took refuge in London and Paris. Unlike Marx, he put food and rural development at the center of a communitarian vision of a better world. Kropotkin was convinced that the mass of poor people could, at last, be made free of the most persistent of human sufferings, hunger, thanks to the immense progress that had been made in the development of agriculture and industry during the previous century. He believed that it could be done through a just order in which the peasantry joined urban workers in a modern but decentralized, cooperative, and self-sufficient society.

To read the book today is to be struck by the sheer idealism of the time and the utopian faith in the revolutionary overthrow of the old order across city and country. How very different from our cynical age, in which free-market capitalism reigns supreme around the world and the American empire seems firm in its throne! Yet Kropotkin's vision came at a bad moment, too. Utopian socialism in Europe was about to expire for good in the wake of the Edwardian boom, interimperial rivalries, and descent into the Great War. Soon after that, the European powers, supreme in their control of the world's destiny, would lose their grip by dint of collective arrogance, stupidity, and bloodletting. Fortunately for Kropotkin, he did not live to see the worst of the wreckage, but he did return to Russia after the October Revolution and was growing increasingly critical of the direction it had taken by the time of his death in 1921.[1]

Kropotkin may seem a distant figure, a Russian revolutionary from the age of the czars, and his *Conquest of Bread* may seem the work of a starry-eyed idealist. But he was, in fact, well immersed in the intellectual debates and sensibilities of the time. He knew Lenin personally and came from the same milieu as Chayanov. He certainly would have known the debates between Lenin, Chayanov, and Kautsky, who are today still considered the grandfathers of agrarian theory and whose works are assigned reading for graduate students at Berkeley. His own investigations into modern and traditional agriculture in England were collected together as *Fields, Factories, and Workshops.*[2]

Why read Chayanov or Kropotkin today? Because the rural transformations taking place in Russia at the turn of the last century are at work today around the world, and the questions of peasant, farmer, and capitalist in the countryside still burn bright. People are still looking for an answer to the cruelties of modernization and expropriation in agriculture. Lenin's variety of utopianism of the vanguard party is not much in favor anymore, even on the far left. Chayanov and Kautsky, who struggled to find a middle road of collectivism with social democracy, seem more nuanced and attractive. They are not at all irrelevant to the aspirations of antiglobalist protestors harassing the great and powerful at the World Trade Organization talks in the fall of 2003 at Cancún over their dreadfully unjust and destructive agrarian policies. The political sensibilities of Kropotkin still have a place in our world, after all.

Looking back at California's triumphal march of agrarian capitalism, one catches a whiff of anarchism, socialism, and utopianism in the air blowing out in the fields and packinghouses. From Henry George to Carolyn Decker, Ricardo Flores Magón to Delores Huerte, Wobblies to National Land for the People, irrigation colonies to organic farmers, there have been visionary Californians trying to bring greater justice and conscience to the groaning table of agribusiness. They have held no less far-fetched hopes of reconciling the unreconcilable than the European radicals of Kropotkin's time. Yet none of these alternative visions has come to pass, either. Capitalism in the California countryside has steamrolled the anarchist communist, the business populist, and the farmworker organizer with equal abandon. It has been absolutely sovereign over the land, as few other agrarian capitalist orders have ever been. That has been the principal secret of its success, far more than beneficent climate, alluvial soils, or well-tamed rivers.

The result has been an astonishing fecundity of agriculture. Capitalist production in California has generated an abundance of things to eat and wear beyond the wildest dreams of Kropotkin's age. Agro-industrialism has driven the growth of the state to levels of affluence that put 19th-century Europe in the shade. How can anyone argue with that? Certainly, before we can take agribusiness to task, we have to know it and appreciate it. That has been the purpose of this book: to see beyond the supermarket shelves and sweeping vistas from Interstate 5, to look deeply into the working of this astonishing system of transforming nature into products for human use, by and for the accumulation of capital. Along the way, we have to marvel at the sheer extent of what has been achieved—beyond all the monuments of pharaohs and emperors, as Marx and Engels so rightly put it—in making the land yield up its gifts, the plants perform miracles, the animals stagger under their own weight, and the waters flow uphill. It is hard not to gawk.

Nonetheless, I come to bury the agrarian Caesar, not to praise him. In performing this anatomy of the historic body of California agribusiness, however, I have tried to follow a path that has not been well trod in critical studies of modern agriculture. I want to portray the California system as distinctive, and not readily comprehended by the tools of agrarian studies cut to fit peasant and family farming orders being invaded from outside by the market and capital. Biologist Richard Lewontin, for example, has argued, following the line of many agrarian theorists, that the natural basis of farming is a barrier to capitalist penetration of agriculture. Agriculture must, in this view, be the terrain of small commodity producers (family farmers or peasants). It is worth quoting him at length, because Lewontin's analytic powers are not to be taken lightly:

> The failure of classical capitalist concentration in farming arises from both financial and physical features of farm production. First, the ownership of farmland is unattractive to capital because it cannot be depreciated, and investment in farmland has very low liquidity as a consequence of the thin farm real estate market. Second, the labor process on very large farms is hard to control because farming operations are spatially extensive. Third, economies of scale are hard to achieve beyond what has already been realized by medium-scale enterprises. Fourth, risks from external natural events like weather, new diseases, and pests are hard to control. [Fifth], the cycle of reproduction of capital cannot be shortened because it is linked to an annual growth cycle in plants, or a fixed reproductive cycle in large animals [except poultry].[3]

The difficulties capital faces in agriculture are not to be gainsaid. Nor is the evidence that in most places, at most times, agrarian production has been the province of small commodity producers. But there's nothing that absolutely necessitates this. We have come up against each of Lewontin's five conditions in the course of this book, and none of them pertains to California agribusiness. The first does hold because the land market here has functioned very well, investment has been brisk, and land can be marketized through leasing and other financial strategies. The second has not held true because mass harvest labor has been mustered to do the job and farm management has overcome many of the difficulties of far-flung, seasonal production. The third point, regarding economies of scale, has to be balanced against economies of scope and the knowledge that capitalism works very well in medium-sized enterprises and by the integration of extensive divisions of labor by a variety of networks. The fourth and fifth pertain to the nature of farm production and, as we've seen, capitalist farmers in California have dealt with these obstacles remarkably well through manipulations of organisms, intensive feeding, pesticide application, and diversification of the whole; indeed, they have made a virtue of seeking out the fecundity of nature rather than merely seeing it as a barrier to stability. So, in the end, Lewontin's formula does not hold up.

Close observers of California agriculture have always recognized its uniquely unrestrained and prolific character, and even gone too far in taking that for granted. The main object of attention and dismay among critics has always been the harvest labor system. In this view, the unchecked power of the growers over field-workers is the crux of the matter, and gross exploitation of the most disenfranchised and ill-treated labor force in America explains the prosperity of farm and factory throughout the great valleys of the Golden State. As if the unorganized state of labor weren't bad enough, the insults are multiplied by the force of racism and the grip of the state on the throats of immigrants and braceros. This labor-centered sensibility has only been compounded by the sheer size of property holdings and the immensity of the land, giving particular weight to the idea of monopoly in landed property as the essential injustice of the agrarian system. The focus on scale and exploitation carries over to the environmental critics of the plumbing system built for the enrichment of agribusiness, even though nature more than workers gets the raw deal in their view.

Nonetheless, the classic attacks on agribusiness, for all their moral force and well-documented revelations of inhumanity, are wrong about what ails California. It is not the imperfect nature of the beast, in its monstrous command of the land and waters or devouring of human beings, that is at issue. After all, few landowners or canning companies here have come close to the bloated size of a Carnegie, Rockefeller, or Bill Gates lording it over their industry and workers. Why so much moral outrage about Bud Antle, by comparison? Nor is California beset by a colonial capitalism dropped in from another world as in many places in the global south. No, capitalism has been there from the beginning of modern California, boring from within, spreading its tentacles smoothly across city and countryside alike, fixing its grip on one and all, workers, rivers, and machines. That emergent production system has grown up big and strong and healthy in the summer sun, achieving a degree of bronzed, agrarian perfection that is hard to find anywhere else on earth in the three hundred years after the English revolution set loose the beast upon the globe.

If the problem is perfection, not failure of laws, governments, organizers, or environmentalists to stop a mutant form of agrarianism, then we have a different kind of opponent on our hands. A problem very close to that faced by antiglobalists camped out on the beaches of Cancún or facing down the police guarding Seattle's trade meetings in 1999 or Milan's in 2002. That is, how to deal with a capitalism without restraint, without a cold war antagonist, without antipode or opposition in an age of market absolutism? No wonder Karl Polanyi, a fine critic of Britain's era of laissez-faire capitalism, is widely cited again today.[4] What we see in California agribusiness is just such an unrestrained, naked form of market society that Polanyi railed against. This has been particularly telling in the ferocious exploitation of harvest labor, but equally has agribusiness drained and poisoned waters; forceably reengineered and fed the plants and animals; and changed forever the landscape of the Golden State. What can we do to stop the juggernaut that cuts through people and countryside alike, while still enjoying the benefits of plenitude from the land? How to halt the conquest and still make enough bread? I don't pretend to have the answer to that, but at least in the world of today we are again free to ask the question as Kropotkin's generation hoped to do. And maybe we'll hit on a solution that has more than a grain of Kropotkin's utopian vision in it.

Notes

Introduction: Unlocking the Secrets of Agribusiness

1. Parsons 1986, 375; income figures, Olmstead & Rhode 1988, 92–93; output data from California Department of Food and Agriculture data for Y2000 and U.S. Census Bureau 1928, 5, for 1925. The total value for 1980 was $13.5 billion; Scheuring 1983, 34.

2. Lenin 1899, Mann 1990, Goodman & Redclift 1991, Friedman 1993, McMichael 1994, Schlosser 2001, Pollan 2002.

3. George 1879, Norris 1901, McWilliams 1939, Worster 1985, Ruiz 1987, and Wells 1996.

4. Jelinek 1982.

5. Smith 1776, Kloppenburg 1988. Alas, Kloppenburg also skates by California.

6. On land and labor, see, e.g., George 1871, Gates 1967, Steinbeck 1939, McWilliams 1939, Levy 1975.

7. For the classic debate, see Lenin 1964 [1899], Kautsky 1988 [1902], Chayanov 1986 [1923]. For a review, see Watts 1996. On growers, see Adams 1921, Stoll 1998; also Daniels 1981, Moses 1995.

8. On nature in production, see Mann and Dickinson 1978, Mann 1990; also Berlan & Lewontin 1986, Goodman & Redclift 1991, Goodman & Watts 1994. On nature and California farming, see Sawyer 1998, Guthman 2004; also Henderson 1999, Worster 1985.

9. Having a natural basis does not distinguish farming from other lines of business, such as metallurgy and chemicals, which have to solve deep puzzles of nature just as surely as does agriculture. In fact, *all* industries are distinctive by virtue of their different natural and social bases, not just agriculture. Agriculture certainly deserves special consideration, but ought not to be relegated to a separate discipline any longer. For further discussion of this point, see Freeman 1982, Storper & Walker 1989.

10. Figures on shares pertain to the entire United States, from Lewontin 1998; around 1900 the on-farm share was two-fifths. On commodity chains, see Friedland et al. 1981,

especially the chart on 16; Friedland 1984, 1994. On the social division of labor in agribusiness, see Goodman et al. 1987; also Kloppenburg 1988, Cronon 1991, and Fine 1994. Frank Norris ought to get credit for his pioneering effort to think through the agrarian system from the fields to the markets in *The Pit* (1906). Agrarian theories were using commodity chain analysis long before it hit the international literature on industrial systems through Gereffi and Korzeniewicz 1994.

11. On the complexity of production chains, see Sayer & Walker 1992, Page & Walker 1991. I have thought a good deal about the question of the social division of labor in modern economics, and this study is a natural extension of that concern. On California's water engineering, see Worster 1985, Reisner 1986. On the drenching with pesticides, see van den Bosch 1978.

12. On California co-ops, see Stoll 1998, Woeste 1998. Examples of contemporary consumption studies are Miller 1995, Fine & Leopold 1992, Frykman & Lofgren 1987, Lears 1988, Cohen 2003. It should also be said that not all food chains end with final consumption, since a good deal of food ends up as decommodified surplus; on this point, see Henderson 2004.

13. For estimates of off-farm agrarian production, see Goodman et al. 1987, 54; Busch et al. 1991, 23. Quote from Jelinek 1982, 61. Kautsky 1988 [1902] was probably the first to bring up the relation of fields and factories.

14. Post 1982; Olmstead & Rhode 1988, 1997. On Midwestern agro-industrialization, see Page & Walker 1991. On California's resource-industrial path of development, see Walker 2001. For a very general discussion of technological change in modernization, see Mokyr 1990.

15. Chandler 1977. On Miller and Lux, see Igler 2001. On recent developments in business organization, see Powell & Dimaggio 1991, Sayer & Walker 1992, Storper 1997, and Chandler 1990.

16. FitzSimmons 1986. For examples of critics trying to apply the corporate model to California agribusiness, see Hightower 1975, Jelinek 1982, Pisani 1984, Burbach & Flynn 1980, Tobey & Wetherell 1995, Moses 1995. On food systems theory, see Friedmann 1993, McMichael 1994, Wrigley & Lowe 1996. It is telling that Goodman et al. 1987, writing in the 1980s, largely ignored the problem of business organization.

17. Henderson 1999.

18. On the theory of class formation, see Przyworski 1976, Sayer & Walker 1992. This is quite different from simply speaking of class *position* in the manner of Wright 1985.

19. On Weberian theories of the state and development, see Skocpol 1979. For Marxist views, see Miliband 1969, Poulantzas 1973. For more current hybrid theories, see Jessop 1990, Evans 1995. On the unique character of the American state, I have a few things to say in Walker & Heiman 1981, Walker 2001. But my greatest education has come from a recent student, Kathy Johnson (2003).

20. Mitchell 1996.

21. George 1871, McWilliams 1939, Worster 1985, Friedland et al. 1981, FitzSimmons 1986, Stoll 1998, Olmstead & Rhode 1997, Vaught 1999, Mitchell 1996, Wells 1996.
22. The term "agribusiness" was invented by Davis and Goldberg (1957) at the Harvard Business School, according to Bill Friedland (personal communication).
23. Scheuring 1983, 2–3.
24. In this I follow Karl Marx in *Capital,* 1863, 1897. Marx was a fine economist, from whom we still can learn, whatever one thinks of his immense and controversial political legacy. For brilliant expositions of Marx's concepts in a more contemporary guise, see Harvey 1982 (2002) and Brenner 1986, 2002.
25. On the question of structure and agency that dogs all social science, of whatever economic, sociological, or political stripe, see Sayer 1992.
26. For recent national comparisons, see Chandler 1990, Herrigel 1986, Coates 1999.
27. On paths of agrarian development, see Moore 1966, Foweraker 1981, Byres 1995, Hart 1999. For an instructive comparison of the logic of capitalist farming with family and slave modes of agrarian production across the United States, see Post 1982, 1995, and 2001.

1: Cornucopia

1. Marx 1864, Smith 1776, Scheuring et al. 1981. On Smith's practical experience, see McNally 1988.
2. Steinbeck 1939. On promotion of California, see Orsi 1973.
3. For a description of the PPIE, see Todd 1921—on which I rely for the following vignettes.
4. On PPIE and San Francisco's imperial pretensions, see Brechin 1999. Alas for San Francisco, Los Angeles was in the process of surpassing it as the biggest metropolis of the Pacific Coast. Walker 1996. On world's fairs in general, see Rydell 1984, Heller 1999.
5. Quote from Marx and Engels 1848. See also Berman 1982, Harvey 1982. On the restoration of the Palace of Fine Arts, see Brechin 1983.
6. Figures from Gates 1967, 62; Hart et al. 1946, 53; Skaggs 1986, 132. On the 19th-century cattle industry, see Eberling 1979, 346; Cleland 1941; Zonlight 1979; Igler 2001. There is no history of sheep raising in California, nor does there seem to be one for the American West as a whole (personal communication with Nathan Sayre), but see Burcham 1957. Ostriches and camels were also tried, to no avail.
7. Figures from Hart et al. 1946, 55; *Pacific Rural Press* 1921; Olmstead & Rhode 1997, 12. There is little on the history of California dairying, but see Hart et al. 1946, 88. California never had many hogs, but Fontana Farms, using Los Angeles's garbage, was for a time the largest hog-feeding operation in the world in the early 20th century. Hart et al. 1946, 95.
8. Data from Olmstead & Rhode 1997, fig. 3, 13; Scheuring 1983, 111ff; Bell 1995.
9. Figures on wheat and barley from Gates 1967; Paul 1968, 1973; Olmstead & Rhode

1997. See also Rothstein 1990. Together barley and wheat in the 1850s and '60s yielded around $70 million to the state's economy. Calculation from figures in Gates 1967, table 3, 52. For typical comments on wheat, see Olmstead & Rhode 1997, 2; Scheuring 1983, 113.

10. Pudup & Watts 1987, 356; Liebman 1983, 89.

11. Fruit data from Adams 1946, 38–39; Stoll 1998, 4; Jelinek 1982, 49–50; Rhode 1995.

12. Mills quote from Stoll 1998, 25. Figures from Vaught 1999, 14; Olmstead & Rhode 1997, 6; Tufts et al. 1946, 155.

13. On the new exotics, see Guthman 2004, Rosenblum 1996. On wine, see Guthey 2004.

14. Tufts et al. 1946, 132–34. I cannot agree with Henderson 1999, 8 (following McWilliams 1939) that sugar beets were "transitional" between wheat and horticulture; they were simply one among many boom crops in California history.

15. Weber 1994, 21; Musoke & Olmstead 1982; Arax & Wartzman 2003.

16. Scheuring 1983, 111, 128.

17. Figures on vegetables (including beans) from Liebman 1983, 48–49, 62; *Pacific Rural Press* 1921; Wickson 1923, 13–15.

18. Figures on tomatoes from Collins et al. 1959, 27; Friedland & Barton 1975; Liebman 1983, 141; Scheuring 1983, 163–64.

19. Figures on strawberries, Wells 1996, 25; on vegetables from Carter & Goldman 1997, 33. On exotic vegetables and organics, see Palerm 1991, Guthman 2004.

20. Figures from Scheuring 1983, 216. On landscape gardening in California, see Padilla 1961, Church 1995, Streatfield 1994.

21. Paul 1973, 19.

22. Fabian 1869, 20.

23. I have compiled this composite of Bay and Delta agriculture from various sources, including Cronise 1868, Baker 1914, Hansen & Miller 1962, Minnick 1969, Paul 1973, Mason & Park 1975, Hornbeck 1983, Jacobsen 1984, Payne 1987, Simone 1990, Vaught 1999, and Chan 1986.

24. Sources on the San Joaquin include Preston 1981, Johnson et al. 1993, Weber 1994, Vaught 1999, and Igler 2001.

25. On Southern California agricultural history in general, see Cleland 1941, Cuthbertson 1955, Fielding 1964, Tobey & Wetherell 1995, Rudy 1995, Palerm 1991, Sawyer 1996, Stoll 1998.

26. On the Sacramento Valley, see Paul 1958, Pudup & Watts 1987, Kelley 1959, 1989, Green 2000.

27. On the Central Coast see Cuthbertson 1955, Modry 1970, Friedland et al. 1981, Palerm 1991, Bardacke 1994.

2: A Landscape of Commodities

1. Marx 1867, ch. 26–30.

2. Nash 1972. On the Gold Rush era, see Paul 1973. Figures from Gates 1967, 62; Walker 2001.

3. On California agriculture given by nature's gifts, see Hodgson 1933; Parsons 1955; Scheuring 1983, 21–95; Musoke & Olmstead 1982. On California's natural setting, see Bakker 1971, Schoenherr et al. 1998, Dallman 1998. Until the middle of the 20th century, the climate of California (especially Southern California) was called "subtropical," but it is properly classified as Mediterranean. The compelling arguments on agricultural origins by Diamond 1997 run into trouble as soon as more complex social orders come into being; it only gets worse as we move into modern times.

4. As Stoll 1998 points out, other states grew many of the same crops as California; so why did this state triumph in national output? Some comparative figures on tree planting are given in Vaught 1999, 105. I remember being startled, when I left California for school in the East, to discover that strawberries, peaches, corn, and other crops grew brilliantly and more deliciously elsewhere. I had been completely sold the myth of California fecundity as a child.

5. On this point, I differ with most Marxist analysis, which leans on productivity growth as *the* logic of capitalist development, as, e.g., Brenner 1986. While one could include product proliferation loosely under the heading of "rising productivity," on closer inspection it must be dealt with as a distinct phenomenon.

6. Mitchell 1996, 58; Crocheron quoted in Scheuring 1988, 24; figure from 1940 census, Hutchinson 1946, vii. There have been some exceptions to California monoculture, such as diversified garden farms near the cities in the 19th century. Tomato growers in 1950 still produced some fifty other crops, usually three or four each in rotation. Collins et al. 1959. Similarly, half the farms still kept cows as late as 1945. Hart et al. 1946, 84.

7. Alston & Pardee 1996, 98. Quote on diversification from Jelinek 1982, 62.

8. On the capitalist logic of overinvestment and overproduction, see Harvey 1982, Walker 1995, Brenner 2002.

9. A big reason for the passing of California's age of wheat was declining prices in Liverpool, as more of the world's grasslands fell to the plow and production glutted the market. Yet nowhere else was wheat abandoned so thoroughly as here. Why? Competition within the Pacific basin put a further squeeze on California wheat growers, first from the Pacific Northwest growers after 1890, then from Chinese millers after 1900, especially with the 1905 Chinese boycott of American goods, which took away roughly half the California flour market. Meisner 1997–98, 91–92. A third reason is soil exhaustion, which was undoubtedly severe; but barley, grown in the same fashion, continued to prosper until horses disappeared. More important was that agriculturalists and investors discovered there was greater gain to be had from fruits than from grain, and the state underwent a dramatic makeover to horticulture.

10. Paul 1973, Stoll 1998, Henderson 1999, Vaught 1999.
11. Olmstead & Rhode 1997, 5; Henderson 1999, 9.
12. 1975 figures from Eberling 1979, 379.
13. Carter & Goldman 1997, 33, 36.
14. Olmstead & Rhode 1997, 4. Similar two-stage models can be found in Nash 1964, Jelinek 1982, Henderson 1999, and Stoll 1998.
15. Henderson 1999, 3.
16. On Indian removal, see Almaguer 1994, Hurtado 1988.
17. On the expropriation of the Californios, see Pitt 1966, Almaguer 1994. Race also played a role, as one local example will suffice to show: ranchero Jose Reyes Berryessa in the San Jose area was lynched on trumped-up charges in a squatter's war; when Berryessa's claims were denied in court in 1869, the family was ruined. None of this happened to Sunnyvale's Murphy family, equally large claimants from the Mexican era who were of Irish origin.
18. Figures from Liebman 1983, 9, 20, 30. On public land disposal, see Gates 1991, Robinson 1948.
19. Paul 1973, 20, 23; Hittell 1863, 162. Paul (1973, 23) contradicts himself by saying, "The curious thing is that so many people recognized from almost the beginning that the wheat ranches were a temporary phenomenon." Paul also leans toward a more cultural explanation than I accept, quoting the *Alta California* from 1855: "[The farmer] is very likely too impatient to plant fruit trees and vines. He can hardly think of waiting for them to grow. No! He must get his returns immediately. He came to California for a fortune, and if he is not making it rapidly he is discontented and eager to try something else." Paul 1973, 20. On Argentina, see Sabato 1988, Johns 1992.
20. Sayre 2002, 52, calls such open range a transitional hybrid from pastoralism to capitalist ranching.
21. On land concentration see George 1871, McWilliams 1949, Zonlight 1979, Liebman 1983, Gates 1991, Rothstein 1995. Farm size from Olmstead & Rhode 1988, 89 (from Historical Statistics of the United States).
22. Quote from Stoll 1998, 33. Farm size figures from Olmstead & Rhode 1988, 89. The sell-off of large holdings is discussed in Liebman 1983, 51–56, and Henderson 1999, 152–60. See also Dumke 1944, McWilliams 1946, Davis 1990, Jacobsen 1984. Recall that the old Mexican regime lasted longer in the South and only broke down completely in the 1860s.
23. On the Sacramento River and the Delta, see Kelley 1959, 1989; Thompson 1957; Minick 1969; Liebman 1983; Chan 1986. On the Tulare Basin and southern San Joaquin, see Villarejo 1980, Preston 1981, Igler 2001, Arax & Wartzman 2003. On the Imperial Valley, see Rudy 1995.
24. On leasing see Villarejo 1980, Pudup & Watts 1987, Wells 1996, Alvarez 2001.
25. Sinclair 1927.
26. For a fuller development of this argument, see Guthman 2004. Even where grow-

ers own their own land, they have usually remortgaged it to raise funds for operations, and they feel the cost of land as interest payments on the loans.

27. One of the weaknesses of the Californios in holding off the Americans was the thinness of the ranchero class and its alienation from Indian workers. On Indian workers, see Almaguer 1994.

28. Quotes by Jelinek 1982, 53; Parsons 1955, 46–47; Henderson 1999, 83. See also Mitchell 1995, ch. 4 on "marked bodies." On labor generally, see Parker 1920, McWilliams 1939, Fuller 1934, Fisher 1953, Daniel 1982, Jelinek 1982, Gonzales 1985, Chan 1986, Martin 1987, Mitchell 1996, Vaught 1999.

29. Polanyi 1944.

30. There is relatively little primary research on 19th-century farm labor, but see Dodge 1885, diag. 5, 13; Chan 1986, Woriol 1992, Vaught 1999, Igler 2001.

31. Wage figures from Gordon 1954, 7. Wage statistics may well be distorted by an undercounting of tramp labor, however.

32. On George's racism, see Mitchell 1996, 93; for the refutation of the coolie theory, see Chan 1986, 318, 323, 328; on Chinese in horticulture generally, see 233, 240–49. Also see Lukes & Okihiro 1985, 12, 20, 48. Chinese earned about two-thirds what whites made, but there is no clear pattern of higher or lower wages being paid depending on farm size.

33. Fisher 1953, 2; Liebman 1983, 63ff, 80; Vaught 1999, 9–10. Vaught is confusing the family labor of the smallest orchardists with the advancing front of horticulture, which had many medium-sized and well-capitalized growers. On the skills of fruit workers, see Vaught 1999, 70–71.

34. On Japanese workers, see Iwata 1962. On anti-Japanese sentiment and laws, see Daniels 1977.

35. On Filipinos, see DeWitt 1980, Crouchett 1982. On Italians, DiLeanardo 1984, Matthews 2003.

36. On this era of labor recruitment and migration, see Mitchell 1996; also Daniel 1982, Liebman 1983. On San Francisco housing, which fed into the lumber industry, construction, shipping, and longshoring, see Groth 1994. On tramp labor in the West, see Woriol 1992.

37. Labor force figures from Mitchell 1996, 59. Early harvest labor estimates are surely undercounts. Three-fifths figure from Henderson 1999, 83. The pioneering studies of harvest labor were Fuller 1934, Taylor & Vesey 1936, and Fisher 1953, complemented by McWilliams 1939 and Steinbeck 1939.

38. Figures on Mexican workers from Jelinek 1982, 68, 70; Mitchell 1996, 91. Figure for cotton pickers in 1925 and percentage of Mexicans in cotton from Weber 1994, 34–35. On special immigration provisions, see Mitchell 1996, 88. On Mexican workers generally, see McWilliams 1961 ed., Galarza 1964, Weber 1994.

39. Immigration figures from Weber 1994, 138, citing the U.S. Congress's Tolan Committee report. Harvest force figure from Adams 1938, 25, cited in Liebman 1983, 110 n.

60. On the Okies generally, see Gregory 1991, Haslam 1974. On Mexican deportations, see Hoffman 1974, Guerin-Gonzales 1994.
40. On the braceros, see Galarza 1964, Calavita 1992. Figures from Calavita 1992, 21, 55, 72, 144. There was a lapse in the bracero program in 1948–51 that was filled by legalization of and failure to pursue undocumented workers by the INS. There was also a mass roundup of illegal immigrants under Operation Wetback, in 1954–55, of perhaps a million people. This barely dented the farm labor supply, which was amply filled by braceros—many of whom were former wetbacks legalized by the INS. Calavita 1992, 25–27, 54–55.
41. Wage data reported in Calavita 1992, 70–71; quote on sprinklers at 58; citations to the two government reports at 44 and 71.
42. Calavita 1992, chap. 6. See also Massey et al. 1987, Nevins 2002.
43. Roybal quote from Calavita 1992, 8. On Operation Gatekeeper and its antecedents, see Nevins 2002. On the agricultural H-2A visas and the possibility of a new bracero program, see Bacon 2002.
44. Sources on the contemporary labor force are Employment Development Department 1992, Zabin et al. 1992, Villarejo & Runston 1993, Martin 1987, Rudy 1995, Wells 1996, Olmstead & Rhode 1997.
45. Fuller 1934 is the classic statement on California farm labor; he was a student, in turn, of Berkeley professor Paul Taylor, the pioneering labor economist and critic of southwestern agriculture. See Taylor & Vesey 1936, Taylor 1979 for his collected essays. See also Fisher 1953, Daniel 1982.
46. Mitchell 1996, 59.
47. Again, see Guthman 2004 for a fuller exegesis of this process. Thanks to Frank Bardacke for pointing out the direct evidence for a land value–wage trade-off in the 1970s.

3: Enter the Grower

1. Quote from McWilliams 1949, 4. His radical contemporaries held the same view, e.g., Taylor 1979, Goldschmidt 1947, Wilson & Clawson 1945.
2. Vaught 1999 goes on to argue that the larger growers eventually "undermined the dream" of the smaller horticulturalists. For a similar attack on McWilliams, see Woeste 1998, 10–11.
3. McWilliams 1949, 1. On the importance of agrarian capitalism in the rise of modern Europe, see Ashton & Philpin 1985, McNally 1988, Jones 1974. On the debates over class and capitalism in the countryside, see Lenin 1964 [1899], Kautsky 1988 [1902], Moore 1966, Watts 1996. As P.K. O'Brien (1977, 168) drily observes, "The relationship between property rights and capital formation in agriculture invites comparisons across systems and countries." California needs to be among those systems compared.
4. On the family farm settlement of the United States, see Smith 1950, Gates 1960, Robbins 1976, Danhof 1979, Cochrane 1979, Page & Walker 1991. On the U.S. agrarian transition of the 1840s, see Post 1995.

5. On these crucial distinctions, see Robbins 1976, Friedmann 1978, Post 1995.
6. Quote on early entrants from Paul 1973, 18. On citrus capitalists, see Stoll 1998, 32; Moses 1995, 26. Moses draws on the observations of Charles Lummis, E.J. Wickson, and Carey McWilliams, among others. The "business on an industrial basis" quote is from Daniel 1982, 43. Giannini quote in James & James 1954, 81, cited by Henderson 1999, 75. SPRR quote from 1910 pamphlet, cited by Rudy 1995, 145.
7. Quote from Hodgson 1933, 354. Given what she documents about Fresno's raisin growers, it is surprising that Woeste 1998 continues to equate them with family farmers across the country.
8. I have taken these examples from Jacobsen 1984, Couchman 1967, Payne 1987, Vaught 1999, Stoll 1998, Moses 1995, Tobey & Wetherell 1995, Taylor 2000. California agrarian novels of the period capture this elision of urbanists and agriculturalists, as shown by Henderson 1999. Davis farm school figures from Scheuring 2001, 31.
9. Chan 1986, 380; Matsumoto 1993; Iwata 1992. Wickson (1923, 11) complained, in the racist mode of the time, that "the competition which the American [*sic*] grower has to encounter [in vegetables] is depressing and discouraging."
10. Number of immigrant farmers from Eberling 1979, 336. On Italian farmers, see Matthews 2003. On Jewish farmers, see Kann 1993. On Armenians, see Woeste 1998, 53–59. On Giannini's policies, see James & James 1954, Nash 1992. Source on present-day immigrants: 1997 agricultural census of California. Out of 74,000 farms, there are 3,400 Southeast Asian, 4,500 Latino, 277 African American, and 524 Native American owners. Half the small farms in Fresno County are owned by nonwhites, mostly immigrants. On the Portuguese, see Graves 2003.
11. Compare Daniel 1982, 17–18. Quotes by Vaught 1999, 53, 4. Vaught contradicts himself again by declaring, "In actuality, horticulture was no more virtuous an understaking than hopculture, wheat growing, or any other type of farming in California." Ibid., 45. On Boswell, see Arax & Wartzman 2003.
12. Figures on California farm profits are scattered through the literature, although no systematic study exists. See, e.g., Cronise 1868, Wickson 1912 and 1923, Adams 1921 and 1948, Stoll 1998, Moses 1995, Payne 1987, Vaught 1999, Woeste 1998.
13. Quote by Daniel 1982, 36. Figures from Mitchell 1996, 116, and U.S. Census of Agriculture. On capital intensity of orchards, see also Tobey & Wetherell 1995, Stoll 1998.
14. Similarly, California would be a state lacking in small towns, as discussed in chapter 7.
15. On economies of scope, see Scott 1988, Storper & Walker 1989, Chandler 1990, Saxenian 1994.
16. Figures on size by county, 1872, in Liebman 1983, 22. See also Preston 1971, Zonlight 1979, Green 2000.
17. On land subdivision, see Liebman 1983. Census figures from Jelinek 1982, 53. On *latifundists*, see Moore 1966.

Santa Clara County farm sizes

YEAR	<100 ACRES	>100 ACRES
1880	721	771
1900	3,057	38
1920	4,390	626

Source: Lukes & Okihiro 1985, 15.

18. Some of this involved legal skulduggery and raw political power. Some of it turned on land prices, higher in California than elsewhere around the West, which may have diverted settlers north to Oregon in the early years. Gates 1967, 1991. See also Jelinek 1982, Liebman 1983, Gates 1991.
19. Norris 1901, Lukes & Okihiro 1985, ch. 4, Woeste 1998, ch. 6.
20. Quote by Daniel 1982, 43. See also Daniel 1982, 37, on the paradox of horticulturalists' criticizing wheat farmers for the same speculative, entrepreneurial behaviors they themselves engaged in. On the national debate, see Smith 1950, Saunders 1999.
21. On the American debate and its Jeffersonian mythology, see Smith 1950. On the second colonization movement, see Requa & Cory 1919, Fiske 1979, Pisani 1984. Henderson (1999) is especially good at capturing the internal condictions of this aspect of California dreaming in agrarian novels of the time. Notably, the rate of tenancy, 15 percent in 1925, was far lower than any other section except the far Northeast. U.S. Bureau of the Census 1928, 5.
22. For example, Daniel 1982, 43, states that "after 1900 family farming survived only as a marginal appendage of a rural economy dominated in fact and in spirit by agribusinessmen as single-minded in the pursuit of profits as the most unwavering urban capitalist." On Boswell and the cotton wars, see Arax & Wartzman 2003.
23. On farm size in California, the most thorough study is Liebman 1988. For a comparison with farm size expansion across the United States in 1939–69, see Cochrane 1979, table 133.
24. On citrus see Stoll 1998, Tobey & Wetherell 1995; on dairies see Gregor 1963, Gilbert & Akor 1988; on rice see Pudup & Watts 1987; on strawberries see Wells 1996; on tomatoes see Collins et al. 1959, Brandt et al. 1978, Smith 2000. On the decline of the ranches, see Skaggs 1986, Starrs 1998.
25. Organic farming is treated exhaustively by Guthman 2004. On wine, see Conoway 1990, Eysberg 1990, Guthey 2004. On Jackson, see Murphy 2004.
26. Guthman 2004, Arax & Wartzman 2003.
27. For a general discussion of farm management, see Cochrane 1979, 382ff. Yet the subject has no historian like Chandler 1962 or Pollard 1968.
28. Liebman 1983, 122. For an example of this in the case of Yolo County, see Vaught 1999.
29. My childhood best friend's grandparents went from being hardscrabble Irish farm-

ers in the Salinas Valley to being wealthy retirees living off their leases in the 1950s and '60s. On early Japanese and Chinese leasing, see Iwata 1962, Chan 1986. Adams deals with leasing in several pamphlets and in Adams 1921, 1952.

30. Iwata 1962, 31, 33.

31. On farm contracting, see Scheuring 1983, 266–67, FitzSimmons 1986. On specialization and interfirm contracting in industry, see Scott 1988, Powell & Dimaggio 1991, Storper & Walker 1989, Storper & Salais 1997.

32. FitzSimmons 1986. See Saxenian 1994, Storper 1996, and Porter 1998 on industrial districts. Guthey 2004 is one of the few to apply industrial district theory to agriculture.

33. Scheuring 1983, 265.

34. On poultry today, see Boyd 2001, Boyd & Watts 1997. For the takeover thesis, see Lewontin 1998.

35. On precision farming, see National Research Council 1997, Pierce & Sadler 1997.

36. California growers were not that different from farmers in the rest of the country in this love of reading and improvement. Wharton 1974. They shared a love of the press similar to their brethren in San Francisco, a famously well-published and well-read city of the 19th century. On the San Jose station, see Arbuckle 1986.

37. On Hilgard see Jenny 1961. On the founding of the agriculture college, see Wickson 1918; Hutchinson 1946; Stadtman 1970, ch. 10; Fiske 1979; Scheuring 2001.

38. California had the second state agricultural research program, after Connecticut. The Farmers' Institute idea began in Wisconsin in the 1860s and diffused through Granger and Farmers' Alliance agitation, getting federal backing in the Second Hatch Act of 1890. Agricultural extension was initiated at Cornell in 1894, secured its first regular support from the California legislature in 1903, and became federal policy in 1914. Kerr 1987, Scheuring 1988.

39. Quote from Tufts et al. 1946, 151. For examples of Wickson's books, see Wickson 1923 [1897], 1912 [1889]. Hilgard has a hall named after him on the Berkeley campus, Wickson one on the Davis campus.

40. Extension publications by the 1910s covered just about every crop and every aspect of farming one could think of. See, e.g., Bioletti 1899, 1911; Adams 1917.

41. On the national history, see Busch & Lacy 1983, Kerr 1987, Alston & Pardey 1996.

42. The county farm agent idea came out of Texas in the 1900s. Crocheron was brought in from the East Coast. On Crocheron, see Scheuring 1988, 2001.

43. On the Davis farm and campus, and much besides, see Scheuring 2001, 50–51.

44. See, e.g., Adams 1921, 1948, 1952. Quote from Adams 1921, 23; comments on farmers, 23; comments on diversification, 51; profit analysis, chap. 12 (the correct method of calculating profit on capital investment was only worked out by Pierre Dupont circa 1923, as shown by Chandler 1962). Like other UC educators on agriculture to come, Adams was a national figure who wrote for a national audience. He makes little direct reference to California, except in his pamphlets for state extension work.

45. Scheuring 1988, 28, confirms the outpouring of work in agricultural economics at the college and by the Giannini Foundation.
46. See, e.g., Scheuring 1983, National Research Council 1997, Pierce & Sadler 1997, Siebert 1997.

4: Down on the Farm

1. On nature and agriculture, see Mann & Dickinson 1978, Goodman et al. 1987, Mann 1990, Goodman & Watts 1994.
2. See especially Mann & Dickinson 1978, Berlan & Lewontin 1986, Lewontin 1998.
3. Kloppenburg 1988.
4. There is no systematic history of California plant and animal breeding that I can find. But see Wickson 1921 and, for ornamental horticulture, see Cuthbertson 1955.
5. On livestock introductions, see Gates 1967; Hart et al. 1946; Jelinek 1982, 24, 33. The new breeds, such as Angus and Hereford, had been developed mostly in Britain in the 18th and early 19th centuries. See Trow-Smith 1959. On cattle introductions to the United States, see Frink et al. 1956, Skaggs 1986, Jordan 1993, Sayre 2002.
6. On the legacy of mission gardens and vineyards, see Hedrick 1950, 364–70. Early Anglo settlers, such as Bidwell, Yount, and Wolfskill, played a part in this.
7. Hedrick 1950, Jacobsen 1984, Tufts et al. 1946, Taylor 2001. Couchman 1967, 10, provides a list of Bay and river area nurseries; San Jose had by far the most. Charles Shinn's *Pacific Rural Handbook* (1879) was the first real guide to horticulture in California.
8. Quotes by Paul 1973, 24, and Wickson 1921, 12. Figures on introductions from Cronise 1868. On the worldwide collection, study, and diffusion of plants, see Brockway 2002. George Vancouver's surgeon collected plants for Kew from California in the late 18th century.
9. Quotes from Wickson 1921, 12; Gates 1967, 73–74. On introductions generally, see Wickson 1921; Tufts et al. 1946, 170–238.
10. Flint quoted in Paul 1973, 25. On state bounties for crops, see Nash 1964. On Haraszthy, see McGinty 1998. On strawberries in California, see Wilhelm & Sagan 1974, 163–232.
11. On oranges see Stoll 1998. On olive introductions, see Taylor 2000, 82, et passim. On avocado introductions, see Charles 2002.
12. See, e.g., Hirabayashi 1989, Rosenblum 1996, Chan 1986.
13. On British and French breeding in the 18th century, see Trow-Smith 1959, Wilhelm & Sagan 1974, Mukerji 1997. On early U.S. plant breeding, see Hedrick 1950, Bailey 1917, Pollan 2001. Americans were especially successful with berries, plums, apples, and grapes, and with tomatoes, corn, and beans—all American natives. There is no modern history of agro-horticulture in the United States, unfortunately.
14. *Pacific Rural Press*, January 1, 1921; Tufts et al. 1946, 211, 171, 181; Wickson 1921, 47; Wihelm & Sagan 1974, 176, 191. On California hybrids more generally, see Scheuring 1983, 142. On citrus breeds, see Webber & Batchelor 1943, Tobey & Wetherell 1995.

15. Thanks to Nathan Sayre, personal communication, for an understanding of early breeding practices. Information on early breeding in California is sketchy, and I have only found mention here and there of individual breeders. Nor is there an overall history of U.S. livestock breeding. Sayre has located a couple of bulletins, to wit, Curtiss 1898, Clay 1899.
16. On Burbank, there are any number of glowing contemporary accounts and later evaluations. See, e.g., Harwood 1907, Burbank 1927, Howard 1945, Dreyer 1975.
17. On the scientific praise, see Wickson 1902, Jordan & Kellogg 1909; ironically, the scientists contributed to Burbank's later loss of face by converting him into a raving eugenicist. On Linnaeus as a Swedish nationalist and royalist, trying to adapt tropical plants to northern climes, see Koerner 1994. On genetics, Dreyer 1975 points out that the rigidly anti-Lamarckian Darwinism of the mid-20th century has given way to a more fluid view of genetic expression, recombination, and mutation that is more forgiving of Burbank's craftwork.
18. Jenny 1962, Amerine 1962, Wilhelm & Sagan 1974, 179.
19. On Hass, see Charles 2002, 145. On the lack of introductions from the UC gardens, see Wickson 1918; Tufts et al. 1946, 156. On canning peaches, see Tufts et al. 1946, 192, 194. On the Davis breeding operation, see Scheuring 2001, 33.
20. Thanks to Nathan Sayre for pointing this out to me. See also Sayre 2002.
21. Hodgson 1933, 352.
22. On the variety of soils, see Jenny et al. 1946, esp. map, 333; Scheuring 1983, 43–64.
23. Jenny 1961, quote at 36. Hilgard also had important insights on soil formation and stratification, the relation of biogeography to soils, and the general relation of climate to the characteristic soils of arid and humid regions. Hilgard was long overlooked in the history of soil science, says Jenny, because he wasn't in the "fertility mainstream." Rossiter 1975 provides a good history of these 19th-century debates on soil, particularly the vitalists versus the mineralists. On Gillet and chlorosis, see Wilhelm & Sagan 1974, 185. On Imperial and UC advisors, see Rudy 1995, 120–21.
24. Hilgard 1884, Storie & Weir 1926. See generally Jenny et al. 1946, 390–93; Scheuring 1983, 268–72.
25. On the Delta, see Chan 1986, 171–75; Chu 1970; Thompson 1957. Before widespread groundwater pumping, periodic saturation by high water tables was a considerable problem, as in the San Joaquin in the years 1914–16. Jenny et al. 1946, 365; Davis 1998, ch. 1.
26. On the Sacramento Valley flood problem, see Kelley 1989. On Japanese rice growers, see Iwata 1962; and on rice in general see Pudup & Watts 1987. On the Tulare Basin, see Preston 1981, Arax & Wartzman 2003.
27. Jenny 1961, Payne 1986, Guthey 2004.
28. On the Asian farmers' contributions, see Iwata 1962, Hirabayashi 1989, Lukes & Okihiro 1985.
29. Soil fertility from Jenny et al. 1946 and interview with Roland Meyer, Agricultural Extension, UC Davis, April 4, 2001. Hilgard quoted in Jenny et al. 370. Hilgard laid

stress on phosphate deficiency, which later research would show to be secondary. Ibid., 383–87. Potassium shortage is even less of a problem. Ibid., 387–89. On grapes, see Lapsley 1996, 46. University of California research in the 1930s showed much less response to fertilizer on deciduous fruits than citrus. It also showed up the myth of low sugar in fertilized beets. Jenny et al. 1946, 377–79. Data for regional fertilizer use from U.S. Census Bureau 1914, plate 358.

30. Worster 1985. See also Pisani 1984 for a more buttoned-down history of water from the bottom up. Hundley 1992, on the other hand, so drains irrigation history of political opinion as to make a scholarly desert of it, despite his prodigious learning. It only shows how worthy Worster's grand errors are, by comparison.

31. On California climate and water resources, see Kahrl 1979. See Davis 1998, ch. 1, on the mismatch of American perceptions and California water realities. On Imperial's problems, see Rudy 1995.

32. This history is laid out in wonderful detail by Whorton (1974), who shows that the world Rachel Carson described had arrived long before DDT. He also notes that honeybees were roundly poisoned intentionally by farmers before entomologists demonstrated their beneficial role in pollination.

33. The citrus story, along with the biological control of the mealy bug, is well told by Sawyer 1996.

34. Jacobsen 1984, 167; Wilson 1998, 47–49.

35. On UC research projects and funding, see Jenny 1961; Scheuring 2001, 15. On Powell and citrus, see Moses 1995, Sawyer 1996. The Davis campus did little entolomology or plant pathology work before World War II, it seems.

36. See generally Smith et al. 1946; Scheuring 1983, 273. On the sulfur ban see Couchman 1967, 35; and on the lead danger on grapes and fruit, see Wharton 1974, 74, 127 et passim. On pyrethrum, see Wharton 1974, 16. Curiously, animal treatment lagged in the state, too; there was no local school of veterinary medicine until 1938.

37. Figure from Kepner et al. 1978, 479. Quote by Weber 1994, 29.

38. See again the table in chapter 2.

39. Fuller 1934, McWilliams 1939, Fisher 1953, Daniel 1982, Mitchell 1996.

40. Fisher 1953, 20, 32–33, is the best source on labor contracting. On internal contracting in steel, see Brody 1960. On the longshore, see Mills 1979. California only began to regulate labor contractors in 1939, and tightened the law in the Farm Labor Contractor Registration Act of 1963, the first of its kind in the United States.

41. Mitchell 1996, Steinbeck 1988.

42. Igler 2001.

43. McWilliams 1939, 60–65; Fisher 1953, 2; Benedict 1946, 404.

44. Chan 1986, Iwata 1992.

45. On the CCIH, see Mitchell 1996. The claim about labor contractors is at 119.

46. Adams 1921, quoted by Henderson 1999, 94. Mitchell 1995, ch. 4. Mitchell portrays this cynicism very well, although he misses what to me is the main point, which is

how little racial doctrine mattered so long as the labor was cheap and abundant. After all, opposition to immigrant labor by the CCIH and nativists was brushed aside by the growers. There is a large literature about the rampant racism in California at the time, e.g., Daniels 1977, Chan 1986.

47. On race views in California cotton, see Weber 1994, 45; on Mexican contractors, see ibid., 69–73; on Imperial Valley's racial division of labor, see Rudy 1995, 160.

48. Mitchell 1996, ch. 3.

49. On the use of machinery against labor, see Friedland & Barton 1975, 1976; Friedland et al. 1981; Hightower 1978. Kepner et al. 1978, 463, attribute tomato mechanization to the shortage of braceros after 1964.

50. Palerm 1991. Thanks also to Frank Bardacke, personal communication, May 2003. For predictions of a mechanical future, see Parsons 1955, Friedland et al. 1981.

51. See generally Employment Development Department 1992, Zabin et al. 1992, Villarejo & Runsten 1993, Rudy 1995, Martin 1997. On cotton, see Musoke & Olmstead 1982.

52. FitzSimmons 1986; Scheuring 1983, 266–67.

53. For the story of these workers, see Nichols 2003.

54. Sources, in order: Collins et al. 1959, 27; Musoke & Olmstead 1982, 387; Hart et al. 1946, 85–86; Wells 1996, 29; Scheuring 1983, 31. See also Olmstead & Rhode 1997, Wickson 1923.

55. Olmstead & Rhode 1997, 7; Wells 1996, 29.

56. Cochrane 1979, 128, 132. American productivity per acre, per person 1880 to 1960 shown in comparison to Japanese in Cochrane 1979, 329–31, based on Hayami & Ruttan. Japanese productivity surged between 1880 and 1930. Goodman et al. 1987, 42–43, also provide a good summary of the petrochemical revolution.

57. Olmstead & Rhode 1997, 25. For a critique of the science-centered view, see Berlan & Lewontin 1986.

5: Industrial Agriculture

1. On the agro-industrialization of agriculture elsewhere, see Thompson 1968, Danhof 1979, Goodman et al. 1987, Page & Walker 1991.

2. Goodman et al. 1987, Lewontin 1998. What these views miss is an adequate theory of the social relations of production in which class struggles make different outcomes possible—rather than one inexorable logic of capital. On the agrarian South, see Wright 1986.

3. Such network organization is the subject of most business organization research today. For a good discussion of the tense but successful balancing of cooperation and competition in network forms, see Saxenian 1994.

4. Olmstead & Rhode 1988, 1997. Also, Fitzgerald 1991. Contrast Bainer's 1975 woeful treatment of the subject, confined only to machinery, or Cochrane's 1979 tendency to start all discussion of farm improvements with machines, e.g., 126 ff.

5. Jelinek 1982, 61ff.

6. Olmstead & Rhode 1988, 1997. For an example of eclecticism in explaining modern technological dynamism, writ large, see Mokyr 1990. For a comparison to the Midwest, see Page & Walker 1991, Post 1982, 1995. On Southern agriculture, see Wright 1986, Post 2003. On Brazil, see Foweraker 1981. On differing agrarian orders and roads to capitalist modernization generally, see Moore 1966.

7. The literature on induced innovation versus forced innovation is voluminous. A starting point is Freeman 1984. Busch et al. (1991) do a good job of trying to provide a balanced view for American agriculture. But they, too, miss the middleman, the agrarian capitalist. For two classic views on the necessary relation between technological progress and capitalism, see Marx 1863 and Schumpeter 1934, discussed in Walker 1995.

8. Estimated number of nurseries from Taylor 2000; he relies on the privately published list of California nurseries, 1850–1900, by Thomas Brown of Petaluma. On Southern California garden nurseries, see Padilla 1961, Cuthbertson 1955.

9. Wickson 1921; Hedrick 1950, 370–85; Wilhelm & Sagan 1974; Taylor 2000. This was a general time of independent nursery founding in the United States; the number rose from a handful to 1,000 in the years 1840–60.

10. On the Santa Clara Valley nurserymen, see Wickson 1921, 40–42; Jacobsen 1984, 70ff; Lukes & Okihiro 1985, 30; Arbuckle 1986, 175ff.

11. Wickson 1912, Hawthorn & Pollard 1954, Cuthbertson 1955, Modry 1970. Santa Clara figures from Lukes & Okihiro 1985, 19. On Sheppard see Padilla 1961. After World War II, seed production began to shift to Central America, thanks to investment by U.S. firms. Modry 1970, 52.

12. On the uptake of Mendel's ideas among breeders in the United States, see Rosenberg 1997.

13. Tufts et al. 1946, 114, 126, 130; Smith et al. 1946, 279; Musoke & Olmstead 1982, 387; Wilhelm & Sagan 1974, 211; Scheuring 2001, 54.

14. On Madson and pure seed, see Scheuring 2001, 54. On postwar UC research expansion, see Scheuring 1988, 36. University of California did almost no work on truck vegetables until after World War II.

15. Kloppenburg 1988, 297. See also Goodman et al. 1987, 37, who hold that "The seed is the bearer of technical progress in plant biology and this discipline . . . has emerged as the privileged locus of industrial appropriation." It's odd that Kloppenburg would take this position, given his sophisticated understanding of the synergies between science and agribusiness. He even quotes Marx and Engels from *The German Ideology* as saying, "[W]here would natural science be without industry and commerce? Even this 'pure' natural science is provided with an aim, as with its material, only through trade and industry." Marx & Engels 1970, 63, quoted in Kloppenburg 1988, 41.

16. Kloppenburg 1988, 99. Tree and other plant propagations were finally protected under the Federal Plant Patent Act of 1930, but seeds were not covered until the 1970 Plant Variety Protection Act.

17. On the tomato story, see Tufts et al. 1946, 144–48; Collins et al. 1959; Friedland & Barton 1975, 1976, 36; Busch et al. 1991, ch. 5; Smith 2000.
18. Lapsley 1996, 177–78; Long 1984; Stuller & Martin 1989, ch. 4; Purdue 1999, ch. 7; Guthey 2004.
19. Thomas & Goldsmith 1945, Wilhelm & Sagan 1974, Welch 1989, Wells 1996, 180–85.
20. On the history of agro-biotechnology, see Busch et al. 1991, ch. 3.
21. Phrase in quotes from Busch et al. 1991, 83. My thanks to Ignacio Chapella for explaining the basic techniques to me. Neither mutation nor interspecies transfer of genes are unique to biotech, since they occur in nature, but this allows humans to introduce such changes willy-nilly, and that is revolutionary. See also Lewontin 2001.
22. On the early development of biotech, see Kenny 1986, Busch et al. 1991, Martineau 2001, Boyd 2003. Thanks to Iain Boal for helping me sort this all out.
23. Alan Rudy and Larry Busch, Michigan State University, personal communication, April 2003.
24. On the corporate move into agro-biotech, see Busch et al. 1991, 15; Tait et al. 2001, Chataway et al. 2002, Boyd 2003; www.etcgroup.org. On the relative eclipse of university agricultural research budgets, see Busch et al. 1991; Alston & Pardee 1996, 38ff; NSF 1996.
25. On corporate strategies, hyperbole, and failures, see *Business Week* 1992, 1999; Powell 1996; Chataway et al. 2002.
26. *Business Week* 1993. On the general problem of biotech's limited ability to get the plants right, see Lewontin 2001. On the widespread reductionist error of seeing life as code instead of as organism and ecology, see Chappela 2003. On the demotion of traditional plant breeders in the biotech age, see Busch et al. 1991, 92.
27. On the controversies, see Busch et al. 1991, Martineau 2001, Chappela 2003.
28. On science in California and Western conquest, militarism, racism, and the rest, see Smith 1987, Brechin 1999.
29. Quotes by Paul 1973, 22–23; Olmsted & Rhode 1997, 14.
30. Olmstead & Rhode 1988 and 1997, 17–18; Rogin 1931; Bainer 1975; Payne 1982.
31. U.S. Census Bureau 1915, 17; Olmstead & Rhode 1997, 14, 16. On the key role of horses in 19th-century British agriculture and economy, see Thompson 1976.
32. Kepner et al. 1978 (3rd ed.), 481.
33. On pumps, see Olmstead & Rhode 1988, 98; 1997, 16.
34. For overviews of field mechanization across the country, see Cochrane 1979, Bainer et al. 1955, Kepner et al. 1971. On ag engineering at Davis, see Scheuring 2001, 57; Bainer 1972. It is important to note the degree of information sharing across the country in this domain, thanks to the work of bodies like the American Society of Agricultural Engineers. Hence leading textbooks written by UC Davis faculty deal with the topic as a national one, without much reference to regional particularities.

35. Pudup & Watts 1987, 346. See also Bainer 1975.
36. On cotton mechnization, see Musoke & Olmstead 1982.
37. On sugar beet mechanization, see Bainer et al. 1955, 456–63; Bainer 1972; Scheuring 2001, 56.
38. On tomato mechanization, see Friedland & Barton 1975, Hightower 1978, Kepner et al. 1976, Busch et al. 1991. Indeed, Runsten & Leveen 1981 argue the exact opposite of Friedland & Hightower: that mechanization of cotton in the 1950s took away critical support for the bracero program by the early 1960s. For a more general discussion of the economics of technological innovation, see Storper & Walker 1989, Walker 1995.
39. Scheuring 1983, 31–32.
40. Thanks to Frank Bardacke for information about failed mechanization projects. Bainer et al. 1955, 4, say quite clearly than mechanical harvesting always reduces quality. It is interesting the first edition of Bainer et al. (1955) has no discussion whatever of fruit and vegetable machinery. The second edition does, but it is overly optimistic about long-term results—and focuses on crops for processing, not the fresh market. Kepner et al. 1971, ch. 21.
41. On precision farming see National Research Council 1997, Pierce & Sadler 1997. For a valuable critique, see Wolf & Buttel 1996, 1271, who call precision farming "a defensive or conservative reaction of an agricultural system struggling to maintain legitimacy in the face of mounting social, economic, and ecological challenges."
42. E.g., Pisani 1984, Worster 1985, Reisner 1986.
43. Olmstead & Rhode 1997, 25, 19, consider irrigation part of the "biological investment" in agriculture, but this makes no sense. External water is not biological in a true sense, even though organisms depend on it. Part of the confusion is that they, like many others, equate land drainage with irrigation because both are about water control and engineering; but drainage is antecedent to irrigation, because it prepares the soil for planting.
44. On the project see Igler 2001, Pisani 1984. The legal battle of *Lux v. Haggin* has been greatly overblown in importance. It settled nothing, and the state continues to this day to have the worst system of water law in the country—and possibly the world—a ridiculous hodgepodge of riparian, appropriative, and beneficial use rights. On California water law, see Hutchins 1956.
45. Igler 2001, Fielding 1964, Gregor 1964, Gilbert & Akor 1988, Storper & Walker 1982. Growing feed grain, often with irrigation, only became general in the Great Plains in the 1920s. Skaggs 1986, 138. For irrigation by county, see Henderson 1999, table 1.2, 15; Kahrl 1979.
46. The best source on irrigation districts, and irrigation development generally, is Pisani 1984.
47. On the Owens story, see Kahrl 1982. The movie *Chinatown* has all the facts distorted, but the truth of the matter emerges intact through a great job of storytelling.

48. On the early years of the Reclamation Service, see LeVeen 1979.
49. Pump figures from Olmstead & Rhode 1997, 19–20. Electric pumps went up from one-sixth to five-sixths of all pumps between 1910 and 1930. Ibid., 20. Thanks to Professor T.R. Narasimhan of UC Berkeley on origins of submersible pumps. Olmstead & Rhode 1997, 20, say one-third of wells had turbine pumps by 1930, two-thirds by the 1940s.
50. The Colorado dam story is recounted in Worster 1985, Reisner 1986, and Hundley 1992, among others. The Imperial Valley's story as a whole is little known, but has been told at great length in a fine dissertation by Alan Rudy (1995).
51. Worster 1985, McWilliams 1939. The interpretation here is entirely mine, and is not made elsewhere, as far as I know. But see Storper & Walker 1982.
52. On the water subsidies, see Storper & Walker 1979, 1982; LeVeen 1979.
53. Reisner 1986, quote at 345. Mike Davis 1998 makes a quite similar argument about environmental hubris in building Los Angeles, but is more attuned to the power of capitalist land developers in the process.
54. On the marginal nature of much of land brought in by the big projects, see Storper & Walker 1982. Reisner 1986, 355, notes that the big projects were mostly to rescue overdrawn groundwater areas, and that "much of the land in need of rescue was second and third class." For a map of irrigated acreage circa 1975, see Donley et al. 1979.
55. LeVeen was denied tenure by the university soon thereafter owing to his temerity in criticizing agribusiness. I can vouch for this, since LeVeen was a colleague and friend of mine; he is a superb economist and social scientist who deserved tenure by any objective measure.
56. On the acreage question, see Villarejo 1980. On the water diversions, see Storper & Walker 1982, Reisner 1986.
57. On inefficiencies, see Teitz & Walker 1982. On the Imperial–San Diego water deal, see Dean Murphy, "California Water War Takes a New Turn," *San Francisco Chronicle,* January 5, 2003, A3. IID currently pays $15.50 per acre-foot of water while San Diego pays $250 per acre-foot.
58. Thompson 1968. The first agricultural revolution of the 17th and 18th centuries in Britain was based on crop rotation, grazing animals on the fallows (ending the long conflict between arable and pasture land), and more intensive use of animals and manure.
59. Haber 1958, 60, 121; Thompson 1968; Scheuring 1983, 268ff.
60. Rossiter 1975; Cochrane 1979, 109, 229–30.
61. On citrus fertilizer use, see U.S. Census Bureau 1914, plate 358. In 1941 the leading California crops demanding fertilization were, in order: citrus, lettuce, potatoes, celery, and deciduous fruits. Over two-thirds of tonnage went to those five crops. Jenny et al. 1946, table 3, 371. On Hilgard's view, see Jenny 1961, 85. On Shell, interview with Roland Meyer, UC Davis, April 4, 2001.
62. Roland Meyer, interview.

63. I am extrapolating from what I've found of such factories in San Francisco and around the Bay Area, such as Pacific Guano and Fertilizer Company in West Berkeley and Union Super-Phosphate Company, subsidiary, at Stege (Richmond). Walker 2001a.
64. Meyer interview; Davis 2002, especially footnote 566 on testimony by John Mills for agriculture asking for more sardine reduction for fertilizer.
65. National figures from Cochrane 1979, 127; California figures from Scheuring 1983, 27. Plant numbers from Meyer, interview. By the 1980s, the number had fallen to two, and today all nitrates come from one plant in Alaska.
66. Guthman 2004, 206. Probably too much credit is given to Green Revolution hybrids in raising productivity and not enough to the sheer mass of cheap fertilizer and better watering. Berlan & Lewontin 1986, Pollan 2002.
67. On the origins of DDT and its spread, see Carson 1962, Wharton 1974, and Russell 2001. The atomic bomb quote is from a public health journal article, cited by van den Bosch 1978, 249.
68. Scheuring 1983, 273, on poundage; van den Bosch 1978, 60, on cost. Van den Bosch 1978, 27, for 5 percent claim.
69. Quote from Pollan 2001, 82. The earliest use of fungicides I have found is for grain shipping in the late 1920s and seed protection in sugar beets in 1940. Scheuring 1988, 27. Monoculture's potential for disaster in other regards, particularly epidemic plant disease that might cause mass famines, led the United Nations to create an International Board for Plant Genetic Resources in 1974.
70. Van den Bosch 1978, vii.
71. On this reaction see Wharton 1974, Rome 2001. I remember well the disappearance of butterflies and birds from my childhood home in the Santa Clara Valley of the 1950s.
72. Van den Bosch 1978 discusses several such incidents of suppression of dissenters, including Robert Rudd (1964). I have my own examples, like the failure to tenure biotech critic Ignacio Chapella in 2003. The hostility was so bad in the 1970s that when van den Bosch died suddenly in 1979, many people feared he had been offed by the powers that be.
73. Paul Feist, "Central Valley pollution bills pass Senate: big farmers lose fight to stay exempt from clean-air rules," *San Francisco Chronicle,* May 3, 2003, A27. On water pollution by pesticides, see Jane Kay, "Suit to End Waiver on Pesticides," *San Francisco Chronicle,* November 28, 2000, A1, A24.
74. On UC cuts, see Scheuring 1988, 50–55. On the national trend, see Busch et al. 1991, Alston & Pardey 1996, National Research Council 1996.
75. Thanks to Jake Kosek for bringing me up to speed on honeybees; personal communication, October 2003.
76. Hawthorn & Pollard 1954, 73–77; Tufts et al. 1946, 115; Scheuring 1988, 42, 47.
77. On the Mississippi see Pollan 2002; on BT persistence see Steinbrecher 1996.
78. Pollan 2001, Wells 1996. Methyl bromide is a major threat to the ozone layer, but California growers have fought its regulation tooth and nail.

79. Quote by Pollan 2001, 82.
80. Ibid. On organic farming as agribusiness, see Guthman 2004.
81. Haber 1958, 1971; Thompson 1968. Also Goodman et al. 1987, 26.
82. On irrigated pasturing by Miller and Lux, see Igler 2001; also Zonlight 1979. New Mexican, Indian, and Mormon irrigators had come to terms with the Western climate earlier than the Yankees and their British investors. It would take Anglo-Americans a long time to figure it out. On the Northern Great Plains disaster, see Frink et al. 1956. On the Southwestern desert ranges, see Sayre 2002.
83. On grass introductions, see Sampson & Chase 1927, Burchell 1981 [1957]. Sampson vies with James Jardine of the U.S. Forest Service as the father of range science. Sampson 1913, 1917, 1919, 1923. For a critique of range science in California, see Hamilton 1997, who also notes that Sampson created the myth of California's natural range species being perennials. I thank Nathan Sayre for teaching me about Sampson and the disputes within range science.
84. The Petaluma story is not well known. But see Hogan 1913; Hart et al. 1946, 96–105; Scheuring 1983, 203–5; Kann 1993; Bell 1995. The rise of Petaluma was helped by the banishment of poultry from San Francisco backyards by hygienic laws aimed at curbing rats as carriers of the plague. Craddock 2000. First use of the term "poultry industry" was apparently by Johnson & Brown 1913, in reference to Petaluma. On the broiler industry in general, see Watts & Kennett 1995, 6–18; Boyd & Watts 1997.
85. On the sardine industry, see McEvoy 1986, Davis 2002. Davis, 170, gives figures of about 50,000 tons a year in 1935 of fishmeal in poultry feed consumed on the West Coast, and 24,000 tons of sardine meal produced in California in 1934, along with the dollar amounts quoted. Hart et al. 1946, 61, cite a larger figure, 100,000 tons, including shark meal.
86. On confinement and vitamin research, see Hart et al. 1946; Scheuring 2001, 54; Boyd 2001.
87. On postwar poultry and its geography, see Boyd & Watts 1997, Watts & Kennett 1995 (esp. table 1), Bell 1995, Boyd 2001. Broilers made a comeback in the state, led by Foster Farms. But California did not adopt the corporate-dominated contracting system popular in the South—an irony, given the long history of contracting in this state.
88. On California poultry, see Moreng 1995; Watts & Kennett 1995; Hart et al. 1946, 103–4.
89. On dairy output and feeding, see Jones 1925; Hart et al. 1946, 58–63; Gregor 1963; Fielding 1964; Gilbert & Akor 1988; Scheuring 1983, 196–97. While dairies across the sunbelt have followed the lead of California feeding practices, the family farm dairy persists in Wisconsin and Vermont.
90. Hart et al. 1946, 73–87; Tufts et al. 1946, 121–25; Scheuring 2001, 53. On UC work with dairy cows, see Hart et al. 1946, 69, 82, 84–86. On stockmen's input, see ibid., 77, 81. For more on nutritional research, see chapter 6.
91. On the rise of feedlots in California (not in the Midwest, as some claim), see Skaggs 1986, 178–79; Scheuring 1983, 190–91. A further move to "boxed beef" precut

at the factory came circa 1970, completing the transformation and wiping out the last of the independent slaughterhouses. Scheuring 1983, 286; Page 1993. The role of the university in feedlot development is unclear to me; it is mentioned in passing in Scheuring 1983, 191.

92. On beef production and consumption, see Schell 1985, Levenstein 1993, Schlosser 2001, Pollan 2002.

93. Possibly the first book to call attention to the consequences of the feeding system was Schell 1985.

94. Goat story from Friedland 2002.

6: The Harvest of Agribusiness

1. Scholars of California agriculture have too often overlooked canning in favor of cooperatives; for example, a check of the index in Vaught 1999 reveals not a single entry for "canneries" or "packers."

2. In the last few years, scholars have taken up the subject of food retailing and restauration with some gusto, as in Levenstein 1993 and Schlosser 2001. They are also looking at the geography of retail supply chains in global food systems, e.g., Wrigley & Lowe 1996. Consumption has also become a hot topic in agrarian studies, shifting the emphasis from agriculture to food systems. There are now some excellent food histories, e.g., Levenstein 1988, 1993; Fine et al. 1996; Belasco 1993; Scranton & Belasco 2001. These have entered California agrarian studies only recently, e.g., Guthmann 2004, Charles 2001.

3. This takes us into the analysis of turnover time of capital, treated by Marx 1893, but relatively neglected in economics until revived under the heading of "throughput" by Alfred Chandler and lean inventory management in Japanese mass production. Chandler 1977, Sayer & Walker 1992, ch. 4.

4. Quote from Tobey & Wetherell 1995, 8. On business integration in general, see Chandler 1977, 1990; Sayer & Walker 1992.

5. Chandler 1962.

6. On wheat see Paul 1958, Rothstein 1990, Meissner 1997/98. See Cronon 1991 on the Chicago grain merchants for a detailed analysis of an evolving commodity trading system. It appears that California wool was mostly shipped to eastern textile companies, but there's no good source on this.

7. Igler 2001.

8. There is no systematic treatment of the agro-merchants and shippers, but see Liebman 1983, Vaught 1999, Teiser 1983.

9. On refrigeration, see Tufts et al 1946, 158–59. The earliest shipper of iced fruit I know of was from Cincinnati in 1843. Wilhelm & Sagan 1974, 156.

10. On co-ops, see MacKay 1925, Kraemer & Erdman 1933, Schoendorf 1947, Tufts et al. 1946, Nash 1964, Couchman 1967, Blackford 1977, Stoll 1996, Woeste 1998, Vaught 1999. Different writers tend to privilege the leadership of different crops, but the action

seems to have shifted back and forth over time and reverberated throughout the state, with each group learning from the others. California did not originate the idea of farmer cooperatives, but it did change it to its modern business form.

11. Stoll 1998, 72. As an example of the severity of the glut, half the pear trees in the state were uprooted. Ibid., 84.

12. Vaught 1999, 35, 106–07. See Chan 1986, 235, on contracting with Chinese. Vaught 1999 has done a service in signaling the importance of Placer County in his study of horticulturalists. It's not clear if the shippers or the co-ops first came up with the idea of standardized grading of fruit.

13. Quote from Woeste 1998, 113. On raisins see Woeste 1998, Vaught 1999.

14. For the history of prunes and Sunsweet, see Couchman 1967.

15. Figures from Morrison 1939, cited in Liebman 1983, 59; Tufts 1946, 166; and Woeste 1998, 30. Ninety percent of the traffic was handled by just eight of the associations (in citrus, deciduous fruit, nuts, raisins, prunes, and avocados). A few co-ops confined their activities to bargaining and services only.

16. Cross quoted in Stoll 1998, 77. Woeste 1998 has provided a thorough analysis of the legal history involved, as well as the disputes among growers, between growers and shippers, and between growers and their consciences as they discarded older notions of cooperation for the new corporate business approach.

17. Nash 1964; Woeste 1998. California agribusiness generally relied on its state umbrella rather than the federal one. Benedict 1946, 421. Some commodities (beef, dairy, avocado, table grapes, etc.) are covered by specific state legislation of their own. On state-aided organization of U.S. agriculture generally, see Cochrane 1979.

18. Emil Horst obituary, *San Francisco Chronicle,* May 25 and November 3, 1940; Benedict 1946, 419; Friedland et al. 1981; Hawthorn & Pollard 1954, 12; Teiser 1983; Wells 1996, 40–42. The federal Agricultural Marketing Act of 1946 extended the marketing order system to fresh produce and enabled the USDA to set federal standards for fresh produce.

19. Levenstein 1993, 27. Iceberg lettuce and other fresh vegetables were grown in central California as well.

20. Ninety-five percent of California tomatoes are processed.

21. Industry ranking from Cleland & Osgood 1928, 266, compiled from U.S. Census figures for 1925. Production of containers including tin cans, boxes, and bags would have been as big as motion pictures. On the overall impact of agriculture and other resource industries, see Walker 2001. The relative decline of food processing among industries nationally is noted by Gabaccia 2002, 188.

22. All from Meissner 1997–98. The figures on wheat exports, cited previously, include flour.

23. Tufts et al. 1946, 132; Liebman 1983, 48–49; Henderson 1999, 221n.

24. On the meatpacking history in general, see Skaggs 1986, Page 1993. On meatpacking plants and tanneries around the Bay Area, see Walker 2001a.

25. Jones 1925; Hart et al. 1946; Gregor 1964; Scheuring 1983, 194.
26. Couchman 1967, Woeste 1998.
27. On machines in packing, see Couchman 1967, 34; on orange packing, see Moses 1995, 34.
28. Pack figures from *Oakland Tribune* 1949, 67–68. There is much confusion about J. Lusk versus A. Lusk company of San Francisco, which seem to have merged together ca. 1890, when A. Lusk disappears from the Sanborn maps of San Francisco. One source is Bunje 1939.
29. Quote from Hampe & Wittenberg 1964, 118. On tomato canning see Collins et al. 1959, 14; Smith 2000. Productivity in all food processing would be almost double the national average in the decade after World War II. Hampe & Wittenberg 1964, 137.
30. Case figures for 1918 and 1928 from *Western Canner & Packer,* cited by Matthews 2003, 34; figure for 1948 from *Oakland Tribune Yearbook 1949,* 67–68. On the history and geography of canning, see Collins et al. 1959, Hampe & Wittenberg 1964, Braznell 1982, Cardellino 1984.
31. On Del Monte, see Burbach & Flynn 1980, Braznell 1982. The first food multinational was Heinz in the late 19th century. Virden and Stewart would fail after 1929.
32. On the contracting system, see Tufts 1946, 168; Collins et al. 1959, 47; Hampe & Wittenberg 1964, 125–26; Jelinek 1982, 64. On the peach campaign, see Hampe & Wittenberg 1964, 126.
33. See figures on contracting in Goodman et al. 1987, 84 (from USDA reports). I have no figures on contracting across California as a whole, but see Wells 1996, 42, and Weber 1994 on strawberries and cotton.
34. Hampe & Wittenberg 1964.
35. Burbach & Flynn 1980, 174–77. These authors make the classic mistake of taking Del Monte, the giant corporation, as the defining figure of California agribusiness.
36. On changes in cannery work, see Matthews 2003, 93, based on a CIO report of 1946. Figures on frozen versus canned foods from FTC 1965, 30.
37. On Tater Tots, see Berelson obituary, *San Francisco Chronicle,* May 15, 1997, A26. On strawberries, see Wells 1996, 33; Wilhelm & Sagan 1974, 211; Bain & Hoos 1963. On UC frozen pack research, see Tufts 1946, 165. On the arrival of frozen foods in general, see Hampe & Wittenberg 1964.
38. See Levenstein 1993, 247, on the Del Monte buyout. On Campbell's, see Smith 2000, 112–13.
39. On the early wine industry, see Carosso 1951; Amerine 1962; Muscatine et al. 1984, 30–48.
40. Nash 1964; Muscatine et al. 1984, 412–47. On CWA and Winehaven, see Adams 1973. The other claimant to "world's largest" was Leland Stanford's 2.5-acre brick winery at Vina in Tehama County in the Sacramento Valley.
41. On the wine revival see Conaway 1990, Eysberg 1990, Lapsley 1996, Guthey 2004.

42. There are, regrettably, no histories of brewing in California. Not all crops were eaten, of course. Cotton and wool have mostly been shipped to eastern textile factories. Wool was sent raw. Cotton was ginned, however—a first-stage form of processing. Gins were a significant element of the landscape of the San Joaquin Valley; there were about 35 gins in 1925 and 120 in 1939, plus cottonseed-oil mills. Cotton growing, ginning, and merchandising were highly integrated by the late 1920s, including a contract system akin to that of the canners. Benedict 1946, 420; Weber 1994; Arax & Wartzman 2003.

43. Figures from Mueller & Garoian 1961, 8; Markin 1963, 14, 16; Matchett 1967, v; Tedlow 1990, 193, 246; Hampe & Wittenberg 1964, 288, 293. On chains, see also Hoffman 1935. On A&P generally, see Tedlow 1990, 190ff. On the prehistory of chains, see Fraser 1981.

44. Figures from Mueller & Garoian 1961, 22; figure 2.4 on 57. On the rise of the supermarket, see Zimmerman 1937, 1955; Mueller & Garoian 1961; Markin 1963; Longstreth 1997, 1999. On opposition to chains, see the same sources and especially Hoffman 1935.

45. Quote from Longstreth 1999, 111, who says that over one-third of all food sales in LA were made by supermarkets in 1936, posing a major threat to the chain stores. See also Zimmerman 1937, 4, 39; Zimmerman 1955, 17–18, 24; Tedlow 1990, 196. The Central Valley was also supermarket country. Zimmerman 1955, 31.

46. Remarkably, there is no history of Safeway, but see Tedlow 1990, 243; Fortune 1940, 61; Markin 1963, 24; Matchett 1967, 32; *Safeway News* 1976. For rankings, see Zimmerman 1955, 138–39, table 11; *Progressive Grocer,* various years (up to 1980). A useful Web site is www.groceteria.net. On Lucky I have found almost nothing except a mention of its merger strategy in Mueller & Garoian 1961, 60.

47. Deutsch 2002. See also Strasser 1989, Cohen 2003.

48. For a thorough discussion of the emergence of modern retailing and retail forms in Los Angeles, with excellent photographs of market interiors and exteriors, see Longstreth 1999. On the experience of shopping in supermarkets, see Willis 1991. Figures on number of items for sale from *Progressive Grocer* 1962, cited in Markin 1963, 60.

49. Quotes by Tedlow 1990, 207; McNair 1931, 31.

50. Tedlow 1990, 207; Markin 1963, 97; Fiske 1955.

51. Fortune 1940, 132; Scheuring 1983, 288.

52. Mueller & Garoian 1961, 72–81, 91; Annual Report of Safeway Stores, Inc. 1954, 6.

53. Zimmerman 1955, 146. I well remember the milkman's deliveries of Toyon brand as a boy in the 1950s, and the disdain my mother had for homogenized milk in cartons.

54. On beef see Zimmerman 1955, 146, 222–28; Walsh 1993, 77ff; Skaggs 1986, Page 1993. Fiske 1955, 18. On California's low standards, see Hart et al. 1946, 57. Wartime shortages forced Safeway into buying out a dozen suppliers to assure its supply.

55. On poultry, see Watts & Kennett 1995, 10–11.
56. Zimmerman 1955, 154. On early failures of prepackaging and its success after 1944, see Hampe & Wittenberg 1964, 180, 182. Quick-frozen meat was developed by Cudahy & Company in 1930, and quick-frozen vegetables by Charles Birdseye in Boston soon thereafter.
57. Zimmerman 1955, 10, 229–34. There's no good data on purchasing of fresh produce, but see Markin 1963, 96–99.
58. Friedland et al. 1981, 52–53, 66.; FitzSimmons 1986, 345; Zimmerman 1955, 147, 232–33.
59. Friedland et al. 1981, 46; figures from FitzSimmons 1986, 342.
60. Quote from Tedlow 1990, 354. The solution depends on economies of scale and scope, on managerial systems, and the nature of the commodity chain. Tedlow, a student of Chandler, divides integration into four types: corporate, contractual, open market, and administered. Ibid., 361.
61. For a history of restaurants and chains, see Pillsbury 1990. On Los Angeles roadside restaurants, see Longstreth 1997, 1999. Figures from Emerson 1990, 30, 19—yearly figures in table 2.1. Levenstein 1993, 233, cites higher figures, up to 38 percent by 1983. California trailed only Hawaii, Alaska, Connecticut, and Massachusetts in per capita restaurant sales.
62. Pillsbury 1990, Schlosser 2001. I have not found a way to trace fast-food suppliers from California, unfortunately.
63. On supermarkets, see Walsh 1993, 9, 52–53. Ward's Business Directory for 2000 shows Ralph's/Food4Less as number one in the United States, American Stores number two, Safeway number three (and again a public company), and Albertson's number six.
64. On fast food's bumps and bruises, see Schlosser 2001.
65. On Wal-Mart and this tumultuous era in retailing, see Ortega 1988—especially 72, 102, 109, 130, 368. Longtime retail leader Sears went into crisis in the 1970s and changed itself into a financial corporation, and many more department store chains and discount chains went under in the 1980s and early 1990s.
66. Pollan quote 2002, 75. Levenstein 1988, 1993 confirm this 200 years of blandness and document the reduction of regional difference in favor of a national cuisine by the early 20th century.
67. It is very hard to sort out exactly when and to what degree California agribusiness has made a difference in national consumption patterns. Since almost all food histories are written from a national perspective (and mostly from the eastern half of the country), California's peculiarties and market clout have usually been minimized. I may make the opposite error of exaggeration, and I hope that future scholars can clarify much that is still murky as I write.
68. Or, as Goodman et al. 1987, 58, put it, "The tendential action of substitutionism is to reduce the rural product to a simple industrial input."

69. Lears 1994, Miller 1995. See Levenstein 1988, ch. 3, on the growth of food processors and food advertising in the late 19th and early 20th centuries. See Agnew 1993 for a critique of cultural consumerist theory. I prefer a dialectical view of the evolution of consumption and production, in the manner exemplified in works such as Forty 1986, Walton 1992, and Scranton 1997. An example of culturalist consumer theory of food is Marsden & Arce 1995.

70. Levenstein 1993, 45, is one of many to make the claim for the three gourmet cities. On Hilgard, see Jenny 1961, 85.

71. On wheat, see Meisner 1997–98. On white sugar, see Levenstein 1988, 33. On olive oil, see Taylor 2000, 43.

72. On the shift toward beef and milk, see Levenstein 1988. On breeding to the British standard of meat, regardless of its effects on land and productivity, see Sayre 2002. In meat and milk, California has always been its own main supplier because of the economics of bulk and freshness—but it has also been a net importer in recent years because of its vast population.

73. On wine consumption and standards, see Amerine 1962, Adams 1973.

74. On apples nationally, see Pollan 2001; on dried fruit, see Couchman 1967; on canners, see Hart et al. 1946, 97; on Tom Foon, see Payne 1987, 96, and Arbuckle 1986, 185–88.

75. On the changing American diet, see the sweeping two-volume history by Levenstein (1988, 1993). On Sunkist, see Stoll 1998.

76. On avocados, see Charles 2002. On prunes see Couchman 1967. On citrus health promotions, see Levinstein 1988, 152–54; 1993, 14. I can testify to the misleading effects of Sunkist's campaign to drink orange and lemon juice to improve teeth and gums: my mother followed this course to excess as a young woman, and it ate the enamel off her teeth! Consumption per capita of fresh fruits and vegetables did rise from 316 pounds in 1919 to 393 in 1939. USDA figures.

77. Hilgard had been working on olive processing for years before olive canning was perfected by Frederic Bioletti and George Colby at UC Berkeley in 1899. Freda Ehmann of Oroville was the first, circa 1905, to popularize the new canned olives—which ended up absorbing 99 percent of olive production in the state. On the canned olive, see Tufts et al. 1946, 231; Taylor 2000. The California Olive Association continued to support Professor Cruess generously. On fruit cocktail, see Payne 1987, 96; Arbuckle 1986, 185–88.

78. On changing 20th-century American food habits, see Levenstein 1993, Goody 1997.

79. On Del Monte see Braznell 1982, Witter 1950. On food advertising, brand names, and General Foods, see Levenstein 1993. On branding in general, see Klein 2000.

80. On vitamin and nutrition fads, see Levenstein 1993, ch. 1 et passim.

81. Figure for 1910 from Produce Marketing Association, www.pma.com; figure for 1940 from Levenstein 1993, citing USDA, handbook #62, *Consumption of Food in the*

U.S., 1909–52, 1953. The change in canned fruits and vegetables was less marked, but still doubled: fresh produce outweighed canned by ten to one in 1919, but only five to one by 1949. USDA figures.

82. Figure from Hempe & Wittenberg 1964, 15.

83. Figures from Hempe & Wittenberg 1964, 21–23.

84. On postwar research funding, and accompanying reorganization of the offices of the USDA, see Scheuring 2001, 52; Alston & Pardee 1996. On food chemistry and additives, see Hampe & Wittenberg 1964; Goodman et al. 1987, 88–89; Schlosser 2001; Hightower 1978. Granny Goose was sold to Del Monte in 1969 and closed in the 1990s.

85. Quotation and Steinbeck paraphrase from Levenstein 1988, 8; 1993, 127. On fast-food restaurants, see Luxenberg 1985, Emerson 1990, Jackle & Skulle 1999, Longstreth 1999, Schlosser 2001. On hygiene and American foodways, see Levenstein 1988, Forty 1986. Quote by Gabaccia 2002, 175.

86. On the midwestern influence in American cuisine in the early 20th century, see Levenstein 1993, 37–38. On Bay Area tastes for wine, fresh produce, Asian foods, and exotics (backed up by an A.D. Nielsen survey), see Michael Bauer and Mona Morgan, "We Eat Chic and Healthy: Local Food Fashions Unlike Any Other," *San Francisco Chronicle,* April 30, 1999, A1, A10.

87. On Hilgard, see Jenny 1961; on the Berkeley study, see Levenstein 1988, 169. Levenstein notes that Californians were spending much less of their income on food than New Yorkers in the 1920s, but fails to note that California foodstuffs were surely cheaper and more abundant by this time. Produce data from USDA, cited in FTC 1965, 30.

88. Data from Produce Marketing Association. The PMA and USDA data are at odds, so take your pick.

89. On organic farming and consumption, see Guthman 2004.

90. On the new cuisine, see Belasco 1993, Tower 2003. California's grew rich faster than in any other state.

91. On olives, see Taylor 2000. On Waters's suppliers, see Saekel 1997.

92. On guacamole see Charles 2002. One might argue that the first "exotic" crop to be grown in large quantity after World War II was garlic, as this quintessentially Italic herb grew in popularity after being shunned like the devil for centuries by Anglo-Americans. Gilroy, south of San Jose, became the garlic capital of the United States. Olive oil, on the other hand, which had been an important food product in California in the 19th century, practically died out in the 20th only to be resusciated in the 1980s.

93. Of course, big food processors frequently buy up small niche producers, but there have been quite a few with staying power.

94. Guthman 2004. Large food processors were equally adept at putting out "natural" lines of foods to capture niche markets without changing their practices very much. Levenstein 1993, 198–99.

95. For a fine discussion of these developments, see Lapsley 1996. On the continued profitability of California premium wines despite the recent downturn, see Carol Emert, "Glut Check," *San Francisco Chronicle,* February 13, 2003, D1, D2.
96. On the microbreweries, see Erickson 1993.
97. Export figures from Scheuring 1983, 297–98; Carter & Goldman 1997. See also California Department of Agriculture's Export Program, www.cdfa.ca.gov/statistics/export.html. On Pacific Rim markets, see Carter 1997. California did not benefit from the huge farm subsidy bill passed by Congress in 2002, either.

7: Capital in the Countryside

1. Henderson 1999, 28.
2. Henderson 1999, 58, 29. Henderson's discussion is a beacon for any treatment of capital in California agriculture.
3. On Murphy, see Lukes & Okihiro 1985. On Reed, see Green 2000. On Miller and Lux, see Igler 2001. On Kern County Land Company, see Zonlight 1979. Alas, no one has come up with figures for this massive flow of capital into land purchases in the 19th century.
4. On 19th-century viticulture, see Carosso 1951, Hutchinson 1984, McGinty 1998. Haraszthy moved on to filibuster in Nicaragua, where he was, according to legend, finished off by alligators. His three sons, however, remained serious winemakers in California. Charles Krug appears to have been backed by banker Charles Moffitt.
5. Henderson 1999, tables and maps at 23–27, 56. On cycles of investment in California, see Gordon 1954.
6. The country banks are generally ignored in the story of agricultural finance; even Henderson 1999 leaps over them. Thanks to Toby Green for bringing their role to my attention. Toward the turn of the century, some of the largest grower-capitalists started putting together regional networks of county banks under their control, prefiguring the branch banking networks of the 20th century. Green 2000. See also Lister 1993. Growth of country assets from Blackford 1977, 96. On lending to Chinese, see Chan 1986.
7. Rhode 1995. Quote from Henderson 1999, 46. Figures from U.S. Census Bureau 1928, 46. See also Benedict 1946, 410. On early-20th-century agricultural prosperity nationally, see Cochrane 1979.
8. On banking development, see Willis 1937, Lister 1993, Doti 1995. On Giannini see James & James 1954, Nash, 1992; on Mercantile Trust see Green 2000. If Sanders 1999 is right, the increased liquidity for agricultural credit was a direct consequence of political intervention by rural interests.
9. As explained by Henderson 1999, 72ff.
10. Henderson 1999, 68–70, has a fine discussion of the financial circuits of the Southern California Fruit Growers Exchange. On contracting and payments, see Collins et al. 1959, Couchman 1967, Vaught 1999, Woeste 1998, Smith 2000.

11. On electric investors, see Issel & Cherny 1986; on water investment, see Pisani 1984.
12. Sobel 1993. The Skaggs brothers also started PayLess Drug Stores and similar chains in the mountain states and Midwest. Robert Magowan ended up on the boards of Del Monte. J.G. Boswell, Bank of California, Southern Pacific, and Fibreboard. Fortune 1940. The Skaggs and Magowans are still major Safeway stockholders and directors.
13. On government credit institutions and figures, see Scheuring 1983, 252, 260–61, and tables at 254. These banking systems are under grower direction, on the cooperative model. Federal subsidies to the big field crops like cotton, wheat, and sugar beets were also beneficial to California growers through the 1970s, but are secondary in this state's agrarian development to credit and investment. On the farm crisis of the 1980s, see Guthman 2004.
14. On feedlots, see Scheuring 1983, 191; on wine, see Muscatine et al. 1984, 50–81; Conaway 1990; Nash 1992; on tomatoes, Smith 2000, 115; on Del Monte, Burbach & Flynn 1981.
15. Henderson 1999, 28–44. On balance in capital accumulation, see Harvey 1982.
16. On fictitious capital and the dynamism of credit, see Harvey 1982.
17. On overaccumulation, finance, and crisis, see Harvey 1982, Brenner 2002. Henderson has the theory right, but puts too much stock in the idea of a one-time transition from wheat to fruit and other intensive crops; rather, the process is cyclic, repetitive, and diverse.
18. On Imperial, see Rudy 1995; on the San Joaquin, see Weber 1994, 20, 33; and Arax & Wartzman 2003.
19. On class formation theory, see Przeworski 1976, Sayer & Walker 1992, ch. 1. Actually, Thompson's (1964) sensibilities about the culture and experience of class are beyond my ability to muster up such a rich portrait of the agrarian class of California. As a starter, however, I recommend Henderson's brilliant discussions of California's agrarian class fantasies portrayed in literature and film.
20. For the discussion of growers' ambivalent class position, see Henderson 1999, 104–14.
21. Goodman et al. 1987, 51. Contrary to Lewontin (1998) southern small farmer degradation is the result of a long history of class dependency and defeat in the southern context. Wright 1986, Post 2003.
22. See generally Adams 1946, Nash 1964, Orsi 1973, Blackford 1977, Liebman 1983.
23. On farmer organization nationally, see Hodgson 1933. On national farmers' movements, see Sanders 1999. On trade associations, see Galambos 1966. On farmer cooperatives, see Woeste 1998.
24. Grower organizing to oppose labor unions will be discussed further below.
25. On Japanese associations, see Iwata 1962, 33; Masumoto 1993.
26. Whatever the term of art, it is essential to avoid the sterile division between struc-

ture and agency that tore apart a fruitful growth of class theory in the 1970s. I refer here to the debate between Nicos Poulantzas and Bill Domhoff, in which the latter's active sense of class formation was marginalized in favor of a rigid functionalism that was itself overtaken by the poststructural turn in social theory in the 1980s.

27. Examples from Green 2000.

28. Quote on co-op domination by Vaught 1999, 96. On Sunsweet, see Couchman 1967. On Fresno and the raisin co-op, see Woeste 1998, 114, table 6.1. See also Stoll 1998. On irrigation and water district politics and power, see Goodall et al. 1978, Pisani 1984.

29. Goldschmidt 1978 [1947]. For a comparison with the Midwest, see Page & Walker 1991. This is to say nothing of worker settlement patterns, which will be touched on later.

30. On Southern Pacific, see Orsi 1975. On the Fresno Colony, see Vaught 1999, 16–25. On Clear Lake schemes, see Green 2000. Cache Creek already had one of the largest of the early irrigation systems—21,000 acres in 1867. On Weinstock, see Woeste 1998.

31. On Imperial Valley, see Worster 1985; Rudy 1995; Henderson 1999, ch. 6. On Los Angeles and Sunkist, see Stoll 1998.

32. On Giannini, see James & James 1964, Nash 1992. On Garden City Bank, see Arbuckle 1986, Green 2000.

33. Green 2000; Safeway annual reports.

34. Examples of local power are rife, e.g., Rudy 1995, 192; Arax & Wartzman 2003. On public-private partnerships, my favorite is Walsh 1978. For reference, Calavita 1992 does an admirable job of rehearsing the theory of the state and capital in her book on California agriculture and working-class rebellion. One can get lost easily in such academic debates, because there is no easy answer to the relative power and relative autonomy of the state under capitalism. It all depends, as they say. It depends, in the American case, most heavily on a highly localized federalist state, a long cultural identification with capitalism, and the resultant hand-shaking relation between business and government on almost every front. Johnson 2003. Unfortunately, most state theory is written by people in Europe, where the state has a quite different history.

35. Nash 1964, Blackford 1977. Phrase in quotes from Vaught 1999, 49, 131. For a fine review of special district government in America, a much-neglected subject, see Dyble 2003. On Miller and Lux, see Igler 2001. On water districts, see Goodall et al. 1978.

36. On farmer-university distrust in general, see Rossiter 1975, 36. On Carr and his suspicions that the university regents were speculating in the college lands allotted under the Morrill Act, see Gates 1961. On Hilgard and the winemakers, see Amerine 1962. On the origins of the Davis university farm, see Scheuring 2001, 13–19. For other examples of university-grower tensions, see Sawyer 1996; Vaught 1999, 49.

37. Fiske 1979, 88. Wickson speech quoted in Scheuring 1988, 9.

38. On Crocheron, see Scheuring 1988 and 2001. On cozy university-grower relations in this era, see Fiske 1979.

39. On prune tariffs, see Couchman 1967, 39. On the avocado limitations, see Charles 2002, 152. On the One Variety Act, see Weber 1994. See also Nash 1964.
40. On this history, see Walker & Storper 1982, Reisner 1986.
41. On the repeated failure of urban unions to grasp the offered hand of farmers and rural workers between the Civil War and World War I, see Sanders 1999. On the NLRA exclusion, see Majka & Majka 1982, Daniel 1982.
42. On Chinese workers see Chan 1986, 332–33. On the anti-Chinese movement, see Saxton 1971.
43. On the Oxnard strike, see Daniel 1982, 77–79; Almaguer 1994, 204. Mitchell 1996 makes the erroneous claim about anti-Japanese sentiment on page 97. On the rural Japanese, see Matsumoto 1993.
44. On labor repression in the 1910s, see Daniel 1982, 89; McWilliams 1939; Mitchell 1996, ch. 3. It was consistent with labor repression during and after World War I all over the country, but the continued degradation of labor moved in the opposite direction of so much "welfare capitalism" of the 1920s around the country.
45. On the Mexican workers and their sources of solidarity and militancy, see Weber 1994, Mitchell 1996. Filipinos and Mexicans generally struck together, although La Unión excluded Filipinos.
46. Daniel 1982; Rudy 1995; Mitchell 1996, 127, 134.
47. On the strikes of the early 1930s, see McWilliams 1939, Daniel 1982, Weber 1994, Mitchell 1996. On the San Francisco General Strike and labor upheaval generally in those years, see Selvin 1996.
48. Daniel 1982, 162–63.
49. McWilliams 1939 is the best source on the Associated Farmers. See also Daniel 1982, Weber 1994. Quote from Steinbeck 1988, 20.
50. Declining wage figures after 1935 calculated by Liebman 1983, 110.
51. On the struggles of 1937–40, see McWilliams 1939, 1942, 1949; Daniel 1982; Weber 1994; Mitchell 1996. On the FSA model housing, see Hise 1997.
52. Mitchell 1996, 154. Weber 1994, 188, hits the nail on the head about the misconception of the farm labor problem. See also Henderson 1999 for an apt critique of John Steinbeck's naive and fatally racist take on the issue. On the range of Okie ideas, see Gregory 1991, Weber 1994, Haslam 1974.
53. Figures from Ruiz 1987, 24–25. Cannery work and life is described by Ruiz 1987, Davis 2002, Matthews 2003.
54. Matthews 2003, 40.
55. Matthews 2003, 56–60.
56. On UCAPAWA organizing in the canneries, see especially Ruiz 1987, Davis 2002, Matthews 2003.
57. This story is told by Matthews 2003, who has a compelling analysis of what the AFL and canners were up to. Daniel 1982 ends his book on a bitter note about the AFL's class treason and the abandonment of the workers by the CIO in 1940, and he fails to see the gains in the plants.

58. Ruiz is highly critical of the Teamsters, while Matthews is more forgiving. Certainly, the decline of several left-leaning unions during the red purges of the late 1940s is to be bemoaned, as it weakened the overall labor movement. On this period generally, see Davis 1986.
59. Quote from Jay 1953, 378.
60. Jay 1953, Zimmerman 1955, Tedlow 1990.
61. Jay 1953, 384–85, 475–92; Selvin 1960. On national unionization of groceries, see also Zimmerman 1955, 285; Tedlow 1990, 221. On the AFL-Teamster perfidy in the Oakland general strike of 1946, see Glass 1996, Rhomberg 2004.
62. Data from the *Progressive Grocer* (April 1983, 94) shows clearly that unionization rates correspond to the size of stores and chains—and are highest in the West. On Chinese supermarkets, see Yee 1999. On the meat cutters, see Skaggs 1987; on packing-shed workers, see Friedland et al. 1981, 66.
63. On fast-food work and organizing, see Leidner 1993, Schlosser 2001, Tannock 2001. On restaurant workers, see Cobble 1991. On Safeway downsizing, see Victor 1989; Mayo 1993, 228–29. On the decline of unions in the 1980s, see Moody 1989.
64. On organizing in the bracero era, see Galarza 1977. On the rise of the UFW and Cesar Chavez, see Levy 1975, Taylor 1975, Jelinek 1982. On the Chicano movement, see Munoz 1989.
65. Bardacke 2002.
66. On race and the American working class in general, see Davis 1986, Katznelson & Zolberg 1986.
67. On California's 19th-century racialization, see Almaguer 1994. For comments on interracial solidarity, see Galarza 1977, Weber 1994, Ruiz 1987. On women and agrarian labor militancy, see Ruiz 1987, Matthews 2003.
68. The gender and racial division of labor is attested to in every study of canning, e.g., Cardellino 1984, Chan 1986, Ruiz 1987, Bardacke 1994, Davis 2002, Matthews 2003.
69. Adams 1921, 519, quoted in Henderson 1999, 92. On Japanese and Chinese farmers, see Chan 1986, Hirabayashi 1989, Iwata 1992, Matsumoto 1993.
70. Mitchell 1996. On strawberry organizing and Mexican aspirations, see Wells 1996. On Mexican grape worker-owners, see Alvarez 2001, Nichols 2003.
71. Weber 1994, 201.

Conclusion: The Conquest of Bread

1. Kropotkin 1892. On the dying of the radical flame, see Clark 1999's treatment of the last works of Camille Pissarro. Kropotkin's influence was widespread and lingering, touching Lewis Mumford, James Joyce, Bertrand Russell, and the Spanish revolutionaries, among others.
2. Kropotkin 1899. On Kropotkin's life and work, see Osofksy 1979 and Purchase 1996. Many thanks to David Hooson for insights into Kropotkin's career. On the agrarian debates in Russia, see Lenin 1964 [1899], Kautsky 1988 [1902], Chayanov 1986 [1923], and the review by Watts 1996.
3. Lewontin 1998, 73.
4. Polanyi 1944.

Bibliography

Adams, Frank. 1946. "The Historical Background of California Agriculture." In Hutchinson, *California Agriculture,* 1–50.

Adams, Leon. 1973. *The Wines of America.* New York: McGraw Hill. Rev. 1978.

Adams, Richard L. 1917. *The Farm Labor Situation in California.* UC College of Agriculture, Agricultural Experiment Station.

Adams, Richard L. 1921. *Farm Management: A Text-Book for Student, Investigator and Investor.* New York: McGraw-Hill. Second edition 1953.

Adams, Richard L. 1948. *Farm Management Crop Manual.* Berkeley: University of California Press.

Adams, Richard L. 1938. *Seasonal Labor Requirements for California Crops.* Berkeley: University of California, Agricultural Experiment Station Bulletin No. 623.

Adams, Richard L. 1952. *Business Management for Western Farms and Ranches.* Danville, IL: Interstates Printers and Publishers.

Agnew, Jean-Christophe. 1993. "Coming Up for Air: Consumer Culture in Historical Perspective." In Brewer, J., and R. Porter, eds., *Consumption and the World of Goods.* London: Routledge, 19–39.

Almaguer, Tomás. 1994. *Racial Fault Lines: The Historical Origins of White Supremacy in California.* Berkeley: University of California Press.

Alston, Julian, and Philip Pardey. 1996. *Making Science Pay: The Economics of Agricultural R&D Policy.* Washington, DC: American Enterprise Institute Press.

Alvarez, Fred. 2001. "Farm Worker to Farmer." *Los Angeles Times,* June 22.

Amerine, Maynard. 1962. "Hilgard and California Viticulture." *Hilgardia* 33, no. 1 (July): 1–23.

Arax, Mark and Rick Wartzman. 2003. *The King of California: J.G. Boswell and the Making of a Secret American Empire.* New York: Public Affairs.

Arbuckle, Clyde. 1986. *Clyde Arbuckle's History of San José.* San Jose: Smith and McKay Printing Company.

Aston, T.H., and C.H.E. Philpin, eds. 1985. *The Brenner Debate: Agrarian Class Structure and Economic Development in Pre-Industrial Europe.* New York: Cambridge University Press.

Azuma, Eiichiro. 1994. "Japanese Immigrant Farmers and California Alien Land Laws: A Study of the Walnut Grove Japanese Community." *California History* 73, no. 1: 14–29.

Bacon, David. 2003. "Is a New Bracero Program in Our Future?" *Z Magazine* 16, no. 10 (October 2003): 25–39.

Bailey, Liberty Hyde. 1917. *Cyclopedia of American Agriculture: A Popular Survey of Agricultural Conditions, Practices and Ideals in the United States.* Revised edition. New York: Macmillan.

Bain, Beatrice, and Sidney Hoos. 1963. *The California Strawberry Industry: Changing Economic and Marketing Relationships.* Giannini Foundation Research Report no. 267. Berkeley: University of California.

Bainer, Roy, R.A. Kepner, and E.L. Barger. 1955. *Principles of Farm Machinery.* New York: Wiley.

Bainer, Roy. 1972. *The Engineering of Abundance.* Oral history transcript. Davis: UC Davis Oral History Project.

Bainer, Roy. 1975. "Science and Technology in Western Agriculture." *Agricultural History* 49, no. 1: 56–72.

Baker, Joseph. ed. 1914. *Past and Present of Alameda County, California.* Chicago: S.J. Clarke Publishing Company.

Bakker, Elna. 1971. *An Island Called California: An Ecological Introduction to Its Natural Communities.* Berkeley: University of California Press.

Bardacke, Frank. 1994. *Good Liberals and Great Blue Herons: Land, Labor and Politics in the Pajaro Valley.* Santa Cruz: Center for Political Ecology.

Bardacke, Frank. 2002. "César Chavez: The Serpent and the Dove." In Clark Davis and David Igler, eds., *The Human Tradition in California.* Wilmington, DE: Scholarly Resources, 209–24.

Belasco, Warren. 1993. *Appetite for Change: How the Counterculture Took on the Food Industry.* Ithaca: Cornell University Press.

Bell, Donald. 1995. "Forces That Have Helped Shape the U.S. Egg Industry: The Last 100 Years." *Poultry Tribune.* Centennial Issue, September, 30–43.

Benedict, M.R. 1946. "The Economic and Social Structure of California Agriculture." In C. Hutchinson, ed. *California Agriculture.* Berkeley: University of California Press, 395–436.

Berlan, Jean-Pierre and R.C. Lewontin. 1986. "The Political Economy of Hybrid Corn." *Monthly Review* 38, no. 3 (July 1986): 35–47.

Berman, Marshall. 1982. *All That Is Solid Melts into Air.* New York: Simon and Schuster.

Bioletti, Frederic. 1899. *Olives.* Agricultural Experiment Station Bulletin No. 123. Berkeley: University of California.

Bioletti, Frederic. 1911. *Principles of Wine-Making.* Agricultural Experiment Station Bulletin No. 213. Berkeley: University of California.

Blackford, Mansel. 1977. *The Politics of Business in California, 1890–1920.* Columbus: Ohio State University Press.

Boyd, William. 2003. "Wonderful Potencies? Deep Structure and the Problem of Monopoly in Agricultural Biotechnology." In Rachel A. Schurman and Dennis Doyle Takahashi Kelso, eds., *Engineering Trouble: Biotechnology and Its Discontents.* Berkeley: University of California Press.

Boyd, William, and Michael Watts. 1997. "Agro-industrial Just-in-time: The Chicken Industry and Postwar American Capitalism." In Watts, Michael, and David Goodman, eds., *Globalising Food: Agrarian Questions and Global Restructuring.* London: Routledge, 192–225.

Braznell, William. 1982. *California's Finest: The History of Del Monte Corporation and the Del Monte Brand.* San Francisco: Del Monte Co.

Brechin, Gray. 1983. "Sailing to Byzantium: The Architecture of the Fair." In Burton Benedict, ed., *The Anthropology of World's Fairs: San Francisco's Panama Pacific International Exposition of 1915.* London and Berkeley: Lowie Museum of Anthropology and Scolar Press, 105–7.

Brechin, Gray. 1999. *Imperial San Francisco: Urban Power, Earthly Ruin.* Berkeley: University of California Press.

Brenner, Robert. 1986. "The Social Basis of Economic Development." In J. Roemer, ed., *Analytical Marxism.* New York: Cambridge University Press, 23–53.

Brenner, Robert. 2002. *The Boom and the Bubble: The U.S. in the World Economy.* London: Verso.

Brockway, Lucille. 2002 [1979]. *Science and Colonial Expansion: The Role of the British Royal Botanic Gardens* [new edition]. New Haven: Yale University Press.

Brody, David. 1960. *Steelworkers in America.* Cambridge: Harvard University Press.

Bunje, E.T.H. 1939. *Oakland Industries, 1848–1938.* Oakland: unpublished manuscript (held by Bancroft Library, UC Berkeley).

Burbach, Roger, and Patricia Flynn. 1980. *Agribusiness in the Americas.* New York: Monthly Review Press.

Burbank, Luther. 1927. *The Harvest of the Years,* with Wilbur Hall. Boston: Houghton Mifflin.

Burcham, L.T. 1981 [1957]. *California Rangeland: A Historico-Ecological Study of the Range Resources of California.* Davis: Center for Archaeological Research at Davis, Publication No. 7 [Sacramento: Department of Natural Resources, Division of Forestry].

Busch, Lawrence, and William Lacy. 1983. *Science, Agriculture and the Politics of Research.* Boulder: Westview Press.

Busch, Lawrence, Willam Lacy, Jeffrey Burkhardt, and Laura Lacy. 1991. *Plants, Power, and Profit: Social, Economic, and Ethical Consequences of the New Biotechnologies.* Oxford: Basil Blackwell.

Business Week. 1992. "Biotech: America's Dream Machine." March 2, 66–67.

Business Week. 1993. "Supertomatoes—the Old-Fashioned Way." October 4, 116–17.

Business Week. 1999. "Fields of Genes." April 2, 62–74.

Byres, Terence. J. 1995. "Political Economy, the Agrarian Question and the Comparative Method." *Journal of Peasant Studies* 22, no. 4: 561–80.

Cardellino, Joan. 1984. *Industrial Location: A Case Study of the California Fruit and Vegetable Canning Industry, 1860 to 1984*. Master's thesis, Department of Geography, University of California, Berkeley.

Carosso, Vincent. 1951. *The California Wine Industry, 1830–1895: A Study of the Formative Years*. Berkeley: University of California Press.

Carson, Rachel. 1962. *Silent Spring*. Boston: Houghton Mifflin.

Carter, Colin. 1997. "International Trade and Pacific Rim Issues." In Jerome Siebert, ed., *California Agriculture: Issues and Challenges*. Berkeley: Giannini Foundation, 231–48.

Carter, Harold, and George Goldman. "The Measure of California Agriculture: Its Significance in the State Economy." 1997. In Jerome Siebert, ed., *California Agriculture: Issues and Challenges*. Berkeley: Giannini Foundation, 29–62.

Caterpillar Tractor Company. 1954. *Fifty Years on Tracks*. Peoria: Caterpillar Tractor.

Chan, Sucheng. 1986. *This Bittersweet Soil*. Berkeley: University of California Press.

Chandler, Alfred. 1962. *Strategy and Structure*. Cambridge: MIT Press.

Chandler, Alfred. 1977. *The Visible Hand*. Cambridge: Harvard University Press.

Chandler, Alfred. 1990. *Scale and Scope*. Cambridge: Harvard University Press.

Chapella, Ignacio. 2003. "The Logic of Biotechnology." Talk given to Department of Geography, University of California, Berkeley, October 15, 2003.

Chataway, Jo, Joyce Tait, and David Wield. 2002. "From Life Sciences to a New Agro-Industry." *Technology Policy Briefs* 1, issue 2, United Nations University, INTECH.

Chayanov, Aleksandr. 1986 (1923). *The Theory of the Peasant Economy*. Edited and Translated by Daniel Thorner, Basile Kerblay, and R.E.F. Smith, with a foreword by Teodor Shanin. Madison: University of Wisconsin Press.

Chu, George. 1970. "Chinatowns in the Delta: The Chinese in the Sacramento–San Joaquin Delta, 1870–1960." *California Historical Society Quarterly* XLIX: 21–37.

Church, Thomas. 1995. *Gardens Are for People*. Edited by Grace Hall and Michael Lurie. 3d edition. Berkeley: University of California Press.

Clark, T.J. 1999. *Farewell to an Idea: Episodes from a History of Modernism*. New Haven: Yale University Press.

Clay, John Jr. 1899. "Work of the Breeder in Improving Live Stock." *Yearbook* of the U.S. Department of Agriculture. Washington, DC: USDA, 627–42

Cleland, Robert, and Osgood Hardy. 1929. *March of Industry*. Los Angeles: Powell Publishing Company.

Cleland, Robert. 1941. *The Cattle of a Thousand Hills: Southern California, 1850–80*. San Marino, CA: The Huntington Library.

Coates, David. 1999. *Models of Capitalism: Growth and Stagnation in the Modern Era*. Oxford: Polity Press.

Cobble, Sue. 1991. *Dishing It Out: Waitresses and Their Unions in the 20th Century*. Urbana: University of Illinois Press.

Cochrane, Willard. 1979. *The Development of American Agriculture*. Minneapolis: University of Minnesota Press.

Cohen, Lizabeth. 2003. *A Consumers' Republic: The Politics of Mass Consumption in Postwar America*. New York: Alfred A. Knopf.

Collins, Norman, Willard Mueller, and Eleanor Birch. 1959. *Grower-Processor Integration: A Study of Vertical Integration between Growers and Processors of Tomatoes in California*. Berkeley: California Agricultural Experiment Station. Bulletin No. 768.

Conaway, James. 1990. *Napa*. New York: Avon Books.

Couchman, Robert. 1967. *The Sunsweet Story: A History of the Establishment of the Dried Tree Fruit Industry in California and of the 50 Years of Service of Sunsweet Growers, Inc.* San Jose, CA: Sunsweet Growers.

Craddock, Susan. 2000. *City of Plagues: Disease, Poverty, and Deviance in San Francisco*. Minneapolis: University of Minnesota Press.

Cronise, Titus. 1868. *The Natural Wealth of California*. San Francisco: Bancroft & Co.

Crouchett, Louise Jacobs. 1982. *Filipinos in California: From the Days of the Galleons to the Present*. El Cerrito, CA: Downey Place Publishing House.

Curtiss, Charles F. 1898. *Some Essentials of Beef Production*. USDA Farmer's Bulletin No. 71. Washington, DC: U.S. Department of Agriculture.

Cuthbertson, Frank. 1955. "History of Vegetable Seed Poduction on the West Coast." Eleven-part series. *Seed World* 76, no. 2–12 (January–June): various pages.

Dallman, Peter. 1998. *Plant Life in the World's Mediterranean Climates*. Berkeley: University of California Press.

Danhof, Clarence. 1979. *Changes in Agriculture: The Northern United States, 1820–1870*. Cambridge: Harvard University Press.

Daniel, Cletus. 1982. *Bitter Harvest: A History of California Farmworkers, 1870–1941*. Berkeley: University of California Press.

Daniels, Roger. 1977. *The Politics of Prejudice: The Anti-Japanese Movement in California and the Struggle for Japanese Exclusion*. 2nd ed. Berkeley: University of California Press.

Davis, John, and Ray Goldberg, 1957. *A Concept of Agribusiness*. Cambridge: Harvard University School of Business Administration.

Davis, Kate. 2002. *Sardine Oil on Troubled Waters: The Boom and Bust of California's Sardine Industry, 1905–1955*. Doctoral dissertation, Department of Geography, University of California, Berkeley.

Davis, Mike. 1986. *Prisoners of the American Dream*. London: Verso.

Davis, Mike. 1998. *Ecology of Fear: Los Angeles and the Imagination of Disaster*. New York: Metropolitan / Henry Holt.

Davis, Mike. 2001. *Late Victorian Holocausts: El Niño Famines and the Making of the Third World.* London: Verso.

DeWitt, Howard. 1980. *Violence in the Fields: California Filipino Farm Labor Unionization During the Great Depression.* Saratoga, CA: Century Twenty-one Publishers.

Di Leonardo, Micaela. 1984. *The Varieties of Ethnic Experience: Kinship, Class, and Gender Among California Italian-Americans.* Ithaca, NY: Cornell University Press.

Dodge, Jacob R. 1885. *Agricultural Graphics: A Report of Exhibits Illustrating Agricultural Statistics at the World's Industrial and Cotton Exposition at New Orleans.* Washington, DC: Government Printing Office.

Donley, Michael, Stuart Allan, Patricia Caro, and Clyde Patton. 1979. *Atlas of California.* Culver City, CA: Pacific Book Center.

Doti, Lynne. 1995. *Banking in an Unregulated Environment: California, 1878–1905.* New York: Garland Press.

Dreyer, Peter. 1975. *A Gardener Touched With Genius: The Life of Luther Burbank.* New York: Coward, McCann & Geoghegan.

Dumke, Glenn. 1944. *The Boom of the Eighties in Southern California.* San Marino, CA: Huntington Library.

Dyble, Louise Nelson. 2003. *Paying the Toll: The Golden Gate Bridge and Highway District and the San Francisco Bay Area, 1919–1971.* Doctoral dissertation, Department of History, University of California, Berkeley.

Ebeling, Walter. 1979. *The Fruited Plain: The Story of American Agriculture.* Berkeley: University of California Press.

Emerson, Robert. 1990. *The New Economics of Fast Food.* New York: Van Nostrand Reinhold.

Employment Development Department. 1992. *Farm Labor Contractors in California.* Sacramento: EDD. *California Agricultural Studies 92-2.*

Erickson, Jack. 1993. *California Brewin'.* Reston, VA: RedBrick Press.

Evans, Peter. 1995. *Embedded Autonomy: States and Industrial Transformation.* New York: Cambridge University Press.

Eysberg, Cees. 1990. *The Californian Wine Economy: Natural Opportunities and Socio-Cultural Constraints—A Regional Geographic Analysis of Its Origins and Perspectives.* Utrecht: Geografisch Instituut, Rijkuniversiteit Utrecht.

Fielding, Gordon. 1964. "The Los Angeles Milkshed: A Study of the Political Factors in Agriculture." *Geographical Review* 54, no. 1: 1–12.

Fine, Ben. 1994. "Toward a Political Economy of Food." *Review of International Political Economy* 1, no. 3: 519–45.

Fine, Ben, and Ellen Leopold. 1992. *The World of Consumption.* London: Routledge.

Fine, Ben, Michael Heasman, and Judith Wright. 1996. *Consumption in the Age of Affluence: The World of Food.* London: Routledge.

Fisher, Lloyd. 1953. *The Harvest Labor Market in California.* Cambridge: Harvard University Press.

Fiske, Donald. 1955. *Evaluation of Vertical Integration in the Chain Food Industry, with Special Emphasis on Safeway Stores, Inc.* MBA Thesis, Haas School of Business, University of California, Berkeley.

Fiske, Emmett. 1979. *The College and Its Constituency: Rural and Community Development at the University of California, 1875–1978.* Doctoral dissertation, Department of Applied Behavioral Sciences, University of California Davis.

Fite, Gilbert. 1966. *The Farmer's Frontier, 1865–1900.* New York: Holt, Rinehart and Winston.

Fitzgerald, Deborah. 1991. "Beyond Tractors: The History of Technology in American Agriculture." *Technology and Culture* 32: 114–26.

FitzSimmons, Margaret. 1986. "The New Industrial Agriculture: The Regional Integration of Specialty Crop Production." *Economic Geography* 62: 334–53.

Fortune. 1940. "Safeway Stores, Inc." *Fortune.* October: 60–134.

Forty, Adian. 1986. *Objects of Desire: Design and Society, 1750–1980.* London: Thames and Hudson / Cameron.

Foweraker, Joseph. 1981. *The Struggle for Land: A Political Economy of the Pioneer Frontier in Brazil from 1930 to the Present Day.* New York: Cambridge University Press.

Foytik, Jerry. 1962. *Agricultural Marketing Orders: Characteristics and Use in California, 1933–1962.* Research Report No. 259. Berkeley: Giannini Foundation.

Fraser, W. Hamish. 1981. *The Coming of the Mass Market, 1850–1914.* London: Macmillan.

Freeman, Christopher. 1982. *The Economics of Industrial Innovation.* 2d ed. London: Frances Pinter.

Friedland, William. 1984. "Commodity Systems Analysis: An Approach to the Sociology of Agriculture." In Harry Schwarzweller, ed., *Research in Rural Sociology and Development.* London: JAI Press, 221–35.

Friedland, William. 1994. "The Global Fresh Fruit and Vegetable System: An Organization Analysis." In Phillip McMichael, ed., *The Global Restructuring of Agro-Food Systems.* Ithaca: Cornell University Press, 173–89.

Friedland, William. 2002. "Agriculture and Rurality: Beginning the Final Separation?" *Rural Sociology* 67, no. 3: 350–71.

Friedland, William, and Amy Barton. 1975. *Destalking the Wily Tomato: A Case Study in Social Consequences in California Agricultural Research.* Research Monograph No. 15. Davis: University of California, Department of Applied Behavioral Sciences.

Friedland, William, and Amy Barton. 1976. "Tomato Technology." *Society* 13, no. 6: 34–42.

Friedland, William, Amy Barton, and Robert Thomas. 1981. *Manufacturing Green Gold.* New York: Cambridge University Press.

Friedmann, Harriet. 1978. "Simple Commodity Production and Wage Labour in the American Plains." *The Journal of Peasant Studies* 6, no. 1: 71–100.

Friedmann, Harriet. 1993. "The Political Economy of Food: a Global Crisis." *New Left Review* 197: 29–57.

Frink, Maurice, W. Turrentine Jackson, and Agnes Wright Spring. 1956. *When Grass Was King.* Boulder: University of Colorado Press.

Frykman, Jonas, and Ovar Lofgren. 1987. *Culture Builders: A Historical Anthropology of Middle-Class Life.* New Brunswick: Rutgers University Press.

Fuller, Varden. 1934. *The Supply of Agricultural Labor as a Factor in the Evolution of Farm Organization in California.* Doctoral dissertation, Department of Economics, University of California, Berkeley.

Gabaccia, Donna. 2002. "As American as Budweiser and Pickles? Nation-Building in American Food Industries." In Warren Belasco and Phillip Scranton, eds., *Food Nations.* London and New York: Routledge, 175–93.

Galambos, Louis. 1966. *Competition and Cooperation: The Emergence of a National Trade Association.* Baltimore: Johns Hopkins University Press.

Galarza, Ernesto. 1964. *Merchants of Labor: The Mexican Bracero Story; An Account of the Managed Migration of Mexican Farm Workers in California, 1942–1960.* Charlotte, NC: McNally & Loftin.

Galarza, Ernesto. 1977. *Farm Workers and Agribusiness in California, 1947–1960.* Notre Dame: University of Notre Dame Press.

Gates, Paul. 1960. *The Farmer's Age: Agriculture 1815–1960.* New York: Holt, Rinehart and Winston.

Gates, Paul. 1961. "California's Agricultural College Lands." *Pacific Historical Review* 30, no. 2 (May): 103–22.

Gates, Paul. 1967. *California Ranchos and Farms, 1846–1862.* Madison: State Historical Society of Wisconsin.

Gates, Paul. 1991. *Land and Law in California: Essays on Land Policies.* Ames: Iowa State University Press.

George, Henry. 1871. *Our Land and Land Policy, National and State.* San Francisco: White and Bauer.

George, Henry. 1879. *Progress and Poverty: An Inquiry into the Cause of Industrial Depressions, and of Increase of Want With Increase of Wealth, the Remedy.* San Francisco: W.M. Hinton.

Gereffi, Gary, and Miguel Korzeniewicz, eds. 1994. *Commodity Chains and Global Capitalism.* Westport, CT: Praeger.

Gilbert, Jess, and Raymond Akor. 1988. "Increasing Structural Divergence in U.S. Dairying: California and Wisconsin since 1950." *Rural Sociology* 53: 56–72.

Glass, Fred. 1996. "We Called It a Work Holiday: The Oakland General Strike of 1946." *Labor's Heritage* 8, no. 2 (fall): 4–25.

Glennie, Paul, and Nigel Thrift. 1993. "Modern Consumption: Theorising Commodities and Consumers." *Society and Space* 11: 603–06.

Goldschmidt, Walter. 1978 [1947]. *As You Sow: Three Studies in the Social Consequences of Agribusiness.* Revised ed. Monclair, NJ: Allanheld, Osmun.

Gonzales, Juan. 1985. *Mexican and Mexican American Farm Workers: The California Agricultural Industry.* New York: Praeger.

Goodall, Merrill, John Sullivan, and Timothy De Young. 1978. *California Water: A New Political Economy.* Montclair, NJ: Allanheld, Osmun.

Goodman, David, Bernardo Sorj, and John Wilkinson. 1987. *From Farming to Biotechnology: A Theory of Agro-Industrial Development.* New York: Basil Blackwell.

Goodman, David, and Michael Redclift. 1991. *Refashioning Natures: Food, Ecology and Culture.* London: Routledge.

Goodman, David, and Michael Watts. 1994. "Reconfiguring the Rural or Fording the Divide? Capitalist Restructuring and the Global Agro-Food System." *Journal of Peasant Studies* 22, no. 1: 1–49.

Gordon, Margaret. 1954. *Employment Expansion and Population Growth.* Berkeley: University of California Press.

Graves, Alvin. 2003. *The Portuguese Californians: Immigrants in Agriculture.* San Jose, CA: Portuguese Heritage Publications of California.

Green, Tobias. 1998. *Persistence and Adaptation in Elite Finance: The Directors of American Trust Company, 1899–1960.* Master's thesis, Department of History, Northern Arizona University.

Green, Toby. 2000. "Unit Banking in California Wheat Country." Unpublished paper, Department of History, UC Davis.

Gregor, Howard. 1963. "Industrialized Drylot Dairying: An Overview." *Economic Geography* 39, no. 4: 299–318.

Gregory, James. 1991. *American Exodus: The Dust Bowl Migration and Okie Culture in California.* New York: Oxford University Press.

Groth, Paul. 1994. *Living Downtown: The History of Residential Hotels in the United States.* Berkeley: University of California Press.

Guerin-Gonzales, Camille. 1994. *Mexican Workers and American Dreams: Immigration, Repatriation, and California Farm Labor, 1900–1939.* New Brunswick: Rutgers University Press.

Guthey, Greig. 2004. *Terroir and the Politics of Agro-Industry in California's North Coast Wine District.* Doctoral dissertation, Geography Department. University of California, Berkeley.

Guthman, Julie. 2004. *Agrarian Dreams: The Paradox of Organic Farming in California.* Berkeley: University of California Press (in press).

Haber, Ludwig F. 1958. *The Chemical Industry During the Nineteenth Century.* Oxford: Clarendon Press.

Haber, Ludwig F. 1971. *Chemical Industry, 1900–1930.* Oxford: Clarendon Press.

Hamilton, J.G. 1997. "Changing Perceptions of Pre-European Grasslands in California." *Madrono* 44: 311–33.

Hansen, Hervey, and Jeanne Miller. 1962. *Wild Oats in Eden: Sonoma County in the 19th Century.* Santa Rosa: n.p.

Hart, George, and collaborators. 1946. "Wealth Pyramiding in the Production of Livestock." In Claude Hutchinson, ed., *California Agriculture.* Berkeley: University of California Press, 51–112.

Hart, Gillian. 1998. "Multiple Trajectories: A Critique of Industrial Restructuring and the New Institutionalism." *Antipode* 30, no. 4: 333–56.

Harvey, David. 1982. *The Limits to Capital.* Oxford: Blackwell.

Harwood, William. 1907. *New Creations in Plant Life: An Authoritative Account of the Life and Work of Luther Burbank.* 2d ed., rev. and enl. New York: Grosset & Dunlap.

Haslam, Gerald. 1974. *Okies: Selected Stories.* San Rafael, CA: New West Publications.

Hawthorn, Leslie, and Leonard Pollard. 1954. *Vegetable and Flower Seed Production.* New York: The Blakiston Company.

Hedrick, Ulysees. 1950. *A History of Horticulture in America to 1860.* New York: Oxford University Press.

Heller, Alfred. 1999. *World's Fairs and the End of Progress: An Insider's View.* Corte Madera: World's Fair, Inc.

Hempe, Edward, and Merle Wittenberg. 1964. *The Lifeline of America: Development of the Food Industry.* New York: McGraw-Hill.

Henderson, George. 1999. *California and the Fictions of Capital.* New York: Oxford University Press.

Henderson, George. 2004. " 'Free Food,' the Local Production of Worth, and the Circuit of Decommodification: A Value Theory of the Surplus." *Environmental and Planning D: Society and Space* 22, no. 4 (2004): 1–28.

Herrigel, Gary. 1996. *Industrial Constructions: The Sources of German Industrial Power.* Cambridge University Press.

Hightower, Jim. 1975. *Eat Your Heart Out.* New York: Crown Publishers.

Hightower, Jim. 1978. *Hard Tomatoes, Hard Times.* Cambridge: Schenkman.

Hilgard, Eugene W. 1884. *Report on the Physical and Agricultural Features of the State of California.* San Francisco: Pacific Rural Press (originally part of a larger *Report on Cotton Production in the United States,* U.S. Census Office, Tenth Census of the United States Washington, D.C. 1884, vol. 6, 649–796).

Hirabayashi, Lane Ryo. 1989. *The Delectable Berry: Japanese American Contributions to the Development of the Strawberry Industry on the West Coast.* Los Angeles: Japanese American National Museum.

Hise, Greg. 1997. *Magnetic Los Angeles: Planning the Twentieth-Century Metropolis.* Baltimore: Johns Hopkins University Press.

Hittell, John. 1863. *The Resources of California, Comprising Agriculture, Mining, Geography, Climate, Commerce, Etc.* San Francisco: A. Roman & Co.

Hodgson, Robert. 1933. "The California Fruit Industry." *Economic Geography* 9 no. 4: 337–55.

Hoffman, A.C. 1935. *Large-Scale Organization in the Food Industries.* Monograph No. 35, Temporary National Economic Committee, Investigation of Concentration of Economic Power. Washington, DC: Government Printing Office.

Hogan, Walter. 1913. *The Call of the Hen.* Petaluma, CA: Petaluma Daily Courier.

Howard, Gregor. 1963. "Industrialized Drylot Dairying." *Economic Geography* 39, no. 4: 298–318.

Howard. W.L. 1945. *Luther Burbank's Plant Contributions.* Berkeley: Agricultural Experiment Station Bulletin No. 691.

Hundley, Norris. 1992. *The Great Thirst: Californians and Water, 1770s–1990s.* Berkeley: University of California Press.

Hurtado, Albert L. 1988. *Indian Survival on the California Frontier.* New Haven: Yale University Press.

Hutchins, Wells. 1956. *California Law of Water Rights.* Sacramento: California State Printing Division.

Hutchinson, Claude, ed. 1946. *California Agriculture.* Berkeley: University of California Press.

Igler, David. 2001. *Industrial Cowboys: Nature, Private Property and the Regional Expansion of Miller & Lux, 1850–1920.* Berkeley: University of California Press.

Iwata, Masakazu. 1992. *Planted in Good Soil: The History of the Issei in United States Agriculture.* Vol. I. Issei Memorial Edition. San Francisco: Peter Lang.

Jackle, John, and Keith A. Sculle. 1999. *Fast Food: Roadside Restaurants in the Automobile Age.* Baltimore, Md.: Johns Hopkins University Press.

Jacobson, Yvonne. 1984. *Passing Farms, Enduring Values: California's Santa Clara Valley.* Los Altos: William Kaufman.

James, Marquis, and Besse Rowland James. 1954. *Biography of a Bank: The Story of the Bank of America, NT & SA.* New York: Harper and Brothers.

Jay, Richard. 1953. *A Case Study in Retail Unionism: The Retail Clerks in the San Francisco East Bay Area.* Doctoral dissertation, Department of Economics, University of California, Berkeley.

Jelinek, Lawrence. 1982 [1979]. *Harvest Empire: A History of California Agriculture.* 2nd ed. San Francisco: Boyd and Fraser.

Jenny, Hans, and collaborators. 1946. "Exploring the Soils of California." In Hutchinson, *California Agriculture.* Berkeley: University of California Press, 317–94.

Jenny, Hans. 1961. *E.W. Hilgard and the Birth of Modern Soil Science.* Pisa: Collana Della Revista Agrochimica.

Jessop, Bob. 1990. *State Theory.* Cambridge: Polity Press.

Johns, Michael. 1992. "The Urbanization of Peripheral Capitalism: Buenos Aires, 1880–1920." *International Journal of Urban and Regional Research* 16, no. 3: 353–74.

Johnson, Kathy. 2003. *Federalism and the Origins of the Urban Crisis: The Geo-Politics of Housing and Highways, 1916–56.* Doctoral dissertation, Department of Geography, University of California, Berkeley.

Johnson, Stephen, Gerald Haslam, and Robert Dawson. 1993. *The Great Central Valley: California's Heartland.* Berkeley: University of California Press.

Johnson, Willis, and George Brown. 1913. *The Poultry Book.* Garden City, NY: Doubleday.

Jones, E.L. 1974. *Agriculture and the Industrial Revolution*. Oxford: Basil Blackwell.

Jones, Robert. 1925. *Dairying in California*. San Francisco: Californians, Inc.

Jordan, David Starr, and Vernor Kellogg. 1909. *The Scientific Aspects of Luther Burbank's Work*. San Francisco: A.M. Robertson.

Jordon, Terry. 1993. *North American Cattle-Ranching Frontiers: Origins, Diffusion and Differentiation*. Albuquerque: University of New Mexico Press.

Kahrl, William. 1982. *Water and Power: The Conflict over Los Angeles' Water Supply in the Owens Valley*. Berkeley: University of California Press.

Kahrl, William, ed. 1979. *The California Water Atlas*. Sacramento: Governor's Office of Planning and Research.

Kann, Kenneth. 1993. *Comrades and Chicken Farmers: The Story of a California Jewish Community*. Ithaca: Cornell University Press.

Katznelson, Ira, and Aristide Zolberg, eds. 1986. *Working Class Formation: Nineteenth Century Patterns in Europe and the United States*. Princeton: Princeton University Press.

Kautsky, Karl. 1988 [1902]. *The Agrarian Question*. Introduction by Hamza Alavi and Teodor Shanin. London: Zwan Publications.

Kelley, Robert. 1959. *Gold versus Grain: The Hydraulic Mining Controversy in California's Sacramento Valley*. Glendale, CA: A.H. Clark.

Kelley, Robert. 1989. *Battling the Inland Sea: American Political Culture, Public Policy, and the Sacramento Valley, 1850–1986*. Berkeley: University of California Press.

Kenney, Martin. 1986. *Biotechnology: The University-Industrial Complex*. New Haven: Yale University Press.

Kepner, Robert A., Roy Bainer, and Edgar L. Barger, 1978 [1971]. *Principles of Farm Machinery*. 3rd ed. Westport, CT: Avi Publishing Company.

Kerr, Norwood. 1987. *The Legacy: A Centennial History of the State Agricultural Experiment Stations, 1887–1987*. Columbia, MO: Missouri Agricultural Experiment Station.

Klein, Naomi. 2000. *No Logo*. Toronto: Knopf.

Kloppenburg, Jack. 1988. *First the Seed: The Political Economy of Plant Biotechnology*. New York: Cambridge University Press.

Koerner, Lisbet. 1994. "Linnaeus' Floral Transplants." *Representations* 47: 144–69.

Kraemer, Erich, and H.E. Erdman. 1933. "History of Cooperation in the Marketing of California Fresh Deciduous Fruits." *Bulletin of the Agriculture Experiment Station*, no 557. Berkeley: Agricultural Experiment Station.

Kropotkin, Peter. 1892. *La Conquète du Pain*. Paris: Stock. Trans. *The Conquest of Bread*. London: Chapman and Hall, 1906.

Kropotkin. 1899. *Fields, Factories and Workshops*. London: Hutchinson and Co.

Lapsley, James. 1996. *Bottled Poetry: Napa Winemaking from Prohibition to the Modern Era*. Berkeley: University of California Press.

Lears, T. Jackson. 1994. *Fables of Abundance: A Cultural History of Advertising in America*. New York: Basic Books.

Leidner, Robin 1993. *Fast Food, Fast Talk: Service Work and the Routinization of Everyday Life*. Berkeley: University of California Press.

Lenin, V.I. 1964 [1899]. *The Development of Capitalism in Russia*. Moscow: Progress Publishers.

LeVeen, Phil. 1979. "Natural Resource Development and State Policy: Origins and Significance of the Crisis in Reclamation." *Antipode* 11, no. 2: 61–80.

Levenstein, Harvey. 1988. *Revolution at the Table: The Transformation of the American Diet*. New York: Oxford University Press.

Levenstein, Harvey. 1993. *Paradox of Plenty: A Social History of Eating in Modern America*. New York: Oxford University Press.

Levy, Jacques E. 1975. *Cesar Chavez: Autobiography of La Causa*. New York: Norton.

Lewontin, Richard. 1998. "The Maturing of Capitalist Agriculture: Farmer as Proletarian." *Monthly Review* 50, no. 3 (July–August): 72–85.

Lewontin, Richard. 2001. "Genes in the Food!" *New York Review of Books*, June 21, 81–84.

Liebman, Ellen. 1983. *California Farmland: A History of Large Agricultural Landholdings*. Totowa, NJ: Rowman and Allanheld.

Lister, Roger. 1993. *Bank Behavior, Regulation and Economic Development: California, 1860–1910*. New York: Garland Publishing.

Livingston, Dewey. 1995. *A Good Life: Dairy Farming in the Olema Valley*. San Francisco: National Park Service.

Longstreth, Richard. 1997. *City Center to Regional Mall: Architecture, the Automobile and Retailing in Los Angeles, 1920–1950*. Cambridge: MIT Press.

Longstreth, Richard. 1999. *The Drive-In, The Supermarket, and the Transformation of Commercial Space in Los Angeles, 1914–1941*. Cambridge: MIT Press.

Lukes, Timothy J., and Gary Y. Okihiro. 1985. *Japanese Legacy: Farming and Community Life in California's Santa Clara Valley*. Cupertino, CA: California History Center, De Anza College.

Luxenberg, Stan. 1985. *Roadside Empires: How the Chains Franchised America*. New York: Viking Penguin.

MacCurdy, Rahno Mabel. 1925. *The History of the California Fruit Growers' Exchange*. Los Angeles: Rice and Sons.

Majka, Linda, and Theo Majka, 1982. *Farmworkers, Agribusiness and the State*. Philadelphia: Temple University Press.

Mann, Susan, and James Dickinson. 1978. "Obstacles to the Creation of a Capitalist Agriculture." *Journal of Peasant Studies* 5, no. 4: 466–81.

Mann, Susan. 1990. *Agrarian Capital in Theory and Practice*. Chapel Hill: University of North Carolina Press.

Markin, Rom J. 1963. *The Supermarket: An Analysis of Growth, Development, and Change*. Pullman: Washington State University Press.

Marsden, Terry, and Alberto Arce. 1995. "Constructing Quality: Emerging Food Networks in the Rural Transition." *Environment and Planning A* 27: 1261–79.

Martin, Philip. 1987. *California's Farm Labor Market.* Davis: University of California, Agricultural Issues Center.

Martineau, Belinda. 2001. *First Fruit.* New York: McGraw-Hill.

Marx, Karl, and Frederick Engels. 1970 [1846]. *The German Ideology.* New York: International Publishers.

Marx, Karl. 1863. *Capital: Volume I.* 1967 edition. New York: International Publishers.

Marx, Karl. 1893. *Capital: Volume II.* 1967 edition. New York: International Publishers.

Mason, Jack, and Helen Van Cleave Park. 1975. *The Making of Marin, 1850–1975.* Inverness: North Shore Books.

Massey, Douglas, Rafael Alarcon, J. Durand, and H. Gonzalez. 1987. *Return to Aztlan: The Social Process of International Migration from Western Mexico.* Berkeley: University of California Press.

Matchett, Robert. 1967. *Departmentalization of the Safeway Organization.* MBA Thesis, School of Business, University of California, Berkeley.

Matsumoto, Valerie. 1993. *Farming the Homeplace.* Ithaca: Cornell University Press.

Matthews, Glenna. 2003. *Silicon Valley, Women, and the California Dream: Gender, Class and Opportunity in the Twentieth Century.* Stanford, CA: Stanford University Press.

Mayo, James. 1993. *The American Grocery Store.* Westport, CT: Greenwood Press.

McGinty, Brian. 1998. *Strong Wine: The Life and Legend of Agoston Haraszthy.* Stanford, CA: Stanford University Press.

McMichael, Philip. ed. 1994 *Global Restructuring of Agro-Food Systems.* Ithaca: Cornell University Press.

McNair, M.P. 1931. "Trends in Large-Scale Retailing." *Harvard Business Review* 10: 30–39.

McNally, David. 1988. *Political Economy and the Rise of Capitalism: A Reinterpretation.* Berkeley: University of California Press.

McWilliams, Carey. 1939. *Factories in the Fields.* Boston: Little Brown. (Also Santa Barbara: Peregrine Smith, 1976.)

McWilliams, Carey. 1942. *Ill Fares the Land; Migrants and Migratory Labor in the United States.* Boston: Little, Brown.

McWilliams, Carey. 1946. *Southern California Country: An Island on the Land.* New York: Duell, Sloan & Pearce. (Santa Barbara: Peregrine Smith, 1973.)

McWilliams, Carey. 1949. *California: The Great Exception.* New York: A.A. Wyn. (Santa Barbara: Peregrine Smith, 1976.)

McWilliams, Carey. 1961. *North from Mexico: The Spanish-Speaking People of the United States.* 2d ed. New York: Monthly Review Press.

Meisner, Daniel. 1997–98. "Bridging the Pacific: California and the China Flour Trade." *California History* 76, no. 4 (winter): 82–93.

Miliband, Ralph. 1969. *The State in Capitalist Society.* London: Weidenfeld and Nicholson.

Miller, Daniel, ed. 1995. *Acknowledging Consumption.* London: Routledge.

Mills, Herb. 1979. "The San Francisco Waterfront: The Social Consequences of Industrial Modernization." In A. Zimbalist, ed., *Case Studies on the Labor Process.* New York: Monthly Review Press, 127–55.

Minick, Roger. 1969. *Delta West: The Land and People of the Sacramento–San Joaquin Delta.* With historical essay by Dave Bohn. Berkeley: Scrimshaw Press.

Mitchell, Don. 1996. *The Lie of the Land: Migrant Workers and the California Landscape.* Minneapolis: University of Minnesota Press.

Modry, Southard. 1970. *The California Flower Seed Industry: A Study in Interrelationships of Commercial, Ecological and Human Factors in the Distribution of a Horticultural Activity.* Doctoral dissertation, Department of Geography, UCLA.

Mokyr, Joel. 1990. *The Lever of Riches: Technological Creativity and Economic Progress.* New York: Oxford University Press.

Moody, Kim. 1989. *An Injury to All: The Decline of American Unionism.* London: Verso.

Moore, Barrington. 1966. *The Social Origins of Dictatorship and Democracy.* Boston: Beacon Press.

Moreng, Robert, 1995. "Development of the Turkey Industry in the United States." *Poultry Tribune.* Centennial issue, September, 19–27.

Moses, Vincent. 1994. *The Flying Wedge of Cooperation: G. Harold Powell, California Orange Growers, and the Corporate Reconstruction of American Agriculture, 1904–1922.* Doctoral dissertation, University of California, Riverside.

Moses, Vincent. 1995. "The Orange Grower Is Not a Farmer: G. Harold Powell, Riverside Orchardists and the Coming of Industrial Agriculture, 1893–1930." *California History* 74: 22–37.

Mueller, Willard, and Leon Garoian. 1961. *Changes in the Market Structure of Grocery Retailing.* Madison: University of Wisconsin Press.

Mukerji, Chandra. 1997. *Territorial Ambitions and the Gardens of Versailles.* Cambridge: Cambridge University Press.

Muñoz, Carlos Jr. 1989. *Youth, Identity, Power: The Chicano Movement.* New York: Verso.

Murphy, Linda. 2004. "Mountain Man." *San Francisco Chronicle,* February 19, D1-D6.

Muscatine, Doris, Maynard Amerine, and Bob Thompson, eds. 1984. *University of California/Sotheby Book of California Wine.* Berkeley: University of California Press.

Musoke, Moses, and Alan Olmstead. 1982. "The Rise of the Cotton Industry in California." *Journal of Economic History* 42, no. 2: 385–412.

Nash, Gerald. 1964. *State Government and Economic Development: A History of Administrative Policies in California, 1849–1933.* Berkeley: Institute of Governmental Studies.

Nash, Gerald. 1972. "Stages of California's Economic Growth, 1870–1970: An Interpretation." *California Historical Quarterly* 51: 315–30.

Nash, Gerald. 1992. *A.P. Giannini and the Bank of America.* Norman: University of Oklahoma Press.

National Research Council, Board on Agriculture. 1996. *Colleges of Agriculture at the Land Grant Universities: Public Service and Public Policy.* Committee on the Future of the Colleges of Agriculture in the Land Grant University System. Washington, DC: National Academy Press.

National Research Council. 1997. *Precision Agriculture in the 21st Century: Geospatial and Information Technologies in Crop Management.* Board on Agriculture, NRC. Washington, DC: National Academy Press.

Nevins, Joseph. 2002. *Operation Gatekeeper: The Rise of the Illegal Alien and the Making of the US-Mexico Boundary.* New York: Routledge.

Nicholls, Sandra. 2003. *Saints, Peaches and Wine: Mexican Migrants and the Transformation of Los Haro, Zacatecas and Napa, California.* Doctoral dissertation, Department of Geography, University of California, Berkeley.

Norris, Frank. 1906. *The Pit: A Story of Chicago.* New York: A. Wessels.

Norris, Frank. 1901. *The Octopus: A Story of California.* New York: P.F. Collier.

O'Brien, Patrick K. 1977. "Agriculture and the Industrial Revolution." *Economic History Review* 30, no. 1: 166–81.

Olmstead, Alan, and Paul Rhode. 1988. "An Overview of California Agricultural Mechanization, 1870–1930." *Agricultural History* 62, no. 3: 86–112.

Olmstead, Alan, and Paul Rhode. 1997. "An Overview of the History of California Agriculture." In Jerome Siebert, ed., *California Agriculture: Issues and Challenges.* University of California, Berkeley: Giannini Foundation, 1–28.

Orsi, Richard. 1973. *Selling the Golden State: A Study of Boosterism in Nineteenth-Century California.* Doctoral dissertation, Department of History, University of Wisconsin.

Orsi, Richard. 1975. "The *Octopus* Reconsidered: The Southern Pacific and Agricultural Modernization in California, 1865–1915." *California Historical Quarterly* 54, no. 3: 197–220.

Ortega, Bob. 1998. *In Sam We Trust: The Untold Story of Sam Walton and How Wal-Mart Is Devouring America.* New York: Times Books.

Osofsky, Stephen. 1979. *Peter Kropotkin.* Boston: Twayne.

Padilla, Victoria. 1961. *Southern California Gardens: An Illustrated History.* Los Angeles: University of California Press.

Page, Brian. 1993. *Agro-Industrialization and Rural Transformation: The Restructuring of Midwestern Meat Production.* Doctoral dissertation, Department of Geography, University of California, Berkeley.

Page, Brian, and Richard Walker. 1991. "From Settlement to Fordism: The Agro-Industrial Revolution in the American Midwest." *Economic Geography* 67, no. 4, 281–315.

Palamountain, J.C. 1955. *The Politics of Distribution.* Cambridge: Harvard University Press.

Palerm, Juan Vicente. 1991. *Farm Labor Needs and Farm Workers in California, 1970 to 1989.* California Agricultural Studies, 91–92. Sacramento: Employment Development Department.

Parker, Carleton. 1920. *The Casual Laborer and Other Essays.* New York: Brace and Howe.

Parsons, James. 1955. "The Uniqueness of California." *American Scholar* 7: 45–55.

Parsons, James. 1986. "A Geographer Looks at the San Joaquin Valley." *Geographical Review* 76: 371–89.

Paul, Rodman. 1958. "The Wheat Trade Between California and the United Kingdom." *Mississippi Valley Historical Review* 45: 391–412.

Paul, Rodman. 1973. "The Beginnings of Agriculture in California: Innovation Versus Continuity." *California Historical Quarterly* 52, no. 1: 16–27.

Payne, Stephen. 1987. *Santa Clara County: Harvest of Change.* Northridge, CA: Windsor (in cooperation with County of Santa Clara Historical Heritage Commission).

Payne, Walter, ed. 1982. *Benjamin Holt: The Story of Caterpillar Tractor.* Holt-Atherton Pacific Center for Western Studies, Monography Series 2, No. 1. Stockton, CA: University of the Pacific.

Pierce, F.J., and E.J. Sadler, eds. 1997. *The State of Site-Specific Management for Agriculture.* Madison, WI: American Society of Agronomy, Crop Science Society of America, and Soil Science Society of America.

Pisani, Donald. 1984. *From the Family Farm to Agribusiness: The Irrigation Crusade in California and the West, 1850–1931.* Berkeley: University of California Press.

Pitt, Leonard. 1966. *The Decline of the Californios: A Social History of the Spanish-Speaking Californians, 1846–1890.* Berkeley: University of California Press.

Polanyi, Karl. 1944. *The Great Transformation.* New York: Rinehart.

Pollan, Michael. 2001. *The Botany of Desire.* New York: Random House.

Pollan, Michael. 2002. "Power Steer." *New York Times Magazine.* March 31, 44–76.

Pollard, Sidney. 1968. *The Genesis of Modern Management.* Harmondsworth: Penguin.

Porter, Michael. 1998. *On Competition.* Boston: Harvard Business School Publishing.

Post, Charles. 1982. "The American Road to Capitalism." *New Left Review* 130: 30–51.

Post, Charles. 1995. "The Agrarian Origins of U.S. Capitalism: The Transformation of the Northern Countryside Before the Civil War." *Journal of Peasant Studies* 22, no. 3: 390–445.

Post, Charles. 2003. "Planter Slavery and Economic Development in the Antebellum Southern United States." *Journal of Agrarian Change* 3, no. 3: 289–332.

Poulantzas, Nicos. 1973. *Political Power and Social Classes.* London: New Left Books.

Powell, Walter, and Paul DiMaggio, eds. 1991. *The New Institutionalism in Organizational Analysis.* Chicago: University of Chicago Press.

Powell, Walter, 1996. "Inter-Organizational Collaboration in the Biotechnology Industry." *Journal of Institutional and Theoretical Economics* 152: 197–215.

Preston, William. 1981. *Vanishing Landscapes: Land and Life in the Tulare Lake Basin.* Berkeley: University of California Press.

Przeworski, Adam. 1976. "Proletariat into Class: The Process of Class Formation from Karl Kautsky's *The Class Struggle* to Recent Controversies." *Politics and Society* 7, no. 4: 343–401.

Pudup, MaryBeth, and Michael Watts. 1987. "Growing Against the Grain: Mechanized Rice Farming in the Sacramento Valley, California." In S. Brush and B. Turner, eds., *Comparative Farming Systems.* New York: Guilford Press, 345–84.

Purchase, Graham. 1996. *Evolution and Revolution: An Introduction to the Life and Thought of Peter Kropotkin.* Petersham, Australia: Jura Books.

Purdue, Lewis. 1999. *The Wrath of Grapes: The Coming Wine Industry Shakeout and How to Take Advantage of It.* New York: Spike.

Reisner, Marc. 1986. *Cadillac Desert: The American West and Its Disappearing Water.* New York: Viking.

Requa, Mark, and Harry Cory. 1919. *The California Irrigated Farm Problem.* Washington, D.C.: U.S. Government Printing Office.

Rhode, Paul. 1995. "Learning, Capital Accumulation, and the Transformation of California Agriculture." *Journal of Economic History* 55, no. 4: 773–800.

Rhomberg, Chris. 2004. *No There There: Race, Class, and Political Community in Oakland.* Berkeley: University of California Press.

Robbins, Roy. 1976. *Our Landed Heritage: The Public Domain, 1776–1970.* 2nd ed. Lincoln: University of Nebraska Press.

Robinson, W.W. 1948. *Land in California, The Story of Mission Lands, Ranchos, Squatters, Mining Claims, Railroad Grants, Land Scrip and Homesteads.* Berkeley: University of California Press.

Rogin, Leo. 1931. *The Introduction of Farm Machinery in Its Relation to the Productivity of Labor in the Agriculture of the United States During the Nineteenth Century.* Publications in Economics, Vol. 9. Berkeley: University of California Press.

Rome, Adam. 2001. *The Bulldozer in the Countryside: Suburban Sprawl and the Rise of the American Environmental Movement.* New York: Cambridge University Press.

Rosenberg, Charles. 1997. *No Other Gods: On Science and American Social Thought.* Baltimore: Johns Hopkins University Press.

Rosenblum, Mort. 1996. *Olives: The Life and Lore of a Noble Fruit.* North Point Press.

Rossiter, Margaret. 1975. *The Emergence of Agricultural Science: Justus Liebig and the Americans, 1840–1880.* New Haven: Yale University Press.

Rothstein, Morton. 1985. *The California Wheat Kings.* Agricultural History Center, Working Paper series No. 22. Davis: University of California Davis

Rothstein, Morton. 1990. *The United States and the United Kingdom as Centers of the World Wheat Trade, 1846–1914.* Working Paper Series No. 60. Davis: Agricultural History Center.

Rudd, Robert. 1964. *Pesticides and the Living Landscape.* Madison: University of Wisconsin Press.

Rudy, Alan. 1995. *Environmental Conditions, Negotiations, and Crises: The Political Economy of Agriculture in the Imperial Valley of California, 1850–1993.* Doctoral dissertation, Department of Sociology, University of California, Santa Cruz.

Ruiz, Vicki. 1987. *Cannery Women, Cannery Lives: Mexican Women, Unionization and the California Food Processing Industry, 1930–1950.* Albuquerque: University of New Mexico Press.

Runsten, David, and Phil LeVeen, 1981. "Mechanization and Mexican Labor in California Agriculture." *Monographs in U.S.-Mexican Studies, no. 6.* University of California, San Diego: U.S.-Mexican Studies Center.

Russell, Edmund. 2001. *War and Nature: Fighting Humans and Insects with Chemicals from World War I to Silent Spring.* New York: Cambridge University Press.

Rydell, Robert. 1984. *All the World's a Fair: Visions of Empire at American International Expositions, 1876–1916.* Chicago: University of Chicago Press.

Sabato, Jorge. 1988. *La Classe Dominant el la Argentina Moderna.* Buenos Aires: Grupo Editor Latinoamericano.

Saekel, Karola. 1997. "The Man Who Gets the Good Stuff." *San Francisco Chronicle,* June 4, F1.

Safeway News. 1976. *Fifty Years of Safeway* 31, no. 1–6 passim. Series dedicated to company history.

Sampson, Arthur, and Agnes Chase. 1927. *Range Grasses of California.* Bulletin No. 430. Berkeley: California Agricultural Experiment Station.

Sampson, Arthur. 1913. *The Reseeding of Depleted Grazing Lands to Cultivated Forage Plants.* Bulletin No. 4. Washington, DC: U.S. Department of Agriculture.

Sampson, Arthur. 1917. *Important Range Plants: Their Life History and Forage Value.* Bulletin No. 545. Washington, DC: U.S. Department of Agriculture.

Sampson, Arthur. 1919. *Plant Succession in Relation to Range Management.* Bulletin No. 791. Washington, DC: U.S. Department of Agriculture.

Sampson, Arthur. 1923. *Range and Pasture Management.* New York: John Wiley & Sons.

Sanders, Elizabeth. 1999. *The Roots of Reform: Farmers, Workers, and the American State, 1877–1917.* Chicago: University of Chicago Press.

Sawyer, Richard. 1996. *To Make a Spotless Orange: Biological Control in California.* Ames: Iowa State University Press.

Saxenian, Annalee. 1994. *Regional Advantage: Silicon Valley and Route 128 in Comparative Perspective.* Cambridge: Harvard University Press.

Saxton, Alexander. 1971. *The Indispensable Enemy: Labor and the Anti-Chinese Movement in California.* Berkeley: University of California Press.

Sayer, Andrew. 1992. *Method in Social Science: A Realist Approach.* 2nd edition. London: Macmillan.

Sayer, Andrew, and Richard Walker. 1992. *The New Social Economy: Reworking the Division of Labor.* Oxford: Blackwell.

Sayre, Nathan. 2002. *Ranching, Endangered Species, and Urbanization in the Southwest: Species of Capital*. Tucson: University of Arizona Press.

Schell, Orville. 1985. *Modern Meat: Antibiotics, Hormones, and the Pharmaceutical Farm*. New York: Vintage.

Scheuring, Ann Foley, ed. 1983. *A Guidebook to California Agriculture*. Berkeley: University of California Press.

Scheuring, Ann, Zoe McCandless-Grossman, and Barry Owen. 1981. *Abalone to Zucchini: A Layman's Guide to California Crops*. Manuscript, prepared for Agricultural Experiment Station, University of California, Davis.

Scheuring, Ann. 1988. *A Sustaining Comradeship: The Story of University of California Cooperative Extension, 1913–1988*. Berkeley: Division of Agriculture and Natural Resources, University of California.

Scheuring, Ann. 2001. *Abundant Harvest: The History of the University of California, Davis*. Davis: UC Davis History Project.

Schlosser, Eric. 2001. *Fast Food Nation: The Dark Side of the All-American Meal*. Boston: Houghton Mifflin.

Schoendorf, A.J. 1947. *Beginnings of Cooperation in the Marketing of California Fresh Deciduous Fruits: History of the California Fruit Exchange*. Sacramento: Inland Press.

Schoenherr, Allan, Robert Fellmeth, and Michael Emerson. 1999. *Natural History of the Islands of California*. Berkeley: University of California Press.

Schumpeter, Joseph. 1934. *The Theory of Economic Development*. Cambridge: Harvard University Press.

Scott, Allen. 1988. *Metropolis: From the Division of Labor to Urban Form*. Berkeley and Los Angeles: University of California Press.

Scranton, Philip. 1997. *Endless Novelty: Specialty Production and American Industrialization, 1865–1925*. Princeton: Princeton University Press.

Scranton, Philip, and Warren Balasco, eds. 2001. *Food Nations: Commerce, Culture and Consumption*. New York: Routledge.

Selvin, David. 1960. *Union Profile: The Fifty Years of the Grocery Clerks Union, Local 648*. San Francisco: Grocery Clerks Local 648.

Siebert, Jerome, ed. 1997. *California Agriculture: Issues and Challenges*. Giannini Foundation, Division of Agriculture and Natural Resources. Berkeley: University of California.

Sinclair, Upton. 1927. *Oil!* New York: A.C. & Boni.

Skaggs, James. 1986. *Prime Cut: Livestock Raising and Meatpacking in the United States, 1607–1983*. College Station, TX: Texas A&M University Press.

Skocpol, Theda. 1979. *States and Social Revolution*. London: Cambridge University Press.

Smith, Adam. 1776. *The Wealth of Nations*. New York: Modern Library Edition, 1937.

Smith, Andrew. 2000. *Souper-Tomatoes: The Story of America's Favorite Food*. New Brunswick: Rutgers University Press.

Smith, Henry Nash. 1950. *Virgin Land: The American West as Symbol and Myth.* Cambridge: Harvard University Press.

Smith, Michael L. 1987. *Pacific Visions: California Scientists and the Environment, 1850–1915.* New Haven: Yale University Press.

Smith, Ralph, and collaborators. 1946. "Protecting Plants from Their Enemies." In C. Hutchinson, ed., *California Agriculture.* Berkeley: University of California Press, 239–316.

Sobel, Robert. 1993. *Dangerous Dreamers: The Financial Innovators from Charles Merrill to Michael Milken.* New York: John Wiley and Sons.

Stadtman, Verne. 1970. *The University of California, 1868–1968.* New York: McGraw-Hill.

Starrs, Paul. 1998. *Let the Cowboy Ride: Cattle Ranching in the American West.* Baltimore: Johns Hopkins University Press.

Steinbeck, John. 1988. *The Harvest Gypsies,* edited by Charles Wollenberg. Berkeley: Heyday Books.

Steinbeck, John. 1939. *The Grapes of Wrath.* New York: Viking.

Steinbrecher, F.A. 1996. "From Green to Gene Revolution: The Environmental Risks of Genetically-Engineered Crops." *The Ecologist* 26, no. 6: 275–83.

Stoll, Steven. 1998. *The Fruits of Natural Advantage: Making the Industrial Countryside in California.* Berkeley: University of California Press.

Storie, R. Earl, and W. W. Weir. 1926. *Generalized Soil Map of California.* Storie Index Bulletin #4, Series 1926.

Storper, Michael, and Richard Walker. 1989. *The Capitalist Imperative: Territory, Technology and Industrial Growth.* New York: Blackwell.

Storper, Michael, and Robert Salais. 1997. *Worlds of Production: The Action Frameworks of the Economy.* Cambridge: Harvard University Press.

Storper, Michael, and Richard Walker. 1982. "The Expanding California Water System." In W.J. Kockelman, T.J. Conomos, and A.E. Leviton, eds., *Use and Protection of the San Francisco Bay System.* San Francisco: Pacific Division, AAAS, 171–90.

Storper, Michael. 1997. *The Regional World.* New York: Guilford Press.

Streatfield, David. 1994. *California Gardens: Creating a New Eden.* New York: Abbeville Press.

Stuller, Jay, and Glen Martin. 1989. *Through the Grapevine: The Real Story Behind America's 8 Billion Dollar Wine Industry.* New York: Wynwood Press.

Tait, Joyce, Joanna Chataway, and David Wield. 2001. *Policy Influences on Technology for Agriculture: Chemicals, Biotechnology and Seeds.* Final Report of the PITA Project, Center for Technology Strategy, Open University, UK. http://technology.open.ac.uk//cts/pita.

Tannock, Stuart. 2001. *Youth at Work: The Unionized Fast Food and Grocery Workplace.* Philadelphia: Temple University Press.

Taylor, Judith. 2000. *The Olive in California: History of a Immigrant Tree.* Berkeley: Ten Speed Press.

Taylor, Paul, and Tom Vasey. 1936. "Contemporary Background of California Farm Labor." *Rural Sociology* 1, no. 4: 281–95.

Taylor, Paul. 1979. *Essays on Land, Water, and the Law in California*. Foreword by Paul W. Gates. New York: Arno Press.

Taylor, Ronald. 1975. *Chavez and the Farm Workers*. Boston: Beacon Press.

Tedlow, Richard. 1990. *New and Improved: The Story of Mass Marketing in America*. New York: Basic Books.

Teiser, Ruth. 1983. *The DiGiorgios: From Fruit Merchants to Corporate Innovators*. Berkeley: Bancroft Library Regional Oral History Office.

Thomas, Harold, and Earl Goldsmith. 1945. *The Shasta, Sierra, Lassen, Tahoe and Donner Strawberries*. Bulletin 690. Berkeley: University of California, Agricultural Experiment Station.

Thompson, E.P. 1964. *The Making of the English Working Class*. London: Penguin.

Thompson, F.M.L. 1968. "The Second Agricultural Revolution." *Economic History Review* 21, no. 1: 62–77.

Thompson, F.M.L. 1976. "Nineteenth Century Horse Sense." *Economic History Review* XXIX: 61–63.

Thompson, John. 1957. *The Settlement Geography of the Sacramento–San Joaquin Delta, California*. Doctoral dissertation, Stanford University.

Tobey, Ronald, and Charles Wetherell. 1995. "The Citrus Industry and the Revolution of Corporate Capitalism in Southern California, 1887–1944." *California History* 74: 6–22.

Todd, Frank. 1921. *The Story of the Exposition: Being the Official History of the International Celebration Held at San Francisco in 1915 to Commemorate the Discovery of the Pacific Ocean and the Construction of the Panama Canal*. Five volumes. For the Panama Canal, Panama-Pacific International Exposition Company. New York: G.P. Putnam's Sons.

Tower, Jeremiah. 2003. *California Dish: What I Saw (and Cooked) at the American Culinary Revolution*. New York: The Free Press.

Trow-Smith, Robert. 1959. *A History of British Livestock Husbandry, 1700–1900*. London: Routledge and Kegan Paul.

Tufts, Warren, and collaborators. 1946. "The Rich Pattern of California Crops." In C. Hutchinson, ed., *California Agriculture*. Berkeley: University of California Press, 113–238.

United States Bureau of Agricultural Economics. 1940. *California Farms*. Bound folio of twelve maps. Division of Land Economics. Berkeley: U.S. Department of Agriculture.

United States Bureau of the Census. 1914. *Statistical Atlas of the United States*. Washington, DC: U.S. Government Printing Office.

United States Bureau of the Census. 1915. *Estimated Valuation of National Wealth, 1850–1912*. Washington, DC: U.S. Government Printing Office.

United States Bureau of the Census. 1928. *1925 Census of Agriculture: Summary Statistics by States.* Washington, DC: U.S. Government Printing Office.

United States Department of Agriculture, 1960. *Consumption of Food in the United States, 1909–52.* Agricultural Handbook No. 62.

Van den Bosch, Robert. 1978. *The Pesticide Conspiracy.* Garden City: Doubleday.

Vaught, David. 1999. *Cultivating California: Growers, Specialty Crops, and Labor, 1875–1920.* Baltimore: Johns Hopkins University Press.

Victor, Kirk. 1989. "What About the Workers?" *National Journal,* February 18, 397–99.

Villarejo, Donald, and David Runsten. 1993. *California's Agricultural Dilemma: Higher Production and Lower Wages.* Monograph. Davis: California Institute for Rural Studies.

Villarejo, Donald. 1980. *Getting Bigger: Large Scale Farming in California.* Davis: California Institute for Rural Studies.

Walker, Richard. 1995. "Regulation and Flexible Specialization as Theories of Capitalist Development: Challengers to Marx and Schumpeter." In Helen Liggett and David Perry, eds., *Spatial Practices: Markets, Politics and Community Life.* Thousand Oaks, CA: Sage, 167–208.

Walker, Richard. 1996. "Another Round of Globalization in San Francisco." *Urban Geography* 17, no. 1: 60–94.

Walker, Richard. 2001. "California's Golden Road to Riches: Natural Resources and Regional Capitalism, 1848–1940." *Annals of the Association of American Geographers* 91, no. 1: 167–99.

Walker, Richard. 2001a. "Industry Builds the City: Industrial Decentralization in the San Francisco Bay Area, 1850–1940." *Journal of Historical Geography* 27, no. 1: 36–57.

Walker, Richard, and Michael Heiman. 1981. "Quiet Revolution for Whom?" *Annals of the Association of American Geographers* 71: 67–83.

Walker, Richard, and Michael Teitz. 1982. "Industry." In E. Engelbert and A. Scheuring, eds., *Competition for California's Water.* Davis: Water Resources Center, 59–75.

Walsh, Annamarie. 1978. *The Public's Business: The Policy and Practice of Government Corporations.* Cambridge: MIT Press.

Walsh, John P. 1993. *Supermarkets Transformed: Understanding Organizational and Technological Innovations.* New Brunswick: Rutgers University Press.

Walton, Whitney. 1992. *France at the Crystal Palace: Bourgeois Taste and Artisan Manufacture in the 19th Century.* Berkeley: University of California Press.

Watts, George, and Connor Kennett. 1995. "The Broiler Industry." *Poultry Tribune,* Centennial Issue, September 1995, 6–18.

Watts, Michael. 1993. "Life Under Contract." In Peter Little and Michael Watts, eds., *Peasants Under Contract.* Madison: University of Wisconsin Press, 21–78.

Watts, Michael. 1996. "Kautsky Redux." *Progress in Human Geography* 20, no. 2: 230–45.

Webber, Herbert, and L.D. Batchelor, eds. 1943. *The Citrus Industry.* Vol. I. *History, Botany, and Breeding.* Berkeley: University of California Press.

Welch, Norman. 1989. *Strawberry Production in California.* Leaflet 2959. Berkeley: University of California Cooperative Extension.

Wells, Miriam. 1996. *Strawberry Fields: Politics, Class and Work in California Agriculture.* Ithaca: Cornell University Press.

West, Charles. 1929. "The Use, Value, and Cost of Credit in Agriculture." *California Agricultural Experiment Station, Bulletin No. 480.* Berkeley: Giannini Foundation of Agricultural Economics, University of California.

Wickson, Edward. 1902. *Luther Burbank: Man, Methods and Achievements.* San Francisco: Southern Pacific Company.

Wickson, Edward. 1912 [1889]. *The California Fruits and How to Grow Them.* San Francisco: Pacific Rural Press. Sixth ed.

Wickson, Edward. 1918. "Beginnings of Agricultural Education and Research in California." In *Annual Report of the Director of the College of Agriculture and Experiment Station, July 1, 1917 to June 30, 1918.*

Wickson, Edward. 1921. *California Nurserymen and the Plant Industry, 1850–1910.* Los Angeles: California Association of Nurserymen.

Wickson, Edward. 1923 [1897]. *California Vegetables in Garden and Field.* San Francisco: Pacific Rural Press. Fifth ed.

Wik, Reynold. 1975. "Some Interpretations of the Mechanization of Agriculture in the Far West." *Agricultural History* 49, no. 1: 73–83.

Wilhelm, Stephen, and James Sagan. 1974. *A History of the Strawberry.* Berkeley: University of California, Division of Agricultural Sciences.

Willis, Paul. 1937. *The Federal Reserve Bank of San Francisco: A Study in American Central Banking.* New York: Columbia University Press.

Willis, Susan. 1991. *A Primer for Daily Life.* London and New York: Routledge.

Wilson, Edwin, and Marion Clawson. 1945. *Agricultural Land Ownership and Operation in the Southern San Joaquin Valley.* Berkeley: U.S. Department of Agriculture.

Wilson, James. 1998. *Terroir.* Berkeley: University of California Press.

Wilson, Simone. 1990. *Sonoma County: The River of Time.* Chatsworth: Windsor Publications.

Woeste, Victoria. 1998. *The Farmer's Benevolent Trust: Law and Agricultural Cooperation in Industrial America, 1865–1945.* Chapel Hill: University of North Carolina Press.

Wolf, Stephen, and Fred Buttel. 1996. "The Political Economy of Precision Farming." *American Journal of Agricultural Economics* 78: 1269–74.

Woriol, George. 1992. *In the Floating Army: F.C. Mills on Itinerant Life in California, 1914.* Urbana: University of Illinois Press.

Worster, Donald. 1985. *Rivers of Empire.* New York: Pantheon.

Wright, Erik O. 1985. *Classes.* London: Verso.

Wright, Gavin. 1986. *Old South, New South: Revolutions in the Southern Economy Since the Civil War.* New York: Basic Books.

Wrigley, Neil, and Michelle Lowe, eds. 1996. *Retailing, Consumption and Capital: Towards the New Retail Geography.* Harlow, England: Longman.

Zabin, Carol, M. Kearney, A. Garcia, D. Runsten, and C. Nagengast. 1993. *Mixtec Migrants in California Agriculture: A New Cycle of Poverty.* Davis: California Institute for Rural Studies.

Zimmerman, Max M. 1937. *Super Market: Spectacular Exponent of Mass Distribution.* New York: Super Market Publishing Co.

Zimmerman, Max M. 1955. *The Super Market: A Revolution in Distribution.* New York: McGraw-Hill Book Company.

Zonlight, Margaret. 1979. *Land, Water and Settlement in Kern County California, 1850–1890.* New York: Arno Press.

Index